Wilderness in Mythology and Religion

Religion and Society

Edited by
Gustavo Benavides, Kocku von Stuckrad
and Winnifred Fallers Sullivan

Volume 55

De Gruyter

Wilderness in Mythology and Religion

Approaching Religious Spatialities, Cosmologies, and Ideas of Wild Nature

Edited by
Laura Feldt

De Gruyter

ISBN 978-1-61451-224-0
e-ISBN 978-1-61451-172-4
ISSN 1437-5370

Library of Congress Cataloging-in-Publication Data

A CIP catalog record for this book has been applied for at the Library of Congress.

Bibliographic information published by the Deutsche Nationalbibliothek

The Deutsche Nationalbibliothek lists this publication in the Deutsche Nationalbibliografie; detailed bibliographic data are available in the Internet at http://dnb.dnb.de.

Printing: Hubert & Co. GmbH & Co. KG, Göttingen
∞ Printed on acid-free paper

Printed in Germany

www.degruyter.com

Contents

1. Wilderness in Mythology and Religion

Laura Feldt*

1. Introduction[1]

Wilderness is often understood as a place where humans have not fully "infected" nature, as uncontaminated by 'civilization' (Cronon 1995; Oehlschlager 1991; Oehlschlager 2005; Garrard 2004,59). Wilderness has been imagined as a part of the solution to humankind's problematic relationship with the non-human world, and as crucial to understandings of nature as alterity. Yet, any understanding of wilderness is of course profoundly cultural. Ideas of natural wildernesses from deep forests, arid deserts, vast steppes and uninhabitable mountains to the depths of the ocean, the rainforest, and the inland ice have played important roles in the history of religions and still do so contemporarily – in religious ideas, narratives, rituals, cosmologies, and in everyday practices. People go to the wilderness to meet themselves, their demons, and their gods; it is simultaneously framed as refuge, paradise, waste land, and hell; it is where you can be lead astray, into idolatry or death, or where you can discover a new subjectivity, where you may find the deepest wisdom or great ignorance. Ideas of the role of various natural environments in diverse religions have also influenced theoretical work on what "religion" is and what it does, from the early days of the scientific study of religion until today. For instance, it might be said that ideas of wilderness played a role in the early studies of 'primitive cultures' and animism, the religion of 'the Semites', over the interest in the Golden Bough (Frazer 1990) and the worship of nature to, perhaps, discussions about the naturalness of religion today.

---

* Laura Feldt gratefully acknowledges a grant from the Danish Council for Independent Research in the Humanities that enabled the research behind this volume.

1 I wish to thank Ingvild S. Gilhus and Jan N. Bremmer warmly for their helpful comments.

Wilderness mythology is probably one of the most abiding creations in the history of religions. It delimits the presence of "the world" and verbalises a natural/imagined frontier. Wilderness in mythology and religion is a space of encounter – between the human self and supernatural others, and between humans and a natural alterity. Wilderness mythology resurfaces in wildlife and nature preservation discourses and still serves as a repository of identity formation today in varying forms of social practice. It is clearly a space of natural, cultural, and social significance. Its interesting multiplicity calls for new approaches that venture beyond previous dichotomizing perspectives. For what are the specific relations between the world's religions and its imagined and real wilderness areas? Analyses of religious conceptions of wilderness and wild nature may throw important light on how humans interact with their 'natural' surroundings and how they respond to the climate debates, as they represent the imagined side of territories and climates. The contributions of this book investigate the territory/climate cluster of wilderness in religions past and present, as well as in the history of the study of religion. It intervenes analytically in the conversation on religious space, climate and nature by analyzing the religious construction/production of wilderness, and the uses, functions, and history of wilderness in diverse contexts.

In spite of its provoking and inspiring hybridity, wilderness has neither previously been subject to detailed scrutiny from a comparative and cross-cultural history of religions perspective, nor has it been subject to extensive study or theorising within the humanistic study of religion. Wilderness has, rather, been investigated as parts of eco-theology (e.g., Deane-Drummond 2004; Eaton 2005; Jasper 2004), by literary critics (e.g., Gersdorf and Mayer 2006; Graulund 2006; Garrard 2004, Heise 2009, et al.), or by philosophers and historians (Cronon 1995; Oelschlager 1991). Wilderness is, in the study of religion, an under-theorized concept and an under-investigated theme. This book addresses the need for humanistic, anthropological and historical investigations of religion and wilderness. As such, it is a contribution to the debate on the use of natural spaces in the study of religion. It contributes to the field by presenting a stimulating variety of theoretical approaches and case studies, thus advancing comparative and cross-cultural research on religion and wilderness.

As Timothy Morton points out, thought, including the ecological thought, has long placed nature 'over yonder', as a thing in the distance (Morton 2007,1–28) and the idea of wilderness has had a role to play in this. Yet, the religious materials analysed in this book clearly demonstrate the wilderness' inevitable culturality, sociality and historicity, its embeddednes in concrete lifeworlds and practices of people (Ingold 2000), and show how the wilderness is at once an imaginary space *and* a concrete-material territory-climate space that cannot be grasped by means of dichotomizing distinctions. The wilderness is a natureculture cluster, the specific conception of which is often consequential for religious identity, orientation and interaction with the natural environment. Wilderness is an epitomical example of the muddiness of the interface of world and humanity. The theoretical perspectives relevant for inquiries into the wilderness and religion nexus testify to the fertility and novelty of the field, as well as to some of the challenges involved in thinking about the interconnections between religion and wilderness, which are clearly more complex than previous understandings suggest. In light of the recent spatial turn in anthropology and the social sciences, space can no longer be seen as 'a substrate for the external imposition of arbitrary cultural form', as a *tabula rasa* or empty container, a preformed surface waiting to be occupied (Ingold 2001,214; Soja 1989, Lefebvre 1991, Crang and Thrift 2004, et al.). Space is not neutral; it requires analysis as much as 'history' or 'sociality'. The idea of wilderness is an epitomical example. Similarly, the understanding of nature and landscape has undergone tremendous developments (e.g., Schama 1996), and 'climate' has become a major research focus (e.g., Beck 2010, Latour 2004, Jasanoff 2010 etc.) Research on cosmology in the general study of religion has been waning (cf. Raudvere 2009) and the interest has shifted to other areas. The crucial motivation for the types of questions asked in this volume, their contemporary relevance, and a direction of interest comes from the budding field of 'religion and nature', while perspectives, theories, and strategies of interest are fruitfully culled from, or entered into dialogue with, research on religious cosmologies and religion and spatiality. Thus, the fruitfulness of wilderness as a research topic for the study of religion stems also from its productive intersection with the existing fields of inquiry of religion and nature, religious cosmologies, and religion and spatiality. The analyses presented here both implicitly and explicitly question standard understandings of the core terms 'wilderness' and wild 'nature' and break new spatial trails in the study of diverse religions.

2. Wilderness and the 'religion and nature' field

Interest in nature and religion has been growing over the last ten years (Ruether and Hessel 2000; Gottlieb 2006, Taylor 2005, Kearns and Keller 2007, Taylor 2010), if not longer, and the contours of a new field is emerging (Jenkins 2009). The renewed interest in religion and nature has called attention to the complexities involved in making the nature/culture distinction, and it has been stressed that 'nature' and 'culture' are invariably interconnected and never appear as distinct entities, but are always as varieties of 'naturecultures' (cf. Haraway 2008). The necessity of new perspective on the world vs. humans interface has been pointed out, and a strong trend in the field suggests that human beings are not opposed to a nature which we do not inhabit, as it were, but are invariably immersed in diverse practical and perceptual engagements with surrounding environments (cf. Ingold 1996; 2000; Taylor 2005). Indeed, wilderness conceptions in religions seem to be significant examples of such naturecul-ture (and textpractice) clusters that cannot be grasped by means of simple nature/culture or cosmos/chaos dichotomies, and which highlight the need for new theoretical approaches. Recent developments in the field of religion and nature are thus relevant for thinking the wilderness-and-religion nexus. Although things are certainly changing (see, e.g., Taylor 2005 and 2005b, Taylor 2010), the field still seems to lack critical, analytical, and comparative work.[2] An implication of this book's research focus is that it represents a turn away from the implicit and explicit assumption of a strong connection between a society's or religion's perception of nature, the nonhuman world, and then a society's environmental behaviour (the "Religious Environmentalist Paradigm", Pedersen 1995); an assumption that has dominated in the study of religion and nature

2 Much previous scholarship on religion and nature has been somewhat activist and/or confessional/ethical in orientation, often with an implicit or explicit theological aim of uncovering the green potential of the world's religions in order to create a common ground for a sustainable future (as in Tucker and Grim (1997, xvi-xix, xxvi), cf. the aim of the *Harvard series Religions of the World and Ecology*; Gottlieb 2006), and assuming that environmental action is conditioned by religious attitudes about nature (Taylor 2005b). The issue of religious/engaged scholarship has been a point of intense debate in the field of religion and nature (Taylor 2005b). The place of political-religious agendas in the academic debate can rightly be discussed and questioned, and certainly the history of the field of religion and nature/ecology calls for greater self-reflexivity with regard to the approaches used (cf. Taylor 2005b) and, I add, the questions asked.

(cf. Kalland 2005) since Lynn White presented his seminal thesis on 'the historical roots of our ecologic crisis' (White 1967). Instead, the research presented here does not proceed from any advance assumption that religions are the source of the ecological crisis – nor that they may therefore also be a solution to it. Instead, the articles analyse the roles that ideas of wilderness have played and play in religions past and present, as a prerequisite for a better understanding of the role that religions may, or may not, play for environmental practice and discussions, and for better understanding of the relations that may pertain between religion, nature and space. The field lacks comparative investigations of the theme of wilderness and religion. This volume, as a comparative study of religion, offers critical, analytical investigations of the nexus of religion and wilderness which reveal some of the complexity of the interactions between humans, wilderness ideas, and religion.

3. Wilderness and the 'religion and spatiality' field

The 'spatial turn' in social sciences and the humanities (Warf and Arias 2009), offers further inspiration for addressing the wilderness as a space. Recent decades have seen a rise of theoretical interest in human conceptions of space and place. The field of critical spatial theory developed in the social sciences, but other academic disciplines such as philosophy, architecture and not least cultural geography have also played important roles. Among the contributors who have sparked the interest in space and spatiality in the humanities are Lefebvre (1974) and Soja (1989) as well as Foucault (1967; 1986), Smith (1987), Harvey (1989), Massey (1994), Serres (1990), Bingham and Thrift (2004) and Thrift and Whatmore (2004). Spatiality theory has also been seminal for new approaches to space within the study of religion (Knott 2005, Corrigan 2009, Lied 2008, and others). What characterises the spatial turn is an interest in how space is produced, imagined and lived, in addition to the conventional interest in space as a geophysical 'reality'. Space is not seen as constituted by fixed positions on a map, but, rather, as a realm of practices, ways of moving about, doing things (cf. de Certau 1984); space is seen as constituted by experiences, relations and forms of engagement. In reaction to a typical modern, 'Western,' notion of space as a given and passively-existing materiality, space is not seen as neutral. Instead, humans create and shape space through their practical engagement with their surroundings. Consequently, it should not be seen as a mere stage or background, but rather as a cultural and social construct. Spatiality theory is thus obviously relevant for investigations of wilderness, and the new spatiality theories offer pertinent input to alternative strategies of analysis or function as apt discussion partners for the analyses of this volume.

4. Wilderness as a cosmological concept

The notion of wilderness irrevocably brings to mind related history of religions concepts such as 'chaos' or 'nature', to which it is, however, not identical. Yet, while distinct from 'chaos' and from 'nature', the idea of wilderness does indeed belong to the same general category of spatial conceptualizations of areas which are seen as opposed to the human 'home space', the world of everyday life, in diverse religions. This does not mean, however, that wilderness can be equated with nature or with chaos, as the analyses of this book demonstrate. The contributions discuss, develop, and complicate the distinct concept of wilderness for the study of religion and investigate its construction, use and functions in religions crossculturally, often in dialogue with previous approaches. In the study of religion, analyses of 'natural' and imagined spatial domains traditionally form part of the subfield of the study of religious cosmologies.

Religious cosmologies are, in the study of religion, investigated as those culturally specific orientations in space (and time) that may be analysed and extracted from religious narratives, images, artefacts, and rituals (Geertz 1999, 471–472; Auffarth 2001). 'Cosmology' is understood both as the general, spatio-temporal view of the cosmos/world embedded in diverse religions, and as the more or less systematized religious knowledge about it (Lincoln 1999, 3–5; Raudvere 2009).[3] In recent decades, the general study of religion has moved away from such classical endeavours.[4] This is probably partly due to how classical studies of cosmology, to some extent, came to be associated with the phenomenology of religion, which was criticized for the grand generalizing, and sometimes religiously

3 It is thus not the same as the study of cosmology in astrophysics. In some quarters, the study of religious cosmologies is associated with the creationism vs. Darwinism-debate in the US (Bolle 1987).

4 This does not mean that it was not and is not still pursued in various area studies and sub-disciplines. Cosmology studies do thrive in some area studies that are sub-disciplines of the study of religion, such as pre-Christian Old Norse religion or in the study of religion in Mesopotamia.

grounded, concepts that it utilized (Jensen 2003; Stausberg 2009). In addition to phenomenology, the study of religious cosmologies has also been associated with structuralist approaches, ever since the discussions about myths and worldviews in the late 1950s by Claude Lévi-Strauss and other structuralists. Myths can supply important materials for fruitful structuralist study of religious cosmologies, but such perspectives are certainly also more applicable to some religions rather than others (Raudvere 2009), just as the extent of systematization and dogmaticism in religious cosmologies is an open, empirical question. The general study of religion has veered away from such classical approaches to cosmologies towards a greater focus on contexts and agency (cf. Jensen 2003, Raudvere 2009, Lincoln 2008,10). For instance, the entrance "cosmology" does not appear in recent guides such as M.C. Taylor's anthology of *Critical Terms in Religious Studies* (1998) or in W. Braun and R.T. McCutcheon's *Guide to the Study of Religion* (2000). As pointed out by Catharina Raudvere, in the post-Foucauldian field of the study of religion, 'cosmology' has been replaced by terms such as mentality, lifeworld, and ideology. This change is related to the influence exerted by related fields such as sociology, anthropology and the history of mentality, and the concomitant greater focus on the contexts of concepts, multi-dimensionality, and visible actors (Raudvere 2009), as well as to the general influence of (post-)poststructuralist studies of religion in the humanities (Stausberg 2009). Yet, this does not mean that there is nothing to gain from an engagement with previous approaches to religious cosmologies. If re-engaged on a critical, contextualised grounding,[5] not only is there something to gain for the general study of religion, but such an endeavour would also participate in the continuous reformulation and advancing of the discipline as a *general* and comparative study of religion. Renewed studies of religious cosmologies would therefore, for instance, move away from the assumption that traditional, pre-modern or indigenous societies share one single cosmological system, just as it cannot be assumed that religious cosmologies are necessarily coherent, systematic, generally shared or uncontested in a society. The status of the cosmological domains, or of the cosmological system as a whole, would also be an open question for investigation, as it can no longer be imagined as necessarily certain and factual, wholly real and convincing for a society's or a

5 Jensen argues that the phenomenology of religion as a subdiscipline within the general study of religions should not be abandoned *per se*. If pursued on the right theoretical grounding it can still be a viable endeavour (Jensen 2003).

group's members.[6] Nor can we assume that any one view expressed in a myth or ritual is regarded as necessarily authoritative in a society; this too would be subject to documentation. The unfortunate assumptions that singular religious statements can be taken as representative of a culture and that the ability to move between differing, sometimes conflicting, understandings of reality, space and world is uniquely modern, while ancient, premodern, and indigenous societies are characterized by a more static and unchanging view, would have to be abandoned (Veyne 1983; Petersen and Schjødt 2007). The ability to pendulate between, or entertain simultaneously, differing representations of reality ('balkanization') is, more plausibly, a general, human capacity. Further, renewed cosmological research would naturally pay much more attention to the interaction between cosmological views and the local climate, materiality and concrete interaction with natural environments, material objects, landscapes, and the political history of the area in question. Large generalizing and dichotomizing concepts such as, for instance, cosmos and chaos, would need qualification, further nuances, and theoretical development, especially in light of advances in spatiality theory and the study of religion and nature, as – for instance – analyses of the wilderness and religion nexus show.

Considering the often striking ambiguity of the wilderness space, a re-consideration or broadening of the classical notion of liminality may be pertinent for cosmological studies. In his study *Rites de passage* (1909), Arnold van Gennep formulated the concept of liminality that Victor Turner later expanded upon (Turner 1974, 1977, 1979). The liminal phase, the middle sequence of a rite of passage, in which the participant is outside everyday life and exposed to 'the transcendent', is seen as a decisive phase of the ritual. In this phase, the transformation in individual or group social, cultural or psychic state takes place. In the liminal phase, one is 'betwixt and between' ordinary cultural classifications. Liminality is, according to Turner, primarily something temporal, a phase in a ritual involving change (Turner 1977, 36–38; 1974, 231–233; 1979; Alexander 1991, 29–33). The liminal phase of rituals as a temporal phase has received much attention in the study of religion (e.g., Turner 1979, Alexander 1991; Brown 2003), while narratively embedded ambiguous spaces have been less theorised. The idea of 'liminal space' (see Endsjø 2000; Schjødt 2008 and 2004; Feldt 2003) may be relevant for

6 Often in cosmological research it has been assumed that cosmology has a world-view-forming function (Bolle 1987).

the analysis of wilderness in mythology. Clearly, 'liminal symbolism' as described by van Gennep and Turner can be found in narrative form (Cohen 1996, Endsjø 2000, Schjødt 2008, Feldt 2011a, Feldt 2011b) and be tied to spatial categories, cf., indeed, the *limen* or threshold. With regard to religious narratives, fantastic-liminal events (miracles) may be understood as the *event*-form of what we find in *agent* form in monsters. Wilderness may, then, be a *spatial* version of liminality. While this may hold true for most religious wilderness constructions, in-depth analyses of the concrete construction of the materiality, climate, and history of a particular wilderness space, its specific uses and functions in the context in which it is found, are still necessary in order to move the discussion beyond what is general to all liminal spaces.

In this way, investigations of the wilderness and religion nexus may offer cause to question and nuance earlier cosmological grids. In classical studies of religious cosmologies, simpler cosmos/chaos, nature/culture, and wilderness/civilization dichotomies have dominated, and sometimes been abandoned or criticized for their inability to capture the diversity of human engagements with, and perceptions of, their surroundings in a cross-cultural perspective. New approaches and understandings are necessary. Further, connections between cosmology, religion and worldview have been assumed, which are now being questioned, as is the explanatory value of the concept of cosmology (Raudvere 2009). Similarly, it is now being acknowledged that spatial conceptualizations are rarely mutually exclusive and often coexist (Ingold 2000, 209–219). Cosmological concepts are often less coherent and structured than earlier approaches allowed for, and general developments in the study of religions point towards a greater focus on actors and contexts. This does, however, not mean that the analysis of spatiality should be abandoned, but it calls for critical re-evaluation of analytical categories in the study of religion.

5. The contributions of this book

In this volume of contributions from the analytical, historical, and anthropological study of religion, each article treats wilderness and religion in a particular case study, and contributes to the elucidation of 1) how the wilderness is constructed in the material examined, and which other domains, places or spaces it interacts with; 2) fruitful theoretical perspectives and strategies of analysis for the notion of 'wilderness' in the study of religion; 3) the religious uses the wilderness is put to, the practices that take place in the wilderness, their significance and meaning, the modes of travelling in and interacting with the wilderness; 4) the functions that are, or may be, ascribed to wilderness in the material or by informants, in emic and/or etic perspectives; 5) tracking the history of the wilderness and whether the construction, uses or functions of the wilderness change over time, and in which ways, whether functional equivalents to wilderness ideas arise, and what may explain any potential absences of wilderness ideas in specific religious contexts.

So while the articles, in the table of contents, have been organized historically and while they will be presented here according to the table of contents, the theoretical connections between the articles move across historical divides. All of the articles of this volume move at the intersections of the fields presented in the previous section, and yet a primary field can be pinpointed in most cases. Situated within the Religion and Nature field we find the articles of Bron Taylor, Anne-Christine Hornborg and Graham Harvey, while the contributions of Jan Bremmer, Jens-Peter Schjødt, and Marianne Fibiger move within the history of religions cosmology field. Thomas Hoffmann's article also dialogues with this field, while his literary and cognitive-semantic studies give it a special twist. Robert A. Segal's article traces parts of the history of wilderness in the study of religion by analyzing the use of the idea of wilderness in W. Robertson Smith's work. Ingvild S. Gilhus, Dag Øistein Endsjø, Morten Axel Pedersen and Laura Feldt engage the religious use of wilderness by means of varieties of spatiality theory.

In the opening article, Jan Bremmer investigates the construction of the mountain wilderness in ancient Greek mythology. He analyses the ways in which human interaction with the mountain wilderness was imagined, focusing on the most prominent mythological inhabitants of the mountains, the Centaurs. Bremmer traces the historical development in this mythology and offers suggestions about its functions and the reasons for the demise of centaur wilderness mythology. The analysis demonstrates how wilderness is never a fixed concept, but one that remains constantly subject to changes in cultural construction and perception. By focusing on the centaurs as mediatory actors Bremmer fruitfully combines the classical cosmological interest with a focus on how the interaction between human society and the wilderness is imagined. The article demonstrates how the Greeks, through the myths about these monstrous figures embodying wilderness, explored the limits of proper hospitality, proper treatment of women and proper drinking of wine. In the characteristics and actions of the Centaurs, and in the ways society is imagined as interacting with the Centaurs, the main aspects of the mountain wilderness are formulated: – they are wild, they are 'before', and they embody reversals. Yet, Bremmer's analysis shows how the Centaurs also possessed positive traits. By combining the negative qualities of the Centaurs with their education of the greatest Greek hero as well as with aspects of the birth of civilisation, the Greeks stressed both the positive and negative qualities of the mountainous wilderness. They recognized that interacting with the wilderness can be dangerous and destructive as well as beneficial and instructive and, moreover, that it can lead to cultural creativity.

Laura Feldt's article treats the idea of wilderness in The Hebrew Bible drawing on perspectives from critical spatiality theory, cultural memory studies, and narratology. Feldt argues that the wilderness is, in the Hebrew Bible, a crucial domain that plays an important role in and for Hebrew Bible religion. The wilderness functions as an arena for reflection on, and negotiation of, central issues and tensions in "Israelite" identity. Contrary to previous research, the article argues that the wilderness of the Hebrew Bible texts treated is presented as irreducibly ambiguous, retaining both positive and negative qualities and combining elements of realistic presentation of geographical space with fantastic-imaginary elements. This means that not only the wilderness, but also its primary spatial counterpart, the land, come to oscillate between positive and negative, realistic and imaginary presentations. Feldt argues that the wilderness of the Pentateuch/Torah texts analysed is characterised more than anything by the strange relation between Yahweh's fantastic marvels and the reiterated emphasis on Israel's rejection, doubt, and apostasy, and suggests that an ethos of religious identity transformation is formulated in the texts. Further, analyses of the subject matter of Yahweh's wilderness marvels show that the marvels communicate that Yahweh is the donor of the fertility of the agricultural land that is celebrated and longed for in the texts. It is therefore suggested that the lived space of the texts is the mixed culture of agriculture and pastoralism, not desert nomadism. Thus, the analysis renders problematic a long-dominant dichotomy in biblical studies between 'nature' and 'history', as well as the equation of 'wilderness' with 'nature'. The wilderness narratives communicate that the desired fertility of the agricultural land stems from Yahweh and that access to it, and to the agricultural land, is contingent on religious transformation. The ethos of religious identity transformation as formulated in these texts deterritorializes Israel's relation to the land while anchoring it solidly in agriculture.

Ingvild Sælid Gilhus' contribution discusses how "the mountain, a desert place" was conceived in the learned Gnostic treatise known as *The Secret Book of John (Ap.John)*, and what its uses and functions were. Drawing on critical spatiality theory, Gilhus examines the spatial categories of the text. The text refers to Timaeus, as well as Genesis, in its descriptions of a cosmological system and its mythological entities. It entails a complete apocalyptic scenario including what belongs to the world above and what belongs to the world below – the world of spiritual beings and the world of humans. Thus *Ap.John* offers a mythological map of the world, which is, like most mythological maps, ideological and strategic. The cosmological narrative has its only possible meeting point with an earthly landscape in the empty space of "the mountain, a desert place", and because this space is stripped, naked and devoid of all life, it is thus opened up for divine revelation. It is a superhuman place of transformation and revelation that has been almost completely taken over by the illuminated vision and the divine voice. The text is written on this space in a way that erases nearly all traces of real geography. In this way the empty space of the frame story becomes a receptacle for the detailed cosmology with all its spiritual beings, and the space of the desert is lifted out of its earthly context and made to reflect the world above. The article thus brings out an ambiguity in the concept of wilderness: The wilderness of *Ap.John* is attractive because it is perceived to be untouched by humans at the same time as it is touched by humans by being constructed as a space for divine revelation and mythology and progressively also as a habitat for the angel-like lives of ascetics.

With Dag Øistein Endsjø's article, we move deeper into the world of Christian asceticism in Late Antiquity. The belief that certain forms of knowledge, or even "truth" itself, may have a special relation to particular geographical areas connects to the most basic ideas about how space is never a neutral entity, but always imbued with various cultural values. Endsjø investigates the idea of wilderness in *Vita Antonii*, Athanasius of Alexandria's biography of Anthony, the first known monk who made the wilderness his home (written briefly after the hermit's death in 355). This material, from the time when Christianity had its breakthrough in the Greco-Roman world, presents itself as a good arena in which to examine how various ideas about knowledge and the wilderness interacted, and how the encounter between traditional Christian and Greek perspectives contributed to the formation of new ideas. Endsjø demonstrates how it was the difference of geography that made the desert fathers stand out from their numerous ascetic predecessors, and so it was, indeed, the wilderness as such that enhanced the entire monastic experience. The wilderness is the key for the understanding of *Vita Antonii*; this spatial dimension is what originally made Anthony what he was – a figure combining boundless wisdom with primordial folly in a way that made him an ultimate Christian role model. Endsjø's paper demonstrates how seminal wilderness ideas have been in the history of religions, and how consequential the idea of wilderness became for Christian thought and practice in Late Antiquity.

'Wilderness' has never been a canonical topic in the curriculum of *ilm al-qur'ān*, the traditional Islamic sciences of the Qur'ān. Neither has the topic 'wilderness' gained access to the standard Islamological encyclopaedias and dictionaries as a entry in own right (as compared to biblical studies). Yet, as Thomas Hoffmann's article shows, 'wilderness' does merit scholarly attention for Qur'ānic studies. Combining textual and literary analyses with observations on the Qur'ān's general ethos and rhetorical make-up, Hoffmann reaches a differentiated conclusion on the provenance, metaphorical conceptualization, use and non-use of wilderness. In cultural memory, the Arabs have been considered the people of 'wilderness' par excellence, both in terms of etymology (since the designation 'Arab' may be connected with various notions of natural wilderness) and in terms of Biblical mythology, namely the references to the Hagarene and Ishmaelite Arabs as 'children of the desert wastelands' and 'wild-asses' (Genesis 16). Moving from the murky etymological and mythological origins, Hoffmann moves to the profile of the Bedouin rider taking off into the desert as depicted in the stock travel section, known as the *rahîl*, of the pre-Islamic Qasidah-poem. From these beginnings, the article examines the traces of the Qur'ānic 'wilderness' (*'arâ'*) idea as a climatic-cum-topographical category associated with aridity, barrenness and general inhospitable surroundings, and addresses other ecological categories that can be associated with wilderness, such as the mysterious Lote-Tree of the Boundary (53:14), and the divinely devastated fields of dry stubble or the wild animals gathering on Judgment Day (Q 81:5). These forays lead to phenomenological and metaphorical categories involving human beings in their engagements with wilderness (and anti-wilderness), e.g. going astray, being naked or staying in paradisical gardens. The article concludes by discussing some strange silences on the wilderness, i.e., a discussion of the lack of some obvious wilderness categories in the Qur'ān.

Jens-Peter Schjødt's article investigates Old Norse mythological understandings of wilderness offering a history of religions cosmological perspective. Wilderness is understood as a term that characterises such domains as are opposed to the ordered world in which we live our everyday lives (cosmos, culture, society, 'this' world). Wilderness may thus be seen as one classificatory construction of a reality which is different from the one we control; it is a subcategory of the 'Other' that points to specific aspects of this 'uncontrollable' reality. Schjødt proposes to see wilderness as that spatial aspect of otherness which is particularly ambivalent in relation to human beings; wilderness is that particular otherworld, known from nearly all cultures, which is characterised by being at the same time – and to a higher degree than other 'otherworlds' – a threat and a potential. In pre-Christian Scandinavian mythology, it is characteristic that the wilderness is part of a cosmological structure which is orientated along both a horizontal and a vertical axis and that the 'threat aspect' is mostly organised along the horizontal axis, whereas the 'potential aspect' is organised along the vertical axis. In the 'human world', on the other hand, the potential aspect may also be acted out along the horizontal axis. This certainly indicates that wilderness is part of a cosmological structure, which is no doubt based on experiences. The cosmological structure is what enables people to cope with experience, by enabling its manipulation through myth as well as ritual.

Marianne Q. Fibiger's article offers two examples of analyses of the idea of wilderness from the Hindu tradition – one drawn from Hindu asceticism and another from goddess worship. Both are seen as key for an understanding of Hindu religion, because the function and use of ideas of wilderness offer a pregnant example of the paradoxicality inherent in Hinduism, and one for which simplistic dualistic analytical models fall short. Fibiger argues that wilderness should, in relation to religion, not always be seen in contradiction to the tamed or ordered world or as something to be avoided, because it brings disorder. It can also function as a necessary feature of the ordered world and be seen as essential for the upkeep of order. In Hindu religion, wilderness is something that has to be encompassed, embraced and lived through, not something that has to be shunned. It is shown how the two examples of asceticism and Goddess worship touch upon the same relation between wilderness and tameness, but with two important differences: 1) While the Goddess is related to the transcendental world but with obligations to both the transcendental world and the profane world (this can be called a top-top-down relation), the ascetic is related to the profane world and with obligations both related to the profane and the transcendental world (this can be called a down-down-top relation). 2) While the ascetic represents the wild or wilderness in relation to the man in the world, the Goddess encompasses both the wild and the tamed. The one and the same Goddess can both be described as wild and furious and as tamed and domesticated. While the ascetic is either or, the Goddess is both and. Therefore, in Hindu religion, the Goddess and the ascetic are, in a sense, mirror images with inverse profane/transcendental indicators, and wilderness is both idealized and sanctioned in ascetic practice as well as in Goddess worship, and assigned a necessary function for the maintenance of order both within the world and in the individual.

With Anne-Christine Hornborg's article we move into the Religion and Nature field. Hornborg's analysis focuses on the encounter between the European newcomers to North America, in the 15th to 17th century, and the Mi'kmaq people in Eastern Canada. We get here an account of the serious impact that strong, religious ideas of wilderness may have in a cultural encounter, and of how an early modern Christian version of the idea of wilderness was used to justify domination of others. The boundary drawn, by the newcomers, between culture and nature, and the use of the idea of wilderness, did not only function as a means of description, it also comprised a power relation and had devastating consequences for the inhabitants of "New France". Wilderness was understood as the condition offered by nature itself, if a human hand had not cultivated the land. The "Indian" was classified as a part of nature, as a savage, a child of the wilderness. Christianity and civilization were seen as the means of their liberation. The article traces, through the texts of the explorers and missionaries, the contours of the encounter between the differing perspectives on landscape, dwelling and personhood of the newcomers and of the Mi'kmaq, whose voices are also brought out in the texts. The article shows clearly that the wilderness idea was used, by the newcomers, as a means of justification in their attempts to force the nomadic Mi'kmaq to be settled and baptized, which, according to the rhetoric of the settlers, rendered them human and cultivated. Yet, the texts also bear witness to the concrete, bodily meetings between Europeans and Mi'kmaq hunters, and demonstrate that this involved a considerable amount of contestation of the European perspective, of negotiation, and, occasionally, it seems, of exchange between the European and the Mi'kmaq perspectives.

Robert A. Segal delves into a decisive section of the history of the use of the concept of wilderness in the study of religion. In his classic *Lectures on the Religion of the Semites* (1889), the great biblicist and Semiticist William Robertson Smith (1846–94) makes two sets of distinctions:

that between the wilderness and civilization and that between the physical and the spiritual, and thus lays out a ground for dichotomizing understandings of wilderness. According to Robertson Smith, on the one hand ancients and "primitives" pit the wilderness against civilization. The wilderness is the domain of demons, of evil, of wild beasts, and of physical danger. Civilization is the domain of gods, of good, of domesticated animals, and of safety. The conquest of an uninhabited spot means the driving out of the resident demons, who are displaced by gods. As decisive as the demarcation of the physical world between wilderness and home is for the ancient Israelites, once the division between material and spiritual starts to emerge, any distinctions within the physical world become pointless, according to Robertson Smith, and the wilderness, if it remains at all, is mental. It is the state of separation from God, the state of sin. Through the focus on the wilderness in his thought, Robert A. Segal's article brings out Roberson Smith's profound impact on the study of religion.

Engaging the distinction between nature and culture, the wild and the tame, Morten Axel Pedersen analyses the encounter between a Buddhist spatial framework and a shamanistic one in Northern Mongolia, and show how this encounter came to have important consequences for understandings of personhood among the Darhad people. Pedersen argues that the spatial contrast between steppe and taiga zones in the Darhad Depression of north-west Mongolia is perceived as an asymmetrical opposition between a homogenous centre (the steppe zone) and a heterogenous margin (the taiga zone). This contrast is replicated in contemporary constructions of Darhad personhood, so that Darhads see themselves to consist of two sides, a Buddhist 'yellow side' and a shamanic 'black side', and these two essential components of the person are homologous to the topographical contrast in question. Unquestionably, the Darhad notion of the 'yellow side' is related to the Mongolian Buddhist domestication interventions made on the Darhads and their land during the 18th and 19th centuries. Yet, according to local narratives collected by Pedersen during his fieldwork, the Mongolian Buddhist church did not bring anything new to the Darhads, but rather brought out a yellowness latent within them – the Darhads were already tame from within. The article brings out the intricate connections between physical landscape, religio-spatial paradigms and personhood that resulted from the encounter between the Buddhist newcomers and the local shamanistic Darhad people. Pedersen's article discusses the continuing relevance of binary distinctions between nature and culture, wild and tame; neither as intrinsic properties nor as arbitrary cultural constructions, but as a relational phenomenon that may be actualized through material forms and social practices.

Graham Harvey investigates the religious use of two types of wilderness among contemporary eco-Pagan activists and environmental educators. His contribution shows how two contrasting notions of what constitutes wilderness are important in understanding the practices and ideas of British Eco-Pagan environmental educators / shamanic practitioners. Harvey demonstrates that the Eco-Pagan environmental educators / shamanic practitioners find employment and direction in the world because their relational participation with the larger-than-human world inspires and requires them to confront the threat posed by a (negative) kind of wilderness: anthropogenic monoculture and its ecocidal consequences. In turn, the threat of monoculture wilderness encourages such people to attempt to disseminate the urgent need for human conversation with the larger-than-human world. Learning the etiquette and protocols for human re-entry into such inter-species conversations (verbal and non-verbal) is perceived to be eased by time spent ritualising in the other kind of wilderness: the less human dominated bio-diverse community of "wild(er)" places. Thus, the friction between these two understandings and experiences of wilderness is what explains the environmentalist, shamanistic, educative and ritualistic ideas and acts of the Eco-Pagans. The article shows the ongoing dynamic tension between these two value-laden uses of the word "wilderness", and demonstrates how these two wildernesses cannot be collapsed into one another. It is the presence of both, and the contrast between them, that inspires and seems to require the kind of things that the Eco-Pagans do and think. Harvey's article thus contributes to the critical debates about alternatives to the modernist separation interposed between "culture" and "nature".

Understanding recent environmental history requires an exploration of competing understandings of wilderness, and disputes over places said to be wilderness. Bron Taylor's article demonstrates how deeply wilderness, and wilderness spirituality, has been rooted in U.S. culture historically, and how it continues to be so. Taylor traces the history of the nexus of wilderness and religion in North America from European contact to Earth Day and speculates on future developments in the relations between religion and wilderness areas. While the establishment of National Parks and wilderness reserves in the United States was not the first time that human beings have set biological systems aside by labelling them sacred, and requiring special and reverent behaviour of humans when present, Taylor shows that in the US, there has, however, been a strong religious dimension to the wilderness and biodiversity preservation movements promoting the establishment and preservation of such reserves. These efforts first bore fruit in the United States, on a large scale. The article traces developments in the rationale for preserving large scale habitats, from being first and foremost for the spiritual benefit and aesthetic pleasures of the human species, to preserving wilderness areas, both terrestrial and marine, because these are the habitats upon which biodiversity, and even the evolutionary process itself, depends. As for future developments in the relations between religion and wilderness areas, we do not at present know whether and to what extent the world's predominant religions *are* greening, or what impact unconventional nature religions will have, including with regard to wilderness preservation, but as Taylor points out, it is important to notice that even as the rationales for wilderness preservation turn more scientific, and even when the arguments insist that saving all parts of nature is the essence of prudence, these arguments are usually infused with religious, or at least, religion-resembling sentiment.

6. Wilderness and Religion at the Crossroads

This book throws comparative light on the complexities of the religion and wilderness interface. As the analyses demonstrate, religious wilderness mythologies and the relations between specific religions and real and imagined wildernesses constitute an important field of research for the academic study of religion. In contemporary parlance, the wilderness is a domain void of humans, opposed to civilization, but the analyses presented here complicate and question this conventional dualism and its implications or corollaries. In the world's religions, wilderness is rarely a domain void of humans, although it may be a domain that is inhospitable or difficult of access for humans. Nor can wilderness be aligned uproblematically with 'nature' or 'chaos', just as previous dichotomizing approaches cannot be used uncritically. Instead, the wilderness appears as an ambiguous boundary region in between the domain of 'home' and the various otherworlds (e. g., heaven, hell, the underworld, 'chaos', etc.) of religious imaginations, while simultaneously being connected to concrete, geographical landscapes, materialities, and climates. Of importance is, rather, how wilderness ideas may influence understandings of religious identity, personhood, religious practice, and spatial practice, as well as the experience of, and interaction with, landscape and environment.

Investigations of the relations between religions and imagined and real natural spaces are vital today for enhancing our understanding of how we as humans situate ourselves in an environment. They also enhance our understanding of human engagement in and interaction with the nonhuman world – especially in relation to those few habitats and species that are still seen as 'wild'. This book offers important pointers to how powerful and seminal the concept of wilderness is and how intimate its ties to religion are. Its analyses and discussions add nuance and complexity to our understanding of the varieties of the interface of religion and environment. In the era of climate change and ecological concern, a deeper, comparative understanding of the role of religion in human interaction with the 'world' is as crucial, and as demanding, an issue as ever.

7. References

Alexander, Bobby C. *Victor Turner Revisited. Ritual as Social Change.* An American Academy of Religion Book, 1991.

Auffarth, Christoph. "Kosmologie". In H.D. Betz, D.S. Browning, B. Janowski, E. Jüngel, eds., *Religion in Geschichte und Gegenwart.* Tübingen: Mohr Siebeck 2001, 1706–1707.

Beck, Ulrich. "Climate for Change, or How to Create a Green Modernity". In *Theory, Culture, Society* 27 (2010), 2–3, 254–266.

Bingham, Nick and Thrift, Nigel. "Some New Instructions for Travellers. The geography of Bruno Latour and Michel Serres". In M. Crang and N. Thrift, *Thinking Space.* Routledge, London, 2004, 281–301.

Bolle, Kees W. "Cosmology". In M. Eliade, *The Encyclopedia of Religion.* New York: Macmillan 1987, 100–107.

Brown, Gavin. "Theorizing Ritual As Performance. Explorations of Ritual Indeterminacy". In *Journal of Ritual Studies* (2003) 17,1,3–18.

Buell, Lawrence. *The Environmental Imagination: Thoreau, Nature Writing and the Formation of American Culture.* Cambridge, MA: The Belknap Press of Harvard University Press, 1996.

Burton, Lloyd. *Worship and Wilderness. Culture, Religion and Law in Public Lands Management.* University of Wisconsin Press, 2002.

Casey, Edward S. *Getting Back into Place: Toward a Renewed Understanding of the Place-World.* Studies in Continental Thought. Bloomington, Ind.: Indiana University Press, 1993.

Casey, Edward S. *The Fate of Place: A Philosophical History.* Berkeley: University of California Press, 1997.

Certeau, Michel de. *The Practice of Everyday Life.* University of California Press: Berkeley, 1984.

Connor, Steven. "Topologies: Michel Serres and the Shapes of Thought". In *Anglistik* (2004) 15, 105–117.

Connor, Steven. "Michel Serres's Milieux". An extended version of a paper given at the ABRALIC (Brazilian Association for Comparative Literature) conference on "Mediations", Belo Horizonte, July 23–26, 2002.

Corrigan, John. "Spatiality and religion". In Warf, B. and Arias, S. 2009,157–172.

Coupe, Laurence. (ed.). *The Green Studies Reader: From Romanticism to Ecocriticism.* Padstow: Routledge, 2000.

Crang, Mike and Thrift, Nigel. (eds.) *Thinking Space.* Bury St Edmunds: Routledge, 2004.

Cresswell, Tim. *In Place, Out of Place. Geography, Ideology and Transgression.* Minneapolis: University of Minnesota Press, 1996.

Cronon, William. *Uncommon Ground. Toward Reinventing Nature.* New York and London: W.W. Norton & Company, 1995.

Deane-Drummond, Celia. *Eco-Theology.* Darton, Longman and Todd, London, 2004.

Duncan, James; Gregory, Derek eds. *Writes of Passage: reading travel writing.* London: Routledge, 1999.

Eaton, Heather. *Introducing Ecofeminist Theologies. Introductions in Feminist Theology,* 12; London and New York: T. & T. Clark Int., 2005.

Endsjø, Dag Øistein. *Primordial Landscapes, Incorruptible Bodies: desert asceticism and the Christian appropriation of Greek ideas on geography, bodies, and immortality.* New York: Peter Lang, 2008.

Endsjø, Dag Øistein. "To Lock Up Eleusis: a question of liminal space". In *Numen* 47 (2000), 351–386.

Feldt, Laura. *The Fantastic in Religious Narrative from Exodus to Elisha.* London: Equinox, 2012.

Feldt, Laura. "Religious Narrative and the Literary Fantastic". In *Religion* (2011) 41,2, 251–283.

Feldt, Laura. "Monstrøsitet som kulturel og religiøs diskurs". In *Religionsvidenskabeligt Tidsskrift* (2003) 42, 43–64.

Foucault, Michel. "Des Espaces Autres," (1967), published in English as "Of Other Spaces," (trans. J. Miskowiec) *Diacritics 16* (1986): 22–27.

Foucault, Michel. *Power/Knowledge: Selected Interviews and Other Writings 1972–1977.* Ed. C. Gordon; trans. C. Gordon et al; Brighton: Harvester Press, 1980, 63–77.

Foucault, Michel. "Other Spaces: the Principles of Heterotopia". *Lotus 48/49* (1986) 9–17.

Frazer, James G. *The Golden Bough. A study in magic and religion.* New York: St Martins Press, 1990 [1911].

Garrard, Greg. *Ecocriticism.* London: Routledge, 2004.

Geertz, Armin W. "Definition as Analytical Strategy in the Study of Religion". In *Historical Reflections* (1999) 25,3, 445–475.

Gersdorf, Catrin and Sylvia Mayer, eds. *Nature in Literary and Cultural Studies. Transatlantic conversations on ecocriticism.* Amsterdam and New York: Rodopi, 2006.

Gilhus, Ingvild S. *Animals, Gods and Humans: Changing Attitudes to Animals in Greek, Roman and Early Christian Ideas.* London: Routledge, 2007.

Goehring, James E. "The Dark Side of landscape: Ideology and Power in the Christian Myth of the Desert", *Journal of Medieval and Early Modern Studies* (2003) 33, 3, 437–451.
Gottlieb, Roger S., ed. *The Oxford Handbook to Religion and Ecology.* Oxford: Oxford University Press, 2006.
Graulund, Rune. "Desert Travel Writing as Indicator Species". In *Studies in Travel Writing* (2006) 10,2, 141–159.
Haraway, Donna. *When Species Meet.* Minneapolis: University of Minnesota Press, 2008.
Heise, Ursula. *Sense of Place and Sense of Planet: The Environmental Imagination of the Global.* Oxford: Oxford University Press, 2009.
Hiebert, Theodore. *The Yahwist's Landscape. Nature and Religion in Early Israel.* Oxford: Oxford University Press, 1996.
Ingold, Tim. "Hunting and Gathering as Ways of Perceiving the Environment", in *Redefining Nature: Ecology, Culture and Domestication,* Roy Ellen and Katsuyoshi Fukui (ed.). Oxford: Berg 1996, pp. 117–155.
Ingold, Tim. *The Perception of the Environment. Essays in livelihood, dwelling and skill.* London: Routledge 2000.
Jasanoff, Sheila. "A New Climate for Society". In *Theory, Culture, Society* 27 (2010) 2–3, 233–253.
Jasper, David. *The Sacred Desert.* Malden, Mass.: Blackwell 2004.
Jenkins, Willis. "Religion and Ecology: A Review Essay on the Field." In *Journal of the American Academy of Religion* (2009), 77, 1, 187–197.
Jensen, Jeppe Sinding. *The Study of Religion in a New Key.* Aarhus: Aarhus University Press 2003.
Journal of Ecocriticism – http://ojs.unbc.ca/index.php/joe
Journal for the Study of Religion, Nature and Culture – http://www.religionandnature.com/journal/
Kalland, Arne. "The Religious Environmentalist Paradigm". In Taylor, ed., *The Encyclopedia of Religion and Nature.* London: Continuum, 2005, s.v.
Kearns, Laurel. and Keller, Catherine eds. *Ecospirit: Religions and Philosophies for the Earth.* New York: Fordham University Press 2007.
Knott, Kim. *The Location of Religion. A spatial analysis.* London: Equinox, 2005.
Landry, Donna. *The Invention of the Countryside: Hunting, Walking, and Ecology in English Literature, 1671–1831.* Basingstoke: Palgrave, 2001.
Latour, Bruno. *The Politics of Nature.* Cambridge, Mass.: Harvard University Press 2004.
Lefebvre, Henri. *The Production of Space.* Oxford: Basil Blackwell. [Translated from *La production de l'espace*, 1974] 1991.
Lied, Liv I. *The Other Lands of Israel. Imaginations of the Land in 2 Baruch.* Leiden: Brill, 2008.
Lincoln, Bruce. *Theorizing Myth.* Chicago: University of Chicago Press, 1999.
Lincoln, Bruce. "Til ære for Chaos". In *Chaos* (2008) 49, 9–28.
Massey, David. *Space, Place and Gender.* Cambridge: Polity Press, 1994.
Morton, Timothy. *Ecology Without Nature. Rethinking Environmental Aesthetics.* Cambridge, MA: Harvard University Press 2007.

Oehlschlager, Max, ed. *The Wilderness Condition. Essays on Environment and Civilisation.* Washington D.C.: Island Press, 1992.

Oelchlager, Max. *The Idea of Wilderness. From Prehistory to the Age of Ecology.* New Haven: Yale University Press, 1991.

Oelschlager, Max. "Wilderness Religion". In Taylor, ed., *Encyclopedia of Religion and Nature,* London: Continuum, 2005, s.v.

Pedersen, Poul. 1995. "Nature, Religion and Cultural Identity: The Religious Environmentalist Paradigm". In *Asian Perceptions of Nature: a critical approach,* edited by O. Bruun and A. Kalland: London: Curzon Press, 258–273.

Petersen, Anders K.P. and Schjødt, Jens-Peter. "Balkanisering og rekonstruktion af kulturer". In *Religionsvidenskabeligt Tidsskrift* (2007) 50, 7–9.

Pomykala, Kenneth E. (ed.), *Israel in the Wilderness. Interpretations of the Biblical Narrative in Jewish and Christian Traditions,* Leiden: Brill, 2008.

Rappaport, Roy. *Ecology, Meaning and Religion.* North Atlantic Books 1979.

Rappaport, Roy. *Ritual and Religion in the Making of Humanity.* Cambridge University Press 2008 [1999].

Raudvere, Catharina. "The Part or the Whole: Cosmology as an Empirical and Analytical Concept". In *Temenos* (2009), no. 45, 7–33.

Ruether, Rosemary. and Hessel, Dieter. eds. *Christianity and Ecology: Seeking the Well-Being of Earth and Humans.* Cambridge, MA: Harvard University Press 2000.

Schama, Simon. *Landscape and Memory.* London: Fontana 1996.

Schjødt, Jens-Peter. *Initiation between Two Worlds. Structure and Symbolism in Pre-Christian Scandinavian Religion. The Viking Collection. Studies in Northern Civilisation.* Odense: The University Press of Southern Denmark. [Translated and revised edition of J.-P. Schjødt, *Initiation, liminalitet og tilegnelse af numinøs viden. En undersøgelse af struktur og symbolik i førkristen nordisk religion.* Doktordisputats, Aarhus 2004] 2008.

Segal, Robert A. Introduction to reprint of the second edition of W. Robertson Smith, *Lectures on the Religion of the Semites,* retitled *Religion of the Semites.* New Brunswick, NJ: Transaction, 2002, pp. vii–xlii.

Serres, Michel. *Le contrat naturel.* Éditions F. Bourin, Paris 1990. In English: *The Natural Contract.* Trsl. by E. MacArthur and W. Paulson. Ann Arbor: The University of Michigan Press, 1995.

Smith, Jonathan Z. *To Take Place. Toward Theory in Ritual.* Chicago: University of Chicago Press, 1987.

Soja, Edward W. *Postmodern Geographies.* New York: Verso 1989.

Stausberg, Michael. "The study of religion(s) in Western Europe III: Further developments after World War II". In *Religion* (2009), no. 39,3, 261–282.

Talmon, Shemaryahu. "The Desert Motif in the Bible and in Qumran Literature". In *Biblical Motifs, Origins and Transformations.* A. Altmann, ed. Cambridge, MA.: Harvard University Press, 1966, 31–63.

Taylor, Bron. *Dark Green Religion: Nature Spirituality and the Planetary Future.* Berkeley: University of California Press, 2010.

Taylor, Bron. *The Encyclopedia of Religion and Nature.* London: Continuum, s.v., 2005.

Taylor, Bron. "Religious Studies and Environmental Concern". In *The Encyclopedia of Religion and Nature.* London: Continuum, s.v. 2005.

Thrift, Nigel. and Whatmore, Susan, eds. *Cultural Geography. Critical Concepts in the Social Sciences.* London: Routledge, 2004.

Tucker, Mary E. and Grim, John. "Series foreword". In M.E. Tucker and D.R. Williams, eds. *Buddhism and Ecology. The Interconnection of Dharma and Deeds.* Cambridge, MA:: Harvard University Press 1997.

Turner, Victor. "Liminality and Communitas" in *The Ritual Process: Structure and Anti-Structure.* Ithaca: Cornell University Press, 1979, 94–130.

Turner, Victor. "Variations on a Theme of Liminality". In S.F. Moore and B.G. Myerhoff, eds. *Secular Ritual.* Assen: Van Gorcum, 1977, 36–52.

Turner, Victor. *Dramas, Fields and Metaphors. Symbolic Action in Human Society.* London: Itacha, 1974.

van Gennep, Arnhold. *Les rites de passage.* Paris: Émile Nourry 1909.

Veyne, Paul. *Les grecs ont-ils cru à leurs mythes?* Paris: Seuil 1983.

Warf, Barney and Arias, Santa. *The Spatial Turn. Interdisciplinary perspectives.* London: Routledge 2009.

White, Lynn, Jr. "The Historical Roots of Our Ecologic Crisis". In *Science* (1967) 155, p. 1203–7.

2. Greek Demons of the Wilderness: the case of the Centaurs

Jan N. Bremmer

1. Introduction

Thinking of wilderness, I inevitably came to recall the books I read in my youth about brave trappers and cruel Red Indians, as they were still called in those days. It was typical of that wilderness that it was in the process of being civilised, as the white settlers built their log huts and gradually developed the land, even though being frequently attacked by the Indians. In short, the wilderness has a strong American connotation for me. Yet the idea of a wilderness, of an area outside, but sometimes close to, civilisation is hardly limited to the USA. On the contrary, we find it already in Archaic Greece, where we have a very typical type of landscape that both in reality and in myth can be seen as wilderness: the mountains. This becomes already apparent in one of the earliest Greek literary texts, Homer's *Odyssey*, where the poet characterises the one-eyed Cyclopes as follows: 'they have neither assemblies for council nor appointed laws, but they inhabit the peaks of high mountains in hollow caves' (9.112–14);[1] in later times the Greek word for mountain, *oros*, was even used to denote the Egyptian desert (Cadell and Rémondon 1967). Now Richard Buxton has very well analysed the mountains and their 'wilderness aspects', both in reality and in the Greek mythological imagination.[2] In his studies he isolated three main aspects of mythical mountains:

1. the mountains are outside and wild,
2. they are 'before',
3. they are places of reversals.

1 For the Greek views of mountains, see Hyde 1915, 1915–1916; Jameson 1989; Meissner 1996. There are many other interesting studies on aspects of Greek mountains in Olshausen and Sonnabend 1996.

2 Buxton 1990, 1992, 1994, 81–96.

Buxton paid hardly any attention, though, to the most prominent mythological inhabitants of the mountains, the Centaurs. That is why I have chosen to look at those beings, half horse, half human or, put differently, humans, usually males and only much later females, first with the barrel and the hindquarters of the horse joined to the buttocks of a full-length man, later males with an equine lower body, who lived on the mountains.[3] By focussing on these mythological mountainous beings, we will inquire to what extent they corroborate Buxton's conclusions or may perhaps somewhat nuance them.

2. The Centaurs: general observations

The Centaurs have regularly been studied, but modern research started only in the 1870 s. It was a time in which nature was seen as *the* key to unlock Greek mythology (Bremmer 2011). Thus the great Wilhelm Mannhardt (1831–1880) admirably collected the then available evidence, made important observations regarding the antiquity of the various motifs, persuasively concluded that the Centaurs went back to a period before Homer, but interpreted the Centaurs as the personifications of tornadoes.[4] Around the same time, the meritorious, if often misguided, Wilhelm Roscher (1845–1923) interpreted them as the personifications of cascading mountain brooks (1890–92, 1058–59). Moreover, Mannhardt had also been impressed by the popular construction of later nineteenth- and earlier twentieth-century etymologists of a connection between the Centaurs and the Indian Gandharvas, which was also taken up by Georges Dumézil (1898–1986) in 'ein luftiges Hypothesengebäude',[5] although 'the names do not correspond by the rules of phonology, and the creatures have virtually nothing in common mythologically' (West 2007, 285 n. 14).

Around 1900, Martin P. Nilsson (1874–1967), a Swedish farmer's son, consolidated the turn away from myth to ritual in the study of Greek religion. He also canonised the agrarian interpretation of ritual in his *Handbuch der griechischen Religion*, the authoritative handbook

---

3 Robert 1920, 426 (still valuable for the literary evidence); more recently, Schiffler 1976; Gantz 1993, 14347; Leventopoulou 1997; Linant de Bellefonds 2009.
4 Mannhardt 1876–77, 2.39–102.
5 Dumézil 1929; note also Carnoy 1936; Nilsson 1967, 231.

of Greek religion in the middle of the twentieth century.[6] Nilsson did not make any progress beyond Mannhardt, although he still paid the necessary attention to the Centaurs. Walter Burkert, on the other hand, in the authoritative handbook of Greek religion of our own times, hardly gives them any consideration.[7] Unlike Nilsson, Burkert concentrates on the sociological and psychological aspects of Greek religion and, in general, is much less interested in its agricultural side. Yet the Centaurs constitute an important part of early Greek mythology, connected as they are with the great mythological heroes Achilles and Heracles. Moreover, as we will see, they proved to be 'good to think with'. As they have been neglected for a long time, there is thus room for a new discussion.

Yet before entering into a discussion of specific episodes, let us first make some general observations on the nature of the Centaurs. In the earliest Greek literature there are already a number of pointers to their wild nature and to their lack of those qualities that characterise civilised mortals. This is immediately clear from the place where they live, as Homer calls them *Phêrsin oreskôioisi*, 'mountain-bred animals'.[8] The place of habitation also reflected itself in their names: in pseudo-Hesiod's *Shield* (186) we find a Centaur named Oureios, 'Mountainous', and on the famous François Vase of about 570 BC one of the Centaurs is called Orosbios, 'Living in the mountains' (Wachter 1991, 99 no. 64).

In Greece, the landscape can easily be divided into the fertile plains and the wild mountains. In a structuralist manner we can see the plain as the civilised area of the *polis* that was in opposition to the sea and the mountains.[9] A Greek curse states 'to the mountain or into the sea' in order to indicate the two places from where no return was possible (Versnel 1978, 41–42), as is also clear from the myth of Oedipus who was exposed either in the mountains or, according to the alternative version, in a basket in the sea (Bremmer 1988, 43–44). Yet mythology, as

6 On Nilsson, see Briggs and Calder 1990, 335–40 (J. Mejer); Calder 1994, 151–78 ('Instinct against Proof: The Correspondence between Ulrich von Wilamowitz-Moellendorff and Martin P. Nilsson on *Religionsgeschichte*').

7 Nilsson 1967, 229–32; Burkert 1985, 173, 209, 232 = Burkert 2011, 267, 320, 353. On Burkert (1931), see most recently Barbu 2007; Bierl and Braungart 2010.

8 Homer, *Il.* 1.267–8, cf. Bacchylides F 66 Maehler: ὀρικοίτας Κένταυρ[ος]; Euripides, *Her.* 364–5: τάν τ' ὀρεινόμον ἀγρίων Κενταύρων ποτὲ γένναν, *IA* 1047: Κενταύρων ἐν ὄρεσι.

9 The plain/mountain contrast is also noted by the Greeks themselves: Theophrast, *Hist. plant.* 1.8.1, 3.11.2; Polybius 18.31; Arrian, *Anab.* 7.9.2.

we have gradually come to realise, not only reflects, but also refracts, clarifies and simplifies reality. That is also clear in the opposition plain/mountain. In reality, there was often an intermediate area of marginal, waste or scrub land on the slopes of the mountains and near the sea, which was exploited in times of economic expansion and abandoned in times of economic decline. This land was called *eschatia.*[10] However, we do not find this marginal land in mythology, which often likes to keep things clear and simple: just plains and mountains.

Furthermore, whereas *phêr* normally means 'wild animal', in this case it looks like an archaic expression that is becoming extinct: apparently, Homer no longer understood the phrase and interpreted *Phêres* as a name for the Centaurs, which suggests an origin for the Centaurs in the Dark Ages.[11] This is all the more likely since the discovery in Euboean Lefkandi in 1969 of a large terracotta Centaur, of the second half of the tenth century BC, and the occurrence of a Centaur on the bottom of a Cypriot plate of the late tenth century as well as their appearance in Attica and Olympia on vases and statuettes in the eighth century. By the mid-seventh century they were even known in West Greece.[12] This widespread occurrence supports the ancient character of the Centaurs, who seem to have been imagined as living in Greece at more places than just at the canonical Mt Pelion (below), witness 'the grave of Nessos and the other Centaurs' on Aetolian Mt Taphiassos (Strabo 9.4.8).

The 'animal type' character of the Centaurs is also expressed by their early epithets *lachnêeis*, 'hairy', *melanchaitês*, 'black-haired', and *lasiauchên*, 'with shaggy neck', an epithet that elsewhere is applied to bears and, not surprisingly, to horses.[13] Yet they are not only hairy like animals, but their diet also runs contrary to that of humans and resembles that of the more dangerous beasts of prey, as they are *ômophagos*, 'raw eating', like lions, wolves, jackals and fish as well as, and not to be forgotten, the Sphinx, itself called an *oureion teras*, 'mountain prodigy', by Euripi-

10 Robert 1969b, 820–21; Hanson 1995, 79–85; Lewis 1997, 291–93; Jameson 2002.

11 Hoekstra 1965, 152–3; Latacz 2000, 108. Hesiod F 209.5 Merkelbach/West already updated the expression by writing *Kentauroisin oreskôioisi*, 'mountainous Centaurs'.

12 For the early, post-Mycenaean Centaurs, see Padgett 2003, 5–14 (with wonderful photos); slightly different, Langdon 2008, 95–98.

13 *Lachnêeis: Iliad* II.743. *Melanchaitês:* Hesiod, *Sc.* 186; Sophocles, *Tr.* 837. *Lasiauchên: Homeric Hymn to Hermes* 224, cf. *Homeric Hymn to Dionysos* 46 (bear); Sophocles, *Ant.* 350 (horse).

des.[14] In fact, an Archaic epic poet succinctly states: 'the Centaurs have no *nous*, 'mind, reason',[15] even though Pindar (*P.* 3.5) says of Cheiron, the most civilised Centaur (below), that he had a '*noos* friendly to men'. It will not be surprising, then, that although many Greeks carried theophoric or herophoric names, there were very few people with the element 'Centaur' in their name.[16] Clearly, Centaurs were not beings a civilised Greek wanted to be associated with.

The Centaurs, then, not only lived in the wilderness, but were also wild themselves, transgressing the norms of civilisation. In the pseudo-Hesiodic, but early, poem *Kaminos* (F 302 Merkelbach/West) they even appear as potential wreckers of human artefacts; this may well be the reason that Centaur names recur as the names of dogs already in the Archaic Age.[17] Although interest in the Centaurs lasted in literature until late antiquity – Centaurs are still mentioned in the ancient Greek and Roman novel, and a Centaur appears in Jerome's biography of Paul, the (probably fictitious) very first Christian hermit[18] – I will limit myself to the older traditions, mainly those attested before the end of the fifth century BC, as from the fourth century onwards Centaurs increasingly figured in contexts in which they did not play a role before (Morawietz 2001, 13); in fact, even female and baby Centaurs may now have appeared on the scene.[19]

Admittedly, references to the Centaurs are rather few and far between in older Greek literature, which shows that their myths were well known and did not need to be spelled out by the poets. Longer, more detailed narratives, often going back to much older sources, can be found only in later mythographers, such as Diodorus Siculus, Hyginus and Apollodorus, all dating from around the turn of the Christian era or well after. These authors operated in a world where knowledge of mythology had become part of people's cultural capital, and thus they fulfilled a social need with their handbooks (Bremmer 2011, 528–9). From them we can isolate five important scenes: after the ancestor of the Centaurs (§ 3)

14 Homer, *Il.* 5.782, 7.256 (lions), 11.479 (jackals), 16.157 (wolves); Ibycus F 40.4 Davies (fish); Segal 1974. Sphinx: Aeschylus, *Sept.* 541; Euripides, *Phoen.* 806.
15 Pisander F 9 Davies = Bernabé, cf. Kassel and Austin on Teleclides F 49.
16 As noted by Robert 1969a, 650–52, who also collects the few examples.
17 Wachter 1991, 97 and 2001, 290.
18 Novel: Apuleius, *Met.* 4.8.22; Lollianus, fr. B 1 verso, 14–15 Henrichs. Jerome: Miller 1996; Wisniewski 2000; Rebenich 2009.
19 But there is good reason to doubt the authenticity of Lucian's description of their occurrence in a painting by Zeuxis, cf. Pretzler 2009.

we will first look at the Centaur Cheiron, the educator of Achilles (§ 4), then the battle between Centaurs and Lapiths (§ 5) and the attempted rape by the Centaur Nessos of the wife of Heracles (§ 6) and, last but not least, Heracles' visit to the Centaur Pholos and the subsequent battle against the Centaurs, which ended in the extinction of the Centaurs (§ 7), an element of their myth that has not received the attention it deserves. Together these five scenes, which we will try to reconstruct from the older sources, will throw light on the nature of the Centaurs and will enable us to sketch a nuanced picture of their embodiment of wilderness.

3. The Birth of the Centaurs

Our oldest source, Pindar, relates how Ixion, the prototypical Greek trickster, tried to rape Hera, but instead he lay with Nephele, a cloud in the shape of 'the foremost heavenly goddess, Kronos' daughter'. As a result, 'that unique mother' bore 'a unique son, who was overbearing and respected neither among men nor in the ways of the gods', who was called Kentauros by his nurse.[20] 'He (Kentauros) mated with Magnesian mares in the foothills of Pelion. And from them issued a wondrous herd of offspring, similar to both parents, with the mother's features below, the father's above', thus reflecting the iconographical development of the sixth century.[21] Martin West has compared the Greek myth with that of the Indian one about the immortal Saranyu, who became pregnant of the Asvins with the mortal Vivasvat and then assumed the shape of a mare, but also ran away leaving her husband with a facsimile of herself. One is also reminded of Poseidon turning himself into a stallion when Demeter tried to escape from his advances in the shape of a mare, thus begetting the very first horse, Arion.[22] We dimly seem to see here the remnants of Indo-European myths connected with the

20 Although in later times other Centaurs also seem to have been considered their offspring, cf. the Hellenistic poem published by Huys 1991, 41 (on line 6); Hyginus, *Fab.* 33.1.

21 Pindar, *P.* 2.33–48, tr. W.H. Race, overlooked by West 2007, 192–3, who compares Scholion (henceforth: Schol.) on *P.Oxy.* 421 and D on Hom. *Il.* 1.263; Nonnos, *D.* 7.125; Eustathius on *Il.* 1.268; add also Aeschylus F *89 Radt. For a possible iconographical identification of Kentauros, see Steinhardt 2009, 182–84.

22 Burkert 1979, 127–28; Doniger O'Flaherty 1980, 166–212 (interesting parallels, improbable interpretations); West 2007, 192 f.

birth of the first horse and other equine beings, which in itself should not surprise us, given the huge importance of the horse for the expansion of the Indo-Europeans. In any case, it is clear that from such an unnatural birth, based on rape, no normal civilised offspring could be expected. And that is of course what we will see.

4. Cheiron

Of all the Centaurs known to us, by far the most famous was Cheiron.[23] As his name shows, there are a few named individuals among the Centaurs as an, in general, anonymous group. This is typical of Greek mythology (Brelich 1956, 325–6), as can also be seen in another group of wild men, the Satyrs and Silenoi, who, later, were often associated with the Centaurs (Appendix). In antiquity, Cheiron's name was already etymologized as 'handy', from Greek *cheir*, 'hand' (Isid. *Orig.* 4.9.12), which is not immediately unconvincing considering his handiness in medicine (*Il.* XI.832) or as *cheirôn*, 'lesser, worse' (*Et. Magnum* 810.37), which is less persuasive. Yet linguistic arguments speak against the connection with 'hand', just as the name 'Centaur' has not been properly explained yet, and these obscure non-Greek etymologies may well be additional arguments for the great antiquity of the Centaurs.[24] Cheiron was the son of Kronos and Philyra, the daughter of Okeanos, although later times confused his father with the just mentioned Ixion (Souidas *FGrH* 602 F 1). Kronos had become a stallion to sire his hybrid son, a myth that was quietly omitted by Homer, who was generally above such unlikely tales.[25] This particular version of the myth cannot be too old, as Kronos entered Greek mythology and religion fairly late (Bremmer 2008, 82), but the motif of the metamorphosis of a god into a horse can also be found in Indo-European mythology as we just saw (§ 3), and probably derives from there. It is clear that the myth had to explain the equine shape of Cheiron, but it also positioned him close to Kronos, the

23 Gisler-Huwiler 1986; Guillaume-Coirier 1995; Aston 2006.

24 Cheiron: Wachter 2001, 263–4. Centaurs: De Angelis 2009 (not really persuasive).

25 *Titanomachia* F 9 Davies = 10 Bernabé; Hesiod, *Th.* 1002; Pindar, *P.* 3.1–3, 4.115, 6.22, 9.30; Pherecydes *FGrH* 3 F 50 = F 50 Fowler; *TrGF* Adesp. *734b. For Homer's minimising of the supernatural, see Griffin 1977, even though he may overstate his case, cf. the sensible comments of Johnston 1999, 32–35.

older divine ruler, who had been defeated by Zeus. As Okeanos also belonged to the oldest generation of gods (Bremmer 2008, 2–4), the message of this genealogy is clear: Cheiron basically belonged to a generation of gods that had to be removed from power before Zeus could institute the present world and the rule of the Olympian gods.[26]

Cheiron lived in Magnesia on Mt Pelion in a cave, the so-called Cheironion,[27] close to the sanctuary of Zeus Akraios,[28] the most important god of the Magnesians, where he even appears on local coins.[29] In the Greek imagination, caves were typical of wild or aboriginal beings. People lived in caves before houses, as did monstrous figures like the Cyclopes (Introduction above), the snake Echidna or the terrible Harpies.[30] Cheiron, then, not only lived on a mountain, but, even there, he did not have a civilised place in which to live.

In the *Iliad* (11.832), Homer mentions Achilles, 'whom Cheiron taught'.[31] The reference is allusive, as we hear no more about it in the passage, which shows that the audience was supposed to know the story. In fact, the *Iliad* is very sparse with information regarding Cheiron, and this has persuasively been connected with the already mentioned minimising of the supernatural by Homer (Mackie 1997, 3–4). Interestingly, the wider passage suggests that a Homeric hero was acquainted with the art of medicine. The medicinal expertise of Cheiron also appears in another Iliadic passage where he is said to have given herbs to Asclepius (4.217–9), which, however, need not already imply his later education by Cheiron as related by Pindar (*P.* 3.5–8), as has often been suggested.[32]

26 For Cheiron as divine, see Pindar, *P.* 4.119; Aeschylus, *Prom.* 1027 (?); Sophocles, *Trach.* 714–15; Apollodorus 2.5.4.

27 Homer, *Il.* 2.744; Hesiod F 40.2 and 204.87 MW; Pindar, *I.* 8.42–3, *P.* 2.45, 3.4, 4.102, 9.30; Euripides, *IA* 705; Nicander, *Ther.* 502; Heraclides, *Periegesis* 2.8 Pfister; Strabo 9.5.19; Diodorus Siculus 4.70; Ovid, *Met.* 11.235–35; Statius, *Ach.* 1.106–08; Schol. *Il.* 1.268, 16.144 (precise location); Eustathius on *Il.* 22.832–34; *SEG* 1.248; Capon 1982, 98 (identification). For interesting descriptions of Mt Pelion see Janssens 1975; Louis 1975.

28 Our source, Heraclides, *Periegesis* 2.8, writes Zeus Aktaios, but this is a later scribal mistake, cf. Graf 2009, 339 note 26; note also *IG* IX 2, 1128 (a dedication of a priest with the second name Kentauros to Zeus Akraios in Demetrias, close to Mt Pelion).

29 Robert 1969a, 651 note 50; Moustaka 1983, 73, Taf. VI.20.

30 Siebert 1990; Buxton 1994, 104–08; Sporn 2010.

31 See also Hesiod F 204 M/W. For a very full account of Achilles' education, which almost certainly goes back to very old sources, see Statius, *Ach.* 2.96–167.

32 *Contra* Mackie 1997, 2.

Still, his Thessalian origin fits well the connection with Mt Pelion: before the beginning of Byzantine times, Cheiron himself is never depicted in a medicinal role.[33]

The allusive character of the Homeric passage about Cheiron's teaching also struck later readers,[34] who knew from other sources that his father Peleus or, as some younger poets reported, his mother Thetis, had handed over young Achilles to the Centaur. The earlier version, however, was always represented on the older vases, which already feature the theme in the later seventh century.[35] Hesiod knew that Cheiron had also educated Jason, the famous hero of the Argonauts, who took to sea from Iolkos, a harbour below Mt Pelion, to fetch the Golden Fleece, just as he had edu cated Medeios, Jason's son.[36] Actaeon was also educated by Cheiron before being torn to pieces by his own dogs, but the latter event took place on Mt Kithaeron, not Mt Pelion, and thus is a clear case of poetic expansion, not age-old tradition. The fact that in later times a whole list could be given of heroes who had been pupils of Cheiron, such as Heracles, Asclepius, Aristaeus and Cyrene (Robbins 1978), is probably an illustration of the same phenomenon.[37]

Later poets also had more to tell about Achilles' education.[38] Pindar (*N.* 3.43–55) relates that he was already hunting at an early age, being extremely nimble and fast-footed, just as the Centaurs themselves were renowned for their speed (Isocrates, *Hel. Enc.* 26). From the fact that Achilles was able to play the lyre in the *Iliad*, subsequent generations concluded, not implausibly, that Cheiron had also taught music to Achilles,[39]

---

33 As observed by Simon 2004, cf. Gisler-Huwiler 1986, no's 104 f.

34 The scholia on Homer, *Il.* 9.486, 489 and 11.832 note that Homer does not know of Achilles being raised by Cheiron.

35 Peleus: Apollodorus 3.13.6; Schol. Homer, *Il.* 18.57a; Schol. Pindar, *N.* 3.76b; Schol. Apollonius Rhodius 1.558. Thetis: Schol. Homer, *Il.* 18.438a. Vases: Gisler-Huwiler 1986, no's 44–64; Kossatz-Deissmann 2009, no's add. 3–8.

36 Jason: Hesiod, F 40.2 MW; Pindar, *N.* 53–4, *P.* 4.102–3; Pherecydes *FGrH* 3 F 59a; *PSI* 10.1173, re-edited by Rossum-Steenbeek 1997, no. 56, fr. 5; Ovid, *Met.* 2.628–30; Apollodorus 3.10.3; Schol. *Od.* 12.69; Schol. Hesiod, *Th.* 993; Schol. Pindar, *P.* 4.135. Medeios: Hesiod, *Th.* 1001–02.

37 Actaeon: Acusilaus *FGrH* 2 F 33 = F 33 Fowler; Apollodorus 3.4.4. List: Xenophon, *Cyn.* 1. Heracles: Schol. Theocritus 13.9b. Aristaeus: Apollonius Rhodius 2.506–10; Asclepius: Pherecydes *FGrH* 3 F 3 = F 3 Fowler; Apollodorus 3.10.2; Mathé 1995.

38 See now Cameron 2009; Stark 2012, 148–163.

39 Homer, *Il.* 9.186 (lyre); Schol. *Il.* 4.219 (Cheiron); Kossatz-Deissmann 1981, no's 51–63.

and a number of later vases do display music-making Centaurs.[40] All this education did not mean that it was for free. The philosopher Antisthenes (F 28: ca. 400 BC) relates that he saw a painting displaying Achilles serving Cheiron. The theme seems old, as it was for a long time normal for young people to be employed as servants, as the European vocabulary – French *garçon*, Old English *cniht*, Dutch *knecht*, German *Knecht* and Dutch *dienstmaagd* – abundantly shows, and menial tasks for young people are well attested in rituals that seem to have a base in former rituals of initiation.[41]

Yet Cheiron apparently also practised other activities. There were whispers of pederastic relations with Heracles in his cave in honour of Pan, himself a pederastic god, and in ancient comedy pederasts who indulged their sexual appetites too often were called 'Centaurs' and 'wild ones'; even the vagina was called 'Centaur' in ancient comedy as it roused sexual lust.[42] Elsewhere in Thessaly, Cheiron was connected with cannibalism, in addition to pederasty.[43] Both motifs, pederasty and cannibalism, are sometimes connected with rites of initiation in ancient Greece and other parts of the world.[44] And indeed, Achilles' stay and education in the wilderness clearly reflect initiatory traditions, as was already seen by Henri Jeanmaire (1884–1960), the pioneering scholar of such traditions in ancient Greece.[45] The initiatory interpretation also fits other characteristics of Achilles' education, such as his change of name and dressing up as a woman just before he joined the Greek expedition against Troy.[46] As the nearby sanctuary of Zeus was the centre of the Magnesian confederacy, and the priest of Zeus Akraios one of its chiefs, this sanctuary must have been the place where once the Magnesians sacrificed to Cheiron and initiated their youths into adulthood.[47] Cheiron's initiatory activities are clearly a reflection of these rites in the Archaic Age, and the

40 Leventopoulou 1997, no's 316–25, probably starting in the late sixth century BC.

41 Bremmer 1999a, 13 (servants); Graf 2000, 262 (menial tasks).

42 Aristophanes, *Nub.* 350, F (dub.) 972 KA; Theopompos F 92 KA (vagina); Eratosthenes, *Cat.* 40 with Jordi Pàmias *ad loc.*; Hesychius κ 225; Photius α 259; Borgeaud 1979, 115–7.

43 Monimos *apud* Clement of Alexandria, *Protr.* 3.42.4 (cannibalism).

44 Pederasty: Bremmer 1980; Davidson 2007. Cannibalism: Bremmer 2002 and 2007, 73.

45 Jeanmaire 1939, 371 and 1949; Gernet 2004, 66.

46 Change of name: Apollodorus 3.13.6. Transvestism: Bremmer 1999b, 188–93.

47 Burkert 1983, 113–4. Priest as chief: Ruge 1930, 470. Sacrifices: Plutarch, *M.* 647 A.

idea that the Centaurs owed their hybrid nature to initiatory adults with horse masks still has a certain plausibility.[48] Yet his genealogy shows that the poets had also adapted older traditions about him to newer mythological developments, and we should claim no more than that initiation was a part of Cheiron's persona.

Education is a civilising activity, and it is therefore not surprising that early traditions knew that Cheiron was properly married to the Naiad Chariclo, who is always represented as completely human on the vases.[49] Both of them appear among the gods on the François Vase, which is another indication of Cheiron's special status compared to that of the other Centaurs.[50] Moreover, Cheiron was also connected with the institution of justice – he was called 'the most just of the Centaurs' by Homer (*Il.* 11.832) –, oaths, sacrifices and the movements of the stars.[51] He thus has the traits of a culture hero, in mythology often indeed a character that is not wholly part of culture himself.[52] In other words, in our oldest sources Cheiron is already a complicated figure, who, despite living in the wilderness, is only partially 'wild' himself. In Victorian language: he is a savage but, importantly, a 'noble' savage.

5. Centaurs versus Lapiths

The third theme regarding the Centaurs was their battle against the Lapiths, a Thessalian tribe. Battling was, so to speak, a generic feature of the Centaurs, and representations on vases of fighting Centaurs start already in the last half of the seventh century. Interestingly, on one of the oldest certain representations, the already mentioned François Vase, the Lapiths are depicted as hoplites, that is, normal soldiers, but the Centaurs use as weapons branches of pine trees and, from the sixth century onwards, pieces of rock. Moreover, on the vase a Lapith receives the name Hoplon,

48 Burkert 1985, 173 = 2011, 267. Note, though, that Dumézil 1929, 258 derived the rites from the myths, whereas Burkert the myths from the rites. For the myth/ritual relation, see Bremmer 2005.
49 Hes. F 42 MW; *Titanomachia* F 6 Davies = 10 Bernabé; Finster-Hotz 1986. Name: Wachter 2001, 253.
50 Gisler-Huwiler 1986, no's 41–42; add Iozzo 2009, 69.
51 *Titanomachia* F 6 Davies = 11 Bernabé.
52 Brelich 1956, 171–85; Long 2005.

'Weapon', whereas a Centaur is called Petraios, 'Rockman',[53] and we find the same opposition in the more or less contemporaneous pseudo-Hesiodic poem *Shield* (178–90). In other words, in their manner of fighting, too, the Centaurs are not really civilised, even though fighting with fir trunks torn up from the ground is a signal of their great strength.[54] Also in this respect, Cheiron was a civilised outsider, as he was famous for having presented a proper ash-wood spear to Peleus, Achilles' father, at the occasion of the latter's wedding.[55]

Our earlier literature, though, focuses on only a few famous battles. The oldest seems to be mentioned in the first book of the *Iliad*, which lists a number of Lapiths and states that 'they used to fight with the strongest, the mountain-inhabiting animals' (1.267–68, cf. above). The iterative form of the imperfect used here strongly suggests a series of fights,[56] and indeed, the fact that Nestor was called up from Pylos in order to help the Lapiths (*Iliad* 1.270) suggests a sequence of previous incidents. Moreover, the resemblances in the Homeric list of Lapiths with that of pseudo-Hesiod's *Shield* (179–82) indicate that both Homer and 'Hesiod' drew on a current tradition of oral poetry, even though those poems have been irretrievably lost (Wachter 1991, 106). These fights eventually came to a climax at a battle that must have taken place at least nine months after the wedding of Peirithoos, the king of the Lapiths, with Hippodameia, as she gave birth to his son on the very day of the battle (*Iliad* 2.742–3).

The ultimate cause of the fighting is probably related in the *Odyssey*, where the brash suitor Antinoos addresses Odysseus in a moralizing tone and compares his behaviour with that of the Centaur Eurytion who 'did bad things' in the house of Peirithoos'. Consequently, 'the heroes', that is, the Lapiths, leapt up, dragged him out of the palace and cut off his nose and ears – a type of mutilation that strongly suggests contemporary Assyrian influence.[57] However, Homer does not tell us which bad things

---

53 Leventopoulou 1997, no's 154–69 at 154 (François Vase) and p. 702; Tomei 2008, 122–23; Dietrich 2010, 79–89 (trees), 89–92 (rocks), 311–12 (names, rocks, trees).

54 As noted by West 2007, 461.

55 *Cypria* F 3 Bernabé = Davies; *Ilias Parva* F 5 Bernabé = Davies; Janko on Homer, *Il.* 16.130–54, 141–44.

56 Latacz *ad loc.*

57 Homer, *Od.* 21.295–304; Ovid, *Met.* 12.219–536; Rollinger 2004, 140 (Assyrian examples). For similar cruelty among the Persians, see Jacobs 2009; Rollinger 2010a, 601–02 (more Assyrian examples in addition to the Persian ones).

Eurytion did, and there is no reason to think of a wedding feast, as he was the only Centaur to have been invited to the feast.[58] The earlier fifth-century poet Bacchylides probably took this line as inspiration for a poem in which he changed the nature of the event and moved the venue of the crime to Elis. Here its king, Dexamenus, says in an only fragmentarily preserved poem:

And he said this: 'I am grieved at heart ...: uninvited (he came) to the lovely feast, the Centaur (Eurytion) whose bed is in the mountain ... and he asks me for my (slender-ankled?) daughter, wishing to take her as his bride to Malea (below): but to me (this is repellent), and since I am unwilling (he threatens me) more harshly ...[59]

The occasion is clearly not a wedding but some kind of feast at which the Centaur arrives uninvited. In the fifth century that was the typical behaviour of parasites, the people who came uninvited to your symposium but whom you could not refuse.[60] In other words, Eurytion also broke the laws of hospitality. Fortunately, Heracles stopped by at the right moment and killed the Centaur. In one way or another, then, it was the behaviour of Eurytion that started the major battle between the Centaurs and the Lapiths.

The wedding seems to have become the scene of the great confrontation only at the beginning of the fifth century, when it is mentioned, for the first time, in a fragment of Pindar (F 166 Maehler), which probably belonged to a poem about Peirithoos' wedding.[61] More or less at the same time we see the wedding battle appear on Attic vases, where it can be read also as a reflection of the battle of the Greeks against the barbarous Persians.[62] It is clear that somebody, not necessarily Pindar, had the idea of combining the wedding and the battles in a way that dramatised the action, although the transfer of the battle against the Centaurs from out in the open to the feast also accentuated the barbarity of the Centaurs who now, in addition, committed a violation of hospitality (Woodford 1974).

58 *Contra* Barron 1972, 25 f.

59 Bacchylides F 66 Maehler, tr. Campbell and note also F 44; Huys 1991, 40, line 7; Diodorus Siculus 4.33.1–2; Ovid, *Ibis* 403–4 with scholion *ad locum*; Hyginus, *Fab.* 31.11, 33.1–2; Pausanias 7.18.1.

60 Asius, fr. 14 West^2; Cratinus F 46, 47 and 182; Aristophanes F 284; Alexis F 213, 259 KA; Timotheus, fr. 1; Apollodorus Carystius, fr. 29 and 31; Lynkeus of Samos in Athenaeus 245a; Lucian, *Demonax* 63; Fehr 1990.

61 Brillante 1998, 53 f. For a detailed description see Ovid, *Met.* 12.210–535 with Bömer *ad loc.*

62 Leventopoulou 1997, no's 171–73, p. 703.

Pindar describes how the Centaurs pushed the milk from the table and started to drink wine 'from silver horns', an interesting detail that suggests the Persian rhyton. Can it be that the poet here too wanted to stress the fact that the Centaurs were not like the Greeks, not even when drinking wine? The Greeks themselves did not drink the milk of cows, and goats' milk was usually drunk where it was produced, as it could not be kept long. On the other hand, the Scythians, and related tribes, who were the typical 'other' of the Greeks, were known as 'mare-milkers', presumably a reference to the disgusting koumiss.[63] The reference to milk drinking Centaurs, then, must have suggested their affinity with the uncivilized nomads of the Black Sea.

As a consequence of their immoderate drinking, the Centaurs started to paw the bride and the Lapith girls but, as has been well noted, 'continuous accounts in literature are surprisingly few'.[64] The vases mainly represent two scenes of battle. In the one, the Centaurs fight the Lapiths outside with no women present, whereas in the other the Centaurs fight the Lapiths in the banquet hall and try to get hold of the women. The two great violations of Greek civilized behaviour, namely raping women and drinking wine immoderately, which come to the fore in these scenes, are further explored in two other scenes in which Centaurs play an important role, namely in the attempted rape of Heracles' wife Deianira by the Centaur Nessos and in Heracles' visit to the Centaur Pholos.

6. Heracles, Nessos and Deianira

The story of Heracles, the Centaur Nessos and Deianira is well known, although it seems to have developed only in the course of the Archaic Age into the well-known version that was used by Sophocles in his *Trachiniae.*[65] In Hesiod's *Theogony* (341), Nessos is just the name of a Thracian river (Kotula 1991), and in the Hesiodic *Catalogue of Women* he is

63 Scythians and other nomads: Homer, *Il.* 13.5; Hesiod F 150, 151 Merkelbach/West; Aeschylus F *198 Radt; Herodotus 1.26.4, 1.216 and 4.2, with Asheri *ad loc.*; Hippocrates, *De aëre, aquis et locis* 18; Shaw 1982–83; S. West 1999.

64 Nisbet and Hubbard on Horace, *C.* 2.12.5, cf. Horace, *C.* 1.18.7, 4.2.14; Vergil, *G.* 2.455, *A.* 7.304.

65 For the story, see especially Robert 1920b, 573–76; Jouan 1983; March 1987, 49–77; Gantz 1993, 431–34; Diez de Velasco 1992; Maass and Kilian-Dirlmeier 1998, 60–61; Peppa-Papaïoannou 2001.

not yet directly connected with the death of Heracles, as here Deianira gives the herald Lichas a garment that she had sprinkled with poison. When Heracles received the garment from Lichas and put it on, death came quickly. In this version, which we can follow from Archilochus to Bacchylides as well as on vase paintings until the beginning of the fifth century, there is not yet a connection between the death of Nessos and that of Heracles.[66] While ferrying her across the river Nessos tried to carry Deianira off, and Heracles killed him in close combat. On black-figure paintings this is virtually always done with a sword, presumably on or near the bank of the river Evenos (Schol. Apoll. Rh. 1.1212). In a poem by, probably, Bacchylides (F 64 dub. Maehler), the poet leaves no doubts about the way he sees the Centaur, even though the text has only been partially recovered. After Nessos '(filled with) Aphrodite's (madness), rushed (at the young woman); she cried out and begged her husband to hurry … (brandishing) his great club in his right hand he (struck) the middle of the savage beast's (head over) the ear and smashed…' The Centaur, then, is called a *phê[r]os agriou*, 'savage beast', the same term we already found in Homer but now explicitly connected with wildness.

However, the more canonical version, in which Heracles kills Nessos with his bow and arrow, is seen in both Sophocles and again, rather surprisingly, in Bacchylides (*Dith.* 16). This can be understood as clear proof of the great impact the tragedian had with his version (March 1987, 62–66). In the *Trachiniae* Deianira relates that Nessos used to carry people in his arms across the river Evenos. As she recalls, 'while he was carrying me upon his shoulders, when I was first accompanying Heracles as bride, after my father had sent me off, while I was in mid-stream he laid lustful hands upon me, and I called out. At once the son of Zeus (Heracles) turned and let fly a feathered arrow,[67] and it went whizzing through his chest into his lungs. As he expired, the monster, *thêr* (= *phêr*), said so much', and then follows Nessos' instruction to Deianira to take the clotted blood from his wound and to use it as a charm so that Heracles will never love another woman instead of her.[68] To introduce her story,

66 Hes. F 25.17–22MW; Archilochus F 286–88 West[2]; Fittschen 1970; March 1987, 52–3.
67 But the arrow is very rare on the vase paintings: Diez de Velasco 1992, 846.
68 Sophocles, *Trach.* 565–81, tr.Lloyd-Jones 1994, 183–85; see also Diodorus Siculus 4.36.4–5; Ovid, *Met.* 9.101–65; Hyginus, *Fab.* 34; Apollodorus 2.151–2; Schol. Lycophron 50.

Deianira starts with: 'I had an ancient gift from an age-old monster, *archaiou … thêros*, hidden in a brazen pot, a thing I received, when I was still a girl, from the blood of the dying Nessos with the hairy chest'.[69] We can see in this myth once again that one of the inhabitants of the wilderness is not only unlike normal people with his 'hairy chest', but also displays an uncontrolled sexual lust. The Centaur really is a 'savage beast'.

7. Heracles, Pholos and the End of the Centaurs

Our last episode brings us back to a more civilised Centaur, Pholos, and the end of these hairy beasts.[70] The scene, which on vase paintings starts to appear in the middle of the seventh century, now shifts to Mt Pholoe in Arcadia, although in his *Heracles* (364–74) Euripides was somewhat casual about its situation and located it in Thessaly, undoubtedly influenced by the traditional battle of the Centaurs against the Thessalian king Peirithoos (§ 3). The best continuous description comes from the late second-century mythographer Apollodorus (2.5.4), who, however, used Archaic and classical sources and whose account can be well illustrated by the vase paintings. It reads as follows:

So passing through Pholoe Heracles was entertained by the Centaur Pholos, the son of Silenos and a Melian nymph. He set roast meat before Heracles while he himself ate his meat raw. When Heracles requested wine, he said he feared to open the jar, which belonged to the Centaurs in common. But Heracles bid him to be of good courage and opened it. Not long afterwards, scenting the smell, the Centaurs arrived at the cave of Pholos, armed with rocks and firs. The first who dared to enter, Anchios and Agrios, were repelled by Heracles with a shower of brands, and the rest of them he shot and pursued as far as Malea.

This part of the mythical cycle of Heracles is confirmed by much older literary sources. Stesichorus (S 19 Davies) already mentions the fact that Pholos put a beaker with wine in front of Heracles. It is clear that he is one of the more civilised Centaurs, as he knows how to drink and to be hospitable, even though he lives in a cave and eats his

69 Sophocles, *Trach.* 555–8, where I have adapted the translation of Lloyd-Jones 1994, 183, as he translates *archaiou* with 'long ago', which would have been *palaiou*, Casevitz 2004; Levy 2007; note also *Trach.* 680: Ἐγὼ γὰρ ὢν ὁ θήρ με Κένταυρος, 1162: Ὅδ' οὖν ὁ θὴρ Κένταυρος.

70 Gantz 1993, 390–92; Leventopoulou 1997, 706–10.

meat raw, which of course normal horses, being vegetarians, never do. This is different from the other Centaurs, who rush out to get drunk and are not armed with proper weapons but with rocks and firs.[71] Evidently, the thematizing of proper, that is moderate, ways of drinking can hardly be separated from the rise of the symposium with its rules in the sixth century.[72] Interestingly, Pholos is also connected with another group of wild beings inhabiting the wilderness, the satyrs, through his birth from Silenos, the patriarchal satyr who was the educator of Dionysos. The detail need not be old, but can still date to the Archaic Age, as the connections of the Centaurs with the satyrs were stressed to an increasing extent from the sixth century onwards.[73] However, the satyrs survived, so to speak, in mythology, because one could perhaps meet them in Dionysiac worship, and of course one would see them in the theatre.

As a result of Heracles' fight against the Centaurs,[74] so Apollodorus notes in his somewhat garbled version, they fled in all directions, and later generations told stories about Centaurs even fleeing to Samos (Strabo 8.3.19). We witness here a theme that has not received any attention in recent discussions. Yet, Homer already states that Peirithoos, the king of the Lapiths, chased his enemies away from Mount Pelion to the Aithikes on Mount Pindos, that is, to the border area of Thessaly.[75] Other narratives relate a more sombre picture. Already around 470 BC, Pindar (*P.* 3.1–3) wishes that Cheiron, *phêr' agroteron*, the 'wild beast', was still alive. Pholos had died immediately after having inadvertently dropped one of Heracles' poisoned arrows on his foot in the aftermath of the latter's fight against the Centaurs,[76] and Nessos was killed by Heracles. As Euripides simply states, 'the mountain-dwelling tribe of fierce Centaurs he laid low, killing them with his winged shafts'.[77].

It is clear from these notices that already from the early Archaic Age onwards the Centaurs were continuously marginalised. First, they were

---

71 Schauenburg 1971; Noël 1983; Valenza Mele 1986; Noël 1999; Verbanck-Piérard 1982.

72 See most recently Murray 1990; Catoni 2010.

73 Leventopoulou 1997, 702, 709; Euripides, *IA* 1058–61; Osborne 1994; Morawietz 2002; see also the Appendix to this article.

74 For a more detailed discussion of the representations of this fight, see Muth 2008, 417–26.

75 Homer, *Il.* 2.742–44; note also Strabo 9.5.12, 9.5.19; Apollodorus 2.5.4 (Cheiron now living at Cape Malea).

76 Diodorus Siculus 4.12.8; Apollodorus 2.5.4.

77 Euripides, *Her.* 364–74, cf. 1272–73, at 364–67, tr. Kovacs 1998, 343.

removed to the 'outskirts' of known civilisation, such as Cape Malea and the Aithikes on Mount Pindos, but later there was no longer any place at all for them, and Greek myth just had them killed or let them die or, to the same effect, related that Poseidon had hidden them in a mountain near Eleusis (Apollodorus 2.5.4).[78] Consequently, when the Augustan traveller Strabo visited the Aithikes he noted about the Centaurs: 'they now tell that they have disappeared' (9.5.12). We witness here the death of a myth (that of the Centaurs) in myth. Evidently, however interested poets and painters were in them, intellectuals could no longer tolerate the thought of such equine-human beings in their own backyard, so to speak. It is a sign of the rising rationality in certain areas of Greek culture to find that already around 500 BC the philosopher Xenophanes stated that he did not want to sing of the Titans, Giants or 'Centaurs, fictions of the earlier generations' (F 1.22 West^2). Evidently, such mythical beings gradually ceased to be believable and acceptable.[79]

It is a corollary of this development that authors of the fifth century BC increasingly noted their anomalous character, and that from about 400 BC their animal character was stressed by giving them the new name Hippokentauroi, 'Horse Centaurs'.[80] And when Sophocles lists six Labours of Heracles in his *Trachiniae* (1095–96), all of which concern the killing of monsters, he mentions as one of these feats the defeating of the Centaurs:

Of double nature, not mingling with others, going on horses' feet, the band of *thêrôn*, 'beasts', wanton, lawless, pre-eminent in might.

Once again they are called 'beasts', but Sophocles now stresses their asocial and lawless character, and the Greek verb used for the killing of the Centaurs, *enairô*, Euripides uses especially for the killing of monsters.[81] It looks as if, in the later fifth century, probably through the discussions of the Sophists about the place of law and civilised behaviour, the anomalous aspects of the Centaurs had come to be perceived as a kind of threat to civilised mankind: they were monsters that had to be completely eradicated.

---

78 For Poseidon's connection with monsters and uncivilised behaviour, see Bremmer 1987.

79 Wilamowitz-Moellendorff 1959, 1.192 on the destruction of the Centaurs: 'weil man an ihr Vorkommen nicht mehr glaubt'.

80 Plato, *Phaedr.* 229d; Xenophon, *Cyr.* 4.3.17.

81 Bond on Euripides, *Her.* 367.

8. Conclusion: the Centaurs as demons of the mountain wilderness

It is time to come to a close and I would like to draw the following conclusions:

1. The Greeks did not have a clear-cut concept and vocabulary of what we call 'wilderness'. This is probably not surprising, as in English the Old Testament seems to have been of paramount influence in creating the popularity of the term. Yet the early Greeks had a notion of an area outside civilisation, which, influenced by their ecology, they located in the mountains. Our analysis suggests that the mountains functioned as a kind of 'liminal space' between the civilised world and the world outside Greek civilisation. As such, this particular space will have to be taken into account in future studies of Greek cosmology. They were not unique in this respect as mountains and wilderness were also closely associated in ancient Mesopotamia and Rome.[82]

2. The Greeks verbalised the 'wild' character of the mountains through myths about the mountains themselves but also through myths of human-equine beings that populated Mt Pelion. The latter myths concentrated less on the actual wilderness than on its 'wild' inhabitants, and verbalised human interaction with the wilderness as interaction with the wild creatures of the mountains (as opposed to interaction with the landscape *per se*). Moreover, through these myths the Greeks explored the limits of proper hospitality, proper treatment of women and proper drinking of wine.

3. In the actions of and stories about the Centaurs we can recognize the main aspects of the mountains as identified by Buxton (§ 1): they are wild in their violent dealings with humans, males and females, they are 'before' by eating raw food and living in pre-civilisation caves, and they embody reversals by mixing with humans. However, and this aspect has to be added, the Centaurs were not only seen in a purely negative light. On the contrary. Both Cheiron and Pholos are relatively civilised, and Cheiron is even a kind of culture-hero.

4. In the course of the Archaic Age (800–500 BC), the Greeks started to explore the world around them, became more urbanised, introduced the symposium with its drinking rules, and initiated more rational ways of thinking. These developments gradually brought it about that the

82 Rollinger 2010b; Acolat 2009.

Centaurs were no longer perceived as more or less human opponents but more and more as monsters that had to be eradicated, even the more civilised ones. The annihilation of the Centaurs shows that, in the rationalising fifth century BC, the ideas of the Greeks about their mountains had considerably changed. They may have remained dangerous territory, but the mountains were now no longer inhabited by animals symbolizing their 'wild' nature. We see here how 'wilderness' is not a fixed concept, but one that remains constantly subject to changes in cultural construction and perception.

5. The ambivalent nature of the Centaurs can hardly be separated from the ambivalent nature of the mountains.[83] However wild and dangerous they were, the mountains were also useful as places for pasturage, the supply of wood, the hunting of game and even the worship of the gods.[84] By combining the negative qualities of the Centaurs with their education of the greatest Greek hero as well as with aspects of the birth of civilisation, the Greeks stressed both the positive and negative qualities of the mountainous wilderness. They recognized that interacting with the wilderness can be dangerous and destructive as well as beneficial and instructive and, moreover, that it can lead to cultural creativity. It is a message that still has something to say to us today.[85]

9. Appendix: unpublished sard-ringstone with young Centaur

Joern Lang

The vertically shaped stone shows a juvenile Centaur standing on a ground line trotting to the left. In his lower right hand he holds a *kantharos*, in the left crook of his arm a *thyrsos* with waving bands.[86]

From a stylistic point of view the stone belongs to a group of late Republican and early Imperial gemstones of the so-called flat bouterolle style (late first century BC – first half of the first century AD). The single forms are engraved in vertically-shaped lively cuts of different sizes (com-

83 This suggestion comes close to Kirk 1970, 160 (in a still worthwhile discussion of the Centaurs).

84 For the useful side of the mountains, see Buxton 1994, 82–86; for the worship of the gods on mountains, add Langdon 2000; Accorinti 2010.

85 I am very grateful to Bob Fowler for his comments, Laura Feldt for her editing and Richard Buxton for correcting my English.

86 Sard-ringstone (Cologne, private Collection): 19 (H) x 16 (W) x 2 mm.

With permission by Kai Scheuermann.
Picture taken by Gisela Geng.

pare for example the details of the horse's tail cut with a very small wheel, or the horse's legs formed by a medium-sized wheel).[87]

Since Geometric times, Centaurs have been a common motif on gemstones.[88] In addition to the usual mythological episodes, such as those featuring Achilles and Cheiron (above § 4),[89] and the familiar depictions of them as wild beasts fighting with trunks or branches (above

87 For the style of the piece, compare Weiß 1996, no. 397; Henig and Whiting 1987, 37 no. 389; Maaskant-Kleibrink 1980, 10 no. 19 (closing date for the findspot: 55 AD); Zwierlein-Diehl 1998, 217 no. 84 (all from the second half of the first century BC – first half of the first century AD). Compare the so called 'Flachperlstil' = flat Bouterolle style of late Republican and early Imperial times: Zwierlein-Diehl 2007, 136; Maaskant-Kleibrink 1978.

88 Boardman 2001, 113 no. 164, 120 no. 230; Scherf 1970, 71 no. 1a Taf. 27. A nice *kottabos*-playing centaur of late archaic times: Zwierlein-Diehl 2002, 74 Cat. 14 Farb-Abb. 1.

89 Schlüter 1975, 193 no. 959 Taf. 127.

§ 5),[90] Centaurs in Roman times were often depicted making music or holding drinking cups[91] (*kantharoi* as well as cups[92] or kraters[93]), which they used even in full movement. On the Sard-ringstone, the *kantharos* held by the juvenile Centaur is an element known from the Centaur Pholos (above, § 7), but it can probably be explained best by the strong association of drinking cups with the *symposion*. The adoption of elements such as *kantharoi* and *thyrsoi* results particularly from the close connection of Centaurs to the Dionysiac *thiasos*, an association which dates to classical times. As early as the sixth century BC, Centaurs and satyrs are depicted on the same vase, and, since the fourth century BC, Centaurs are seen as participants in the *thiasos* itself. From that period, Centaurs are increasingly shown with Dionysiac attributes and as comasts,[94] a relation predestined by the frequently documented idea that these hybrid creatures fostered a lust for wine. In the present case, although the Centaur's posture does not suggest drunkenness, the over-turned drinking cup and the *thyrsos* can be seen as subtle references to the inebriating influence of wine.[95]

10. References

Accorinti, Domenico. "La montagna e il sacro nel mondo greco," in: A. Grossato, ed., *La montagna cosmica*. Milan: Medusa, 2010, 17–42.

Delphine Acolat, "Le modèle de la montagne sous le haut-empire: vrai "locus horridus" ou prétexte littéraire?", in: *Les Etudes Classiques* 77 (2009), 61–77.

Angelo de Angelis, "Tra dati linguistic e fonti letterarie: per un'etimologia del gr. Kéntauros 'divoratore di viscere'," in: *Glotta* 85 (2009), 59–74.

Aston, Emma. "The Absence of Chiron," in: *Classical Quarterly* 56 (2006), 349–62.

Barbu, Daniel, "Entretien avec Walter Burkert," in: *Asdiwal* (Geneva) 2 (2007), 7–15.

John P. Barron, "New Light on Old Walls: Murals of the Theseion," in: *Journal of Hellenic Studies* 92 (1972), 20–45.

---

90 Schlüter 1975, 298 no. 1623 Taf. 216.

91 Platz-Horster 1987, 14 no. 24 Taf. 4, 103 no. 180 Taf. 37; Brand 1972, 147 no. 3066. 3067 Taf. 299 (playing diaulos); Scherf 1970, 50 no. 174 Taf. 21.

92 Brand 1968, 89 no. 500 Taf. 54; Weiß 1996, 143 no. 390 Taf. 54.

93 Platz-Horster 1987, 14 no. 25 Taf. 5.

94 Morawietz 2000, 59–62; for the close connections of Centaurs and satyrs in the sixth and fifth centuries BC, see Osborne 1994.

95 I thank Suzanne Lye for kindly correcting my English.

Bierl, Anton and Wolfgang Braungart, eds, *Gewalt und Opfer: im Dialog mit Walter Burkert.* Berlin and New York: de Gruyter, 2010.
Boardman, John. *Greek Gems and Finger Rings. Early Bronze Age to Late Classical.* London: Thames and Hudson, 2001.
Borgeaud, Philippe. *Recherches sur le dieu Pan.* Rome: Institut suisse de Rome, 1979.
Brelich, Angelo. *Gli eroi greci.* Rome: Ateneo, 1956.
Brand, Elfriede *et al. Antike Gemmen in Deutschen Sammlungen I-1 Staatliche Münzslg. München.* Wiesbaden: Steiner, 1968.
Brand, Elfriede, *Antike Gemmen in Deutschen Sammlungen I-3 Staatliche Münzslg. München.* Wiesbaden: Steiner, 1972.
Bremmer, Jan N., "An Enigmatic Indo-European Rite: Paederasty," in: *Arethusa* 13 (1980), 279–98.
Bremmer, Jan N., "'Effigies Dei' in ancient Greece: Poseidon," in: D. v. d. Plas, ed., *Effigies Dei. Essays on the History of Religions.* Leiden: Brill, 1987, 35–41.
Bremmer, Jan N., "Oedipus and the Greek Oedipus Complex," in: J.N. Bremmer, ed., *Interpretations of Greek Mythology.* London: Routledge, 1988², 41–59.
Bremmer, Jan N., "Fosterage, Kinship and the Circulation of Children in Ancient Greece," in: *Dialogos* 6 (1999a), 1–20.
Bremmer, Jan N., "Transvestite Dionysos," in: *The Bucknell Review* 43 (1999b), 183–200.
Bremmer, Jan N., "Odysseus versus the Cyclops," in: S. des Bouvrie, ed., *Myth and Symbol* I. Athens: The Norwegian Institute, 2002, 135–52.
Bremmer, Jan N., "Myth and Ritual in Ancient Greece: Observations on a Difficult Relationship," in: R. von Haehling, ed., *Griechische Mythologie und Frühchristentum.* Darmstadt: Wissenschaftliche Buchgesellschaft, 2005, 21–43.
Bremmer, Jan N., "Myth and Ritual in Greek Human Sacrifice: Lykaon, Polyxena and the Case of the Rhodian Criminal," in: J.N. Bremmer, ed., *The Strange World of Human Sacrifice.* Leuven: Peeters, 2007, 55–79.
Bremmer, Jan N., *Greek Religion and Culture, the Bible and the Ancient Near East.* Leiden: Brill, 2008.
Bremmer, Jan N., "A Brief History of the Study of Greek Mythology," in: K. Dowden and N. Livingstone, eds., *Blackwell Companion to Greek Mythology.* Oxford: Blackwell, 2011, 527–47.
Briggs, Ward W. and William M. Calder III, eds., *Classical Scholarship. A Biographical Encyclopedia.* New York and London: Garland, 1990.
Brillante, Carlo, "Ixion, Peirithoos e la stirpe dei centauri," in: *Materiali e Discussioni* 40 (1998) 41–76.
Burkert, Walter. *Structure and History in Greek Mythology and Ritual.* Berkeley, Los Angeles, London: University of California Press, 1979.
Burkert, Walter, *Homo necans.* Berkeley, Los Angeles, London: University of California Press, 1983.
Burkert, Walter, *Greek Religion.* Cambridge Mass.: Harvard University Press, 1985.

Burkert, Walter, *Griechische Religion der archaischen und klassischen Epoche*, 2nd ed., Stuttgart: Kohlhammer, 2011.
Buxton, Richard, "Montagnes mythiques, montagnes tragiques," in: *Ktema* 15 (1990 [1994]), 151–72.
Buxton, Richard, "Imaginary Greek Mountains," in: *Journal of Hellenic Studies* 112 (1992), 1–15.
Buxton, Richard, *Imaginary Greece.* Cambridge: Cambridge University Press, 1994.
Cadell, Hélène and Roger Rémondon, "Sens et emplois de *to oros* dans les documents papyrologiques," in: Revue des Études Grecques 80 (1967), 343–49.
Calder III, William M., *Further Letters of Ulrich von Wilamowitz-Moellendorff.* Hildesheim: Olms, 1994.
Cameron, Alan, "Young Achilles in the Roman World," *Journal of Roman Studies* 99 (2009), 1–22.
Capon, Moses and Rachel. *Magnesia, a Story of a Civilisation.* Athens: M. and R. Capon, 1982.
Carnoy, Albert, "Le concept mythologique du gandharva et du centaure," in: *Muséon* 49 (1936), 99–113.
Casevitz, Michel, "Remarques sur le sens de ἀρχαῖος et de παλαιός," in: *Metis* NS 2 (2004), 125–36.
Catoni, Maria Luisa. *Bere Vino Puro: Immagini del Simposio.* Milan: Feltrinelli, 2010.
Davidson, James. *The Greeks & Greek Love.* London: Weidenfeld & Nicolson, 2007.
Dietrich, Nikolaus. *Figur ohne Raum?* Rome: de Gruyter, 2010.
Diez de Velasco, Francisco. "Nessos," in: *LIMC* VI.1 (1992), 838–47.
Doniger O'Flaherty, Wendy. *Women, Androgynes, and Other Mythical Beasts.* Chicago and London: University of Chicago Press, 1980.
Dumézil, Georges, *Le problème des Centaures.* Paris: Geuthner, 1929.
Fehr, Burkhard. "Entertainers at the *Symposion:* The *Akletoi* in the Archaic Period," in: O. Murray, ed., *Sympotica.* Oxford: Oxford University Press, 1990, 185–95.
Finster-Hotz, Ursula. "Chariclo," in: *LIMC* III.1 (1986), 189–91.
Fittschen, Klaus. "Zur Herakles-Nessos-Sage," in: *Gymnasium* 77 (1970), 161–71.
Gantz, Timothy. *Early Greek Myth.* Baltimore: Johns Hopkins, 1993.
Gernet, Louis. *Polyvalence des images. Testi e frammenti sulla leggenda greca.* Pisa: ETS, 2004.
Gisler-Huwiler, M. "Cheiron," in: *LIMC* III.1 (1986), 237–48.
Germaine Guillaume-Coirier, "Chiron Phillyride," in: *Kernos* 8 (1995), 113–22.
Graf, Fritz. "The Locrian maidens," in: R. Buxton, ed., *Oxford Readings in Greek Religion.* Oxford: Oxford University Press, 2000, 250–70.
—, "Zeus and his *Parhedroi* in Halikarnassos. A Study on Religion and Inscriptions," in: A. Martinez Fernandez, ed., *Estudios de Epigrafia Griega.* La Laguna, Universidad de La Laguna, 2009, 333–48.
Griffin, Jasper. "The Epic Cycle and the Uniqueness of Homer," in: *Journal of Hellenic Studies* 97 (1977) 39–53, reprinted with minor revisions in D.

Cairns, ed., *Oxford Readings in Homer's Iliad.* Oxford: Oxford University Press, 2001, 365–84.
Hanson, Victor D., *The Other Greeks.* New York: Free Press, 1995.
Henig, Martin and Mary Whiting. *Engraved Gems from Gadara in Jordan. The Sa'd Collection of Intaglios and Cameos.* Oxford: University Committee for Archaeology, 1987.
Hoekstra, Arie. *Homeric Modifications of Formulaic Prototypes.* Amsterdam: Noord-Hollandsche Uitgevers Maatschappij, 1965.
Huys, Marc. *Le poème élégiaque P.Brux.inv.E.8934 et P.Sorb.inv.2254: édition, commentaire et analyse stylistique.* Brussels: Musées Royaux d'Art et d'Histoire, 1991.
Hyde, Walter W. "The Mountains of Greece," in: *Bulletin of the Geographical Society of Philadelphia* 13 (1915), 1–16, 47–64, 110–26, summarized in: *Classical Weekly* 12.13 (1919), 97–99.
Hyde, Walter W., "The Ancient Appreciation of Mountain Scenery, in: *Classical Journal* 11 (1915–1916), 70–84.
Iozzo, Mario. "Un nuovo *dinos* da Chiusi con le nozze di Peleus e Thetis," in: E. Moormann and V. Stissi, eds, *Shapes and Images. Studies …Herman A.G. Brijder.* Leuven: Peeters, 2009, 63–85.
Jacobs, Bruno. "Grausame Hinrichtungen – friedliche Bilder. Zum Verhältnis der politischen Realität zu den Darstellungs-szenarien der achämenidischen Kunst," in: M. Zimmermann, ed., *Extreme Formen von Gewalt in Bild und Text des Altertums.* Munich: Herbert Utz Verlag, 2009, 121–53;
Jameson, Michael H. "Mountains and the Greek City States," in: J.-F. Bergier, ed., *Montagnes, fleuves, forêts dans l'histoire.* St. Katharinen: Scripta Mercaturea Verlag, 1989, 7–17.
Jameson, Michael H., "Attic *Eschatia*," in: K. Ascani *et al.*, eds, *Ancient History Matters.* Rome: L"Erma" di Bretschneider , 2002, 61–69.
Janssens, Emile. "Le Pélion, le Centaure Chiron et la sagesse archaïque," in: J. Bingen *et al.*, eds, *Le monde grec … Hommages à Claire Préaux.* Brussels: Editions de l'Université de Bruxelles, 1975, 325–37.
Jeanmaire, Henri. *Couroi et courètes.* Lille: Bibliothèque universitaire, 1939.
Jeanmaire, Henri, "Chiron," in: *Annuaire de l'Institut de Philologie et d'Histoire Orientales de l'Université Libre* 9 (1949), 255–65.
Johnston, Sarah I. *Restless Dead.* Berkeley, Los Angeles, London: University of California Press, 1999.
Jouan, François. "Déjanire, Héraclès et le Centaure Nessos. Le cheminement d'un mythe," in: H. Limet and J. Ries, eds, *Le mythe, son langage et son message.* Louvain-la-Neuve: Centre d'histoire des religions, 1983, 225–43.
Kirk, Geoffrey S. *Myth.* Cambridge University Press, 1970.
Kossatz-Deissmann, Anneliese. "Achilleus," in: *LIMC* I.1 (1981), 37–200.
Kossatz-Deissmann, Anneliese. "Achilleus," in *LIMC, Supplementum 2009*, I, 1–15.
Kotula, Tadeusa. "Νέσσος et Νάϊσσος: problème topographique et historique des campagnes de Gallien et de Claude II contre les Goths," in: *Eos* 79 (1991), 237–43.

Kovacs, David. *Euripides III.* Cambridge Mass. and London: Harvard University Press, 1998.
Langdon, Merle. "Mountains in Greek Religion," in: *Classical World* 93 (2000), 461–70.
Langdon, Susan. *Art and Identity in Dark Age Greece, 1100–700 B.C.E.* Cambridge: Cambridge University Press, 2008.
Latacz, Joachim *et al. Homers Ilias. Gesamtkommentar* I.2. Leipzig: Saur, 2000.
Leventopoulou, Maria *et al.*, "Kentauroi et Kentaurides," in: *LIMC* VIII.1. Zürich: Artemis, 1997, 671–721.
Lévy, Edmond, "'Αρχαῖος et παλαιός chez Hérodote," in: *Ktema* 32 (2007), 497–510.
Lewis, David M. *Selected Papers in Greek and Near Eastern History*, ed. P.J. Rhodes. Cambridge: Cambridge University Press, 1997.
Linant de Bellefonds, Pascale, "Kentauroi et Kentaurides," in: *LIMC, Supplementum 2009.* Düsseldorf: Artemis, 2009, 1.306–10.
Lloyd-Jones, Hugh. *Sophocles II.* Cambridge Mass.: Harvard University Press, 1994.
Long, Jerome H. "Culture Heroes," in: L. Jones (ed.), *Encyclopaedia of Religion* 3. Detroit: MacMillan, 2005[2], 2090–93.
Louis, Pierre. "Monstres et monstruosités dans la biologie d'Aristote," in: J. Bingen *et al.*, eds, *Le monde grec … Hommages à Claire Préaux.* Brussels: Editions de l'Université de Bruxelles, 1975, 277–84.
Maaskant-Kleibrink, Marietta. *Catalogue of the Engraved Gems in the Royal Coin Cabinet The Hague.* Wiesbaden: Steiner, 1978.
Maaskant-Kelibrink, Marietta. "The Velsen Gems," in: *Bulletin Antieke Beschaving* 55 (1980), 1–28.
Maass, Michael and Imma Kilian-Dirlmeier, "Aegina, Aphaia-Tempel. 18, Bronzefunde ausser Waffen," in: *Archaeologische Anzeiger* 1998, 57–104.
Mackie, Chris J. "Achilles' Teachers in the Iliad," in: *Greece & Rome* 44 (1997), 1–10.
Mannhardt, Wilhelm. *Wald- und Feldkulte*, 2 vols. Berlin: Weidmann, 1876–77.
March, Jennifer. *The Creative Poet.* London: Institute of Classical Studies, 1987.
Mathé, Suzanne. "Les enfances chez Chiron," in: D. Auger, ed., *Enfants et enfances dans les mythologies.* Paris: Les Belles Lettres, 1995, 45–62
Meissner, Burkhard. "Vorstellungen der Griechen von den Bergen," in: E. Olshausen and H. Sonnabend, eds, *Stuttgarter Kolloquium zur historischen Geographie des Altertums 5, 1993.* Amsterdam: Hakkert, 1996, 50–69.
Miller, Patricia C. "Jerome's centaur: a hyper-icon of the desert," in: *Journal of Early Christian Studies* 4 (1996), 209–33
Morawietz, Georg. *Der gezähmte Kentaur: Bedeutungsveränderungen der Kentaurenbilder in der Antike.* Munich: Biering & Brinkmann, 2002.
Moustaka, Aliki. *Kulte und Mythen auf thessalischen Münzen.* Würzburg: Konrad Triltsch, 1983.
Murray, Oswyn, ed., *Sympotica.* Oxford: Oxford University Press, 1990.
Muth, Susanne. *Gewalt im Bild.* Berlin and New York: de Gruyter, 2008.

Nilsson, Martin P., *Geschichte der griechischen Religion I.* Munich: Beck, 1941^{1}, 1967^{3}.
Noël, Daniel. "Du vin pour Héraklès!," in: F. Lissarrague, ed., *Image et céramique grecque.* Rouen: Université, 1983, 141–50
Noël, Daniel. "Centaures en société autour d'une jarre de vin," in: *Dialogues d'Histoire Ancienne* 25 (1999), 55–82.
Olshausen, Eckart and Holger Sonnabend, eds, *Stuttgarter Kolloquium zur historischen Geographie des Altertums 5, 1993.* Amsterdam: Hakkert, 1996.
Osborne, Robin. "Framing the Centaur: Reading Fifth-Century Architectural Sculpture," in S. Goldhill and R. Osborne, eds, *Art and Text in Ancient Greek Culture.* Cambridge: Cambridge University Press, 1994, 52–84.
Padgett, J. Michael, "Centaurs and Satyrs in Early Greek Art," in: idem (ed.), *The Centaur's Smile.* New Haven and London: Yale University Press, 2003, 3–46.
Peppa-Papaïoannou, Irini, "Un relief mélien trouvé à Trézène," in: *Bulletin de Correspondance Hellénique* 125 (2001), 109–31.
Platz-Horster, Gertrud. *Die antiken Gemmen aus Xanten* I. Bonn: Rheinland Verlag, 1987.
Pretzler, Maria. "Form over Substance? Deconstructing Ecphrasis in Lucian's *Zeuxis* and *Eikones*," in: A. Bartley, ed., *A Lucian for our Times.* Cambridge: Cambridge Scholars Publishing, 2009, 157–71
Rebenich, Stefan. "Inventing an Ascetic Hero. Jerome's *Life of Paul the First Hermit*," in: A.J. Cain and J. Lössl (eds), *Jerome of Stridon. His Life, Writings and Legacy.* Farnham: Ashgate, 2009, 13–27.
Robbins, Emmett. "Cyrene and Cheiron," in: *Phoenix* 32 (1978), 91–104.
Robert, Carl. *Die griechische Heldensage I.* Berlin: Weidmann, 1920a^{4}.
Robert, Carl. *Die griechische Heldensage II.1.* Berlin: Weidmann, 1920b^{4}.
Robert, Louis. *Opera Omnia Selecta 1.* Amsterdam: Hakkert, 1969a.
Robert, Louis. *Opera Omnia Selecta 2.* Amsterdam: Hakkert, 1969b.
Rollinger, Robert. "Herodotus, Human Violence and the Ancient Near East," in: V. Karageorghis and I. Taifacos, eds, *The World of Herodotus.* Nicosia: Foundation Anastasios G. Leventis, 2004, 121–50.
Rollinger, Robert. "Extreme Gewalt und Strafgericht. Ktesias und Herodot als Zeugnisse für den Achaimenidenhof", in: B. Jacobs and R. Rollinger, eds, *Der Achämenidenhof.* Wiesbaden: Harrassowitz, 2010a, 559–666.
Rollinger, Robert. "Berg und Gebirge aus altorientalischer Perspektive," in: W. Kofler *et al.* (eds), *Gipfel der Zeit. Berge im Texte aus fünf Jahrhunderte.* Freiburg: Rombach, 2010b, 11–52.
Roscher, Wilhelm H., 'Kentauren', in: idem, ed., *Ausführliches Lexikon der griechischen und römischen Mythologie* II. Leipzig: Teubner, 1890–92, 1032–88.
Rossum-Steenbeek, Monique van. *Greek Readers' Digests?* Leiden: Brill, 1997.
Ruge, Walter. "Magnesia," in: *RE* XIV (1930), 459–73.
Schauenburg, Konrad. "Herakles bei Pholos," in: Athenische Mitteilungen 86 (1971), 43–54.
Scherf, Volker *et al. Antike Gemmen in Deutschen Sammlungen III.* Wiesbaden: Steiner, 1970.

Schiffler, Birgitt. *Die Typologie des Kentauren in der antiken Kunst vom 10. bis zum Ende des 4. Jhs. v. Chr.* Frankfurt: Peter Lang, 1976.

Schlüter, Margildis *et al. Antike Gemmen in Deutschen Sammlungen IV. Hannover und Hamburg.* Wiesbaden: Steiner, 1975.

Segal, Charles. "The Raw and the Cooked in Greek Literature: Structure, Values, Metaphor," in: *Classical Journal* 69 (1974), 289–308.

Shaw, Brent D. "'Eaters of Flesh, Drinkers of Milk': the Ancient Mediterranean Ideology of the Pastoral Nomad," in: *Ancient Society* 13–14 (1982–83), 5–31.

Siebert, Georges. "Imaginaires et images de la grotte dans la Grèce archaïque et classique," in: *Ktema* 15 (1990 [1994]), 151–61.

Simon, Erika. "Heilende Heroen," in: *Archiv für Religionsgeschichte* 6 (2004), 39–43.

Sporn, Katja. "Espace naturel et paysages religieux: les grottes dans le monde grec," in: *Revue de l'Histoire des Religions* 227 (2010), 553–71.

Stark, Michaela. *Göttliche Kinder. Ikonographische Untersuchung zu den Darstellungskonzeptionen von Gott und Kind bzw. Gott und Mensch in der griechischen Kunst.* Stuttgart: Steiner, 2012.

Steinhardt, Matthias. "Demeter Lernaia und Kentauros", in: E. Moormann and V. Stissi, eds, *Shapes and Images. Studies ...Herman A.G. Brijder.* Leuven: Peeters, 2009, 179–85.

Tomei, Dolores. "I kantharoi greci a figure nere e rosse," in: Ostraka 17 (2008), 111–80.

Valenza Mele, Nazarena. "Il ruolo der Centauri e di Herakles. Polis, banchetto e simposio," in: C. Annequin, ed., *Les grandes figures religieuses.* Paris: Les Belles Lettres, 1986, 333–70.

Verbanck-Piérard, Annie. "La rencontre d'Hérakles et de Pholos," in: L. Hadermann-Misguich and G. Raepsaet, eds, *Rayonnement grec: Hommages à Charles Delvoye.* Brussels: Editions de l'Université de Bruxelles, 1982, 143–54.

Versnel, Hendrik S. "Polycrates and his ring," in: *Studi Storico-religiosi* 1 (1978), 17–46.

Wachter, Rudolf. "The inscriptions on the François Vase," in: *Museum Helveticum* 48 (1991), 86–113.

Wachter, Rudolf. *Non-Attic Greek Vase Inscriptions.* Oxford: Oxford University Press, 2001.

Weiß, Carina. *Die antiken Gemmen der Sammlung Friedrich Julius Rudolf Bergau im Germanischen Nationalmuseum. AGD Nürnberg.* Nürnberg: Verlag des Germanischen Nationalmuseums 1996.

West, Martin L., *Indo-European Poetry and Myth.* Oxford: Oxford University Press, 2007.

West, Stephanie. "Introducing the Scythians: Herodotus on koumiss (4.2)," in: *Museum Helveticum* 56 (1999), 76–86.

Wilamowitz-Moellendorff, Ulrich von. *Die Glaube der Hellenen*, 2 vols. Darmstadt: Wissenschaftliche Buchgesellschaft, 1959[3].

Wisniewski, Robert. "*Bestiae Christum loquuntur* ou des habitants du désert et de la ville dans la *Vita Pauli* de saint Jérôme," in: *Augustinianum* 40 (2000), 105–44.

Woodford, Susan. "More Light on Old Walls. The Theseus of the Centauromachy in the Theseion," in: *Journal of Hellenic Studies* 94 (1974), 158–65.

Zwierlein-Diehl, Erika. *Die Gemmen und Kameen des Dreikönigenschreines.* Cologne: Verlag Kölner Dom. 1998.

Zwierlein-Diehl, Erika. *Siegel und Abdruck.* Bonn: Köllen Druck und Verlag, 2002.

Zwierlein-Diehl, Erika. *Antike Gemmen und ihr Nachleben.* Berlin: de Gruyter, 2007.

3. Wilderness and Hebrew Bible Religion – fertility, apostasy and religious transformation in the Pentateuch

Laura Feldt

1. Introduction[1]

This paper investigates the construction and function of 'wilderness' in Hebrew Bible religion.[2] In Hebrew Bible research history, the desert wilderness has been seen as ancient Israel's formative environment and as decisive for the development of its distinctive religion.[3] The idea that "Israelite religion" was unique in its ancient Near Eastern context because of its formative, desert wilderness experience has also played a seminal role in the history of religions (e. g., Segal this vol., Partridge 2004,8 – 11), just as it has appeared in recent discussions about religion and nature.[4] In order to approach the question of the construction and function of wilderness in Hebrew Bible religion, I analyse here a selection of epitomical wilderness narratives from the Pentateuch / Torah by means of a perspective that draws on the religion and nature field and critical spatiality theory. I argue that the wilderness is, in these texts, primarily used to establish Yahweh as the donor of agricultural fertility, and to communicate an ethos of religious identity transformation, the aim

1 I wish to thank Catherina Raudvere, Iram Asif, and Kirsten Nielsen for fruitful comments.

2 The term "Hebrew Bible religion" is explained in section 2. This study assumes that *midbār* in the Torah, the area that Israel travels through on the journey from Egypt to the promised land, can usefully be analysed as 'wilderness'.

3 See Hiebert 1996, 9 – 12. The idea that Israel's religion could only be understood in the context of desert nomadism, and its "historical consciousness" as forged in "the furnace of the desert", goes back, at least to A. Alt (1925, 1934), but was popularized in Frankfort (1949); see also de Vaux 1968. The related "desert ideal hypothesis" has a long history (K. Budde 1895, J.W. Flight 1923, Bright 1975, et al.); references in Cohn 1981 and Talmon1966.

4 The debates in the wake of the Lynn White thesis (White 1967), see (Kalland 2005).

of which is to ensure Israel's continued access to the fruits of the agricultural land. These two uses of the wilderness space have implications for traditional accounts of Hebrew Bible religion, as it has traditionally, in biblical studies, been characterised as a religion which attributes little significance to the nonhuman world (including agriculture cf. Hiebert 1996),[5] as well as for the understanding of the role of wilderness, which has predominantly been seen as negative domain. Previous treatments have not paid much attention to the impact of the wilderness narratives on the process of reception or its functions for religious identity formation.

2. Approaching Wilderness and Hebrew Bible Religion

Biblical studies have focused on discussing whether attitudes towards the wilderness in the Hebrew Bible are positive or negative, and on the wilderness as a concrete, geophysical space and historical period. Contrariwise, I argue that the wilderness is presented in an ambiguous way and that there is a need to move away from seeing it as an either predominantly negative (e.g., Talmon 1966) or predominantly positive space (the "nomadic ideal" theory in exegesis, e.g., Budde; de Vaux; references in Talmon 1966), or in terms of a simple cosmos/chaos dichotomy (Pedersen 1958–1960; Brueggemann 1977, 29[6]). To properly un-

5 Contrariwise, environmentalist literature has seen strong ties between Judaism, Christianity and Islam and agriculture, and viewed agriculture as inherently expansionist and imperialist (Nash 2001; Oelschlaeger 1991; cf. Taylor this volume). In environmental writing and ecocriticism, wilderness is seen as a notion based on an agricultural economy (Oelschlaeger 1991, 28; Garrard 2004,60). According to this view, the implication of the agricultural construction of 'home' vs. 'wilderness' is that agricultural life becomes a struggle *against nature*, and so the transition from hunter-gatherer to agriculture is a fall from a primal ecological grace. In this way, agriculture becomes the cause and the symptom of an anthropocentric alienation from the earth which was completed by monotheistic religion and eventually modern science (Garrard 2004,60–61). This view is in some ways problematic (Garrard 2004,61).

6 An exception is Jensen (2000), who notes the ambiguity of the desert, but does not investigate the wilderness per se. Talmon has rightly criticized the theory of a "desert ideal" or "nomadic ideal" in the Hebrew Bible (Talmon 1966). Instead, he argues that the view of the wilderness is predominantly negative (Talmon 1966, 48; Talmon in TWAT s.v. *midbār*; he is followed by Cohn 1981). I think there is reason to ask why, then, is the wilderness the primary arena for the encounter of deity and people and for communication between deity and

derstand the religious construction and function of the wilderness of the Hebrew Bible, it does not suffice to look at its historical-political[7] or geographical aspects only. On the other hand, seeing the wilderness as a literary motif (Schofield 2008 and Talmon 1966) or as an imagined space (as in a cosmological approach) only,[8] also leaves important aspects of these texts out of sight: the wilderness as a concrete, geophysical area, its climate and materiality. While newer studies document more diversity in the biblical attitudes to wilderness and a broader range of scholarly interests (Leal 2006, 55–56; Lee 2008; Schofield 2008[9]), Cohn's 1981 study is still the only study that investigates the religious use of the wilderness and engages perspectives from the study of religion. He demonstrates that the wilderness period fits van Gennep's triparte model as a liminal space, and he models Israel's journey from Egypt, through the wilderness, to the promised land as a *rite de pas-*

people in the Pentateuch? I also question Talmon's evaluation of the wilderness chronotope as having an "untergeordnete Stellung", a "beschränkten Stellenwert", in the Hebrew Bible (Talmon TWAT 686) – considering that almost all of Ex, Lev, Num and Dt are set in the wilderness, as well as his conclusion that since Israel spent only 40 years in the desert, then it is of little importance (Talmon TWAT 686). This seems not to do justice to the great amount of textual attention spent on the wilderness in the Hebrew Bible.Brueggemann sees wilderness as the "historical form" of chaos, as formless and lifeless (Brueggemann 1977, 29). Pedersen sees the wilderness as a negative, chaotic domain of curse (1958–1960); his view has been criticised by Addinall (1981).

7 T. Dozeman's insightful book *God at War* (1996) treats many wilderness texts, arguing that the "rise of the wilderness tradition in the Pentateuch is an exilic innovation by deuteronomistic tradents" (Dozeman 1996, 95–96). Yet, he does not engage the wilderness as a space or the importance of the nonhuman world in the wilderness traditions, but relates them to ideas of historiography and salvation history. In his assessment, however, the "rise of the wilderness" is, in the Pentateuch, "its most significant addition for constructing theology" (Dozeman 1996,179), thus attributing space social significance (cf. also Dozeman 1989).

8 Cohn 1981 veers in this direction.

9 These authors suggest that neither the negative nor the positive view of Israel's desert experience is adequate, and yet they do not treat the religious function of the wilderness, nor the spatial and material aspects of the Hebrew Bible wilderness and its focus on the non-human world. Lee (2008) understands the wilderness narratives through the lens of Yahweh's "punishment and forgiveness", focusing on the theological message. R.B. Leal has recently, in a work of eco-theology, demonstrated the diversity of the attitudes to the wilderness in the Hebrew Bible (2006). Yet, he understands the wilderness as a real space and does not offer detailed analyses of the texts or the wilderness' religious functions.

sage/paysage. This valuable analysis, however, does not address the concrete-material aspects of the wilderness space and the images of the nonhuman world used,[10] or its religious instrumentality beyond what is general to (all) liminal spaces.

What is so interesting about the Hebrew Bible desert wilderness is exactly that it oscillates between a real and a fantasmatic presentation, in between cosmology, literary motif, spatial practice and geography. It retains references to and elements of realistic geographical space while also being the primary site of fantastic marvels[11] in the Hebrew Bible. The basic motivation for this paper comes from recent developments in the field of religion and nature which guide my interest in the presentation of the materiality, climate and modes of interaction with the nonhuman world, rather than previous dichotomizing ideas of culture vs. nature.[12] A view of humanity and culture as inseparable from the nonhuman world, as *not* distinct from 'nature', but as implicated and involved in, always interacting with, its environment, has become commonplace in the 'religion and nature' field (Taylor 2005) as in other current investigations of climate and environment (Jasanoff 2010, Ingold 2000, et al.) The decisive theoretical inspiration and methodology for this study comes, however, from the larger field of "the spatial turn in the social sciences" and the application of this theoretical perspective by scholars of religion and anthropologists. Critical spatiality theory has, in the last decade, become an important influence in many fields, including biblical studies (Camp 2002, Berquist and Camp 2008, Lied 2008, George 2009). The major point of critical spatiality theory is that space, like history and society, is always construct-

10 For Cohn, "history" is still the organizing principle for religion in the Hebrew Bible (Cohn 1981,1).

11 I here use the words marvels or fantastic events to designate those elements of the texts traditionally designated 'miracles'. For arguments in favour of a change of vocabulary, see Feldt 2011a.

12 Wilderness does not equate with the idea of nature. It is conventionally seen as those parts of nature that are largely free of human influence (Oelschlaeger 2005). I use the term 'non-human world' in line with the insight that humans are always involved and engaged in relations with the nonhuman world; that humans do not belong to a realm of 'culture' that is separable from 'nature'. The concept of 'nature' has played a tremendous role in Hebrew Bible research vis-à-vis the concept of 'history' (cf. Simkins 1994, Hiebert 1996,4–22), and for that reason I sometimes use also the term 'nature' to communicate that my results may have implications for previous research involving this dichotomy.

ed[13] and produced in social practice and that it must therefore be theorized. Like many other areas of scholarly inquiry, the field of biblical studies has been dominated by an interest in history and sociality, while the dimension of space has received less attention. The work of philosopher Henri Lefebvre and geographer Edward W. Soja demonstrates that space is not a mental construct alone, it is also always material, and representations of space in social thought and practice cannot be understood as projections of modes of thinking independent of socio-material conditions (Soja 1989,124–126; Lefebvre 1991). Soja, building upon the work of Lefebvre, introduces a useful tripartite strategy for the analysis of social space (Soja 1996), which, I contend, will benefit this investigation of the religious instrumentality of the wilderness in the Hebrew Bible.

Soja's theory hinges on a tripartite approach to space as *Firstspace*, *Secondspace* and *Thirdspace*, a distinction which is indebted to Lefebvre's distinctions between perceived space, conceived space and lived space. Firstspace concerns the concrete materiality of spatial forms, the geophysical realities as empirically observable, Secondspace involves imagined space, ideas and representations of space in mental or cognitive forms, while Thirdspace designates the lived experience of social space, the lived realities as practiced.[14] Thirdspace is a creative recombination that builds on both Firstspace and Secondspace and, as such, it moves beyond the alternatives of describing spatiality only as material, physical, or as an imagined, mental construct (Soja 1996; Camp 2002; Lied 2005). Thirdspace is a term pointing to space as practiced in the lifeworlds of human beings; the field of social spatiality that uses and incorporates Firstspace and Secondspace aspects, but which cannot be reduced to the combination of the two (Lied 2005,108). In actual, human practice, these aspects cannot be separated. For purposes of analysis, however, they may be usefully distinguished.

This theory formation offers inspiration for approaching the construction of the wilderness space in the Hebrew Bible, but because the wilderness space is a narrated space, in which all three kinds of spatial aspects are intermingled, I supplement critical spatiality theory with

13 That space is culturally constructed is of course a commonplace insight. Many anthropologists (and narratologists, cf. Bal 1997) use the helpful distinction between place and space (e. g. Kirby 2009), which I do not engage here.

14 As mentioned by C. Camp, Jim Flanagan has helpfully dubbed these aspects material space, designed space and lived space (Camp 2002).

insights from the narratology of Mieke Bal (1997).[15] For the purposes of the present analysis, I approach *Firstspace* as the concrete, geophysical and geographical materiality of the land areas or spaces mentioned in the texts.[16] As for *Secondspace*, it is not correct simply to equate written texts, especially of the canonised sort, with Secondspace, on the assumption that "if it is written, it must be conceptualized" and accordingly Secondspace, as Claudia Camp notes, for the boundaries between the three aspects of spatiality collapse in any actual text. Narrative literature "may create "worlds", in which recipients and personae may "live"" (Camp 2002); narrative literature may supply models (which may come to function as Secondspatial paradigms) for thinking thirdspatially and be a site of Thirdspace experience, but may also incorporate thirdspatial aspects. In narrative (as elsewhere), spatialities are always intervowen in complex ways, and narratives may be important sites of negotiation and struggle between spatial aspects.[17] The texts' secondspace aspects will, for these reasons, here be treated by focusing on those spatial models and conceptual, geographical 'systems' (itineraries, maps) used to explain and think space that are drawn upon in the texts (George 2009, 25–27; Bal 1997, 214–216) for instance ideas of spatial oppositions between city-land, land-desert, the sown-unsown, political oppositions like hostile vs. friendly that may be ascribed to them, and other ways of speaking of, thinking about, and measuring space; or spatial paradigms, for instance of mountains or "Egypt", as represented in the texts.[18] Since Secondspace aspects represent a discourse of power (Soja 1996, Camp 2002 and 2008), a dominating appropriation of space, I combine the above with attention to the story-level utterances

---

15 Thus, I do not adhere strictly to Soja's theory, but use his tripartite approach to space strategically, akin to Bal's tripartite approach to narrative (text, story and fabula) – as different aspects that never appear separate in any text, but which may be distinguished for the purpose of analysis.

16 As a combination of archaeological and other knowledge of the area with information supplied in the text. This is far from simple – cf. Lied 2005, 108 note 30.

17 The Hebrew Bible is a good example of a text collection that bears the stamp of official religion, while nevertheless expressing many heterogenous voices – this is recognized both by the Documentary Hypothesis in standard biblical studies, and by Bakhtinian approaches in literary approaches to the biblical texts.

18 Bal distinguishes analytically between fabula, story, and text in the analysis of narrative (Bal 1997); for an extensive introduction to Bal's narratology, see Feldt 2011c. Both the fabula's construction of location/place, and the 'story'-layer's construction of 'space' are relevant in analysis (Bal 1997, 132–141 and 214–216).

of narrator and deity about the wilderness space as important indicators about the texts' surface ideology and spatial ideals,[19] on the assumption that the deity and narrator represent authoritative voices in these religious texts. I attend to the texts' *thirdspatial* aspects by looking at expressions of lived experience, emotion, sensation, opinion, and the like in the embedded character reactions and utterances, which I take to point towards (models of and models for) the reception and use of the wilderness in lived religious space.[20] My focus is here on how the characters experience (in sight, hearing, touch, etc.) and perceive space as expressed in their utterances and reactions. Further, I approach the texts with a narratology-based assumption that the events happening in a particular place, and how the characters react to them, are just as important in a spatial analysis of narrative as are straightforward descriptions of space (cf. Bal 1997, 134–137), an assumption which is in accord with critical spatiality theory (Soja 1996, 65–70, Camp 2002). Therefore, I hope to bring out what the events of the wilderness, as well as the personae reactions and utterances, communicate about the wilderness and its religious function. My assumption is that this approach may illumine how the texts aim to influence the process of reception and the lived space of their intended recipients, as well as bear witness to the social space of their author-redactors – the spaces in which, from which, these seminal wilderness texts were produced and used.

As for the material, the Pentateuch is the composite result of multiple writings, rewritings, redactions etc., and I read it here in the Masoretic edition,[21] as a composite, redacted text, and I take it to speak of the

19 Importantly, these may be qualified, contested, questioned or backed up in other discourse.

20 In making such a distinction, and assuming that the reactions of Israel reflect and model the texts' (intended) recipients, it is of course important not to lose sight of the fact that "Israel" and the reactions of the "Israelites" in the texts are textual constructions. Importantly, the secondspace aspects may be qualified, contested, questioned or backed up in other discourse, cf. in a narratological perspective, Bal 1997,25, 34, 43–50, 52–61.

21 The Pentateuch/Torah is a collectivity of text types – mythology, ritual text, wisdom, song, poetry, etc. – and in all likelihood put together from a diversity of sources. Further, as for manuscripts, there is no "the text", only copies of copies of copies, but it is clear that *a* final edited form appeared in the Common Era (while elements of the texts of course represent older traditions). The Masoretic text forms the basis for this study, even if it reflects exegesis and does not present "the original text"; the absence of original texts (the composite text as the condi-

late author/redactors and recipients of this version, on the assumption that the Pentateuch as a "collectivity … invites interpretation" for the study of religion (quote in Leach 1987, 581; Hendel 2005,96–98), while recognizing its basic multifariousness (cf. Hendel 1997). My theoretical object is the use of 'wilderness' in Hebrew Bible religion. 'Hebrew Bible religion' is a shorthand term for the various idealized practices and ditto theologies expressed in the anthology of texts known as the Hebrew Bible,[22] and the role that the wilderness plays in and for it. While wilderness texts are found elsewhere in the Hebrew Bible, the Pentateuch represents the richest and most complete development of wilderness traditions in the Hebrew Bible. I focus here on sections of texts which frame and epitomize the wilderness experience, namely the stories of Israel's entry into the wilderness and Yahweh's revelation on Mt. Sinai, Ex 15,22-Ex 19, as well as the stories that relate Israel's departure from Sinai and first glimpses of the promised land, Num 11–14.

tion for the process of reception) is a feature Hebrew Bible study shares with the study of cuneiform literature (Feldt 2006a; Feldt 2012,87; Carr 2010). The composite, redacted version is also worth investigating in terms of its effects on the process of reception. It has been demonstrated that Hebrew Bible narrative is textual-literary in character (Kawashima 2004, van der Toorn 2007) and the ancient use of scrolls of "books" is well-documented (Propp 1999,39–54). On the textualization of religion in the late 1st millennium, see Schaper 2009, van der Toorn 2007, Veldhuis 2003.

22 "Hebrew Bible religion" differs from "Israelite religion", which is the hypothetical, reconstructed religion of the ancient "Israelites". The sources for Israelite religion are not only the Hebrew Bible texts but also archaeological material from the relevant epochs, but the reconstruction of "Israelite religion" or the" history of religion of Israel" is, unfortunately, often based on a more or less uncritical acceptance of the image given in the Hebrew Bible (Jensen 1998, 13). As H.J.L. Jensen explains, it is debatable whether the Hebrew Bible refers to any actually existing ethnic group by the name of Israel, or whether "Israel" is primarily constructed by the Hebrew Bible (Jensen 1998, 17). The Hebrew Bible is a religious anthology aiming to promote and maintain proper worship of the deity Yahweh. It addresses itself to the group "Israel" which worships the deity Yahweh, or which ought to do so. The texts postulate that this group possesses a common identity genealogically (common ancestors), historically (common experiences), territorially (common land), and religiously (common worship) (Jensen 1998). The authors/redactors stand at a considerable distance from the time when the narrated events purportedly took place, and the texts in their composite form therefore speak of the world of their relatively late author/redactors and their situatedness in early Judaism.

3. Analyses

First, I analyse the desert wilderness as represented in terms of firstspatial, geophysical materiality and human ways of interacting with it in the selected texts in combination with archaeological and geographical information. Afterwards, I attend to the events, actions and experiences that characterise the wilderness.

3. a. The desert wilderness – firstspace aspects

As table 1 demonstrates, the desert wilderness (*midbār*) is, in the selected texts, presented in primarily realistic terms, as a vast, waterless expanse where the sun burns. The desert dryness is interrupted by occasional oases.

Table 1. Verbalization

Terms used	*midbār* (passim)
1. Climate, materiality, nature	– Burning sun (Ex 16,3) – Vast expanse (Ex 15,22), large distances (e.g., Ex 16,1) – An oasis of 12 springs and 70 palms at Elim (Ex 15,27) – Uninhabited (wilderness vs. settled land (*ʾereṣ nôšāvet* (*yšb*)) Ex 16,35) – Dryness and lack of water, bitter water (Ex 15,22–26) – A mountain in the wilderness of Sinai (Ex 19,2.4) – Lack of (varied) food (Num 11), an alimentary lack of variety: it lacks grain, figs, grapes, pomegranates, water, miserable food (Num 11,5; 20,4–5) – Dew falls on the camp in the morning (Num 11,9) – Yahweh descends in a cloud (Num 11,25) – A storm from Yahweh with quails from the sea (Num 11,31) – Snakes (Num 20,6–9)

Table 1. Verbalization *(Continued)*

Terms used	*midbār* (passim)
2. 'Geographical' interaction with the wilderness / ways of moving about	– Movement from specified place to specified place (*passim*) (e.g., Yam-Suf (Ex 15,22) to *midbār Šur* (Ex 15,22) to Mara (Ex 15,23), to Elim (Ex 15,22), to *midbār* Sin (Ex 16,1); Massa and Meriba (Ex 17,7); Num 9,15–23; Num 10,11–13; Num 11,35; Num 12,16)); Tabera (Num 11,3); Kibrot-HaTa'ava (Num 11,34–35); Haserot (Num 11,35); *midbār Pā'rān* (Num 12,16; 13,3); Kadesh (Num 13,26); *hammidbār derek yam-sûp* the wilderness by the way to *yam-sûp* (Num 14,25) – Camping in the wilderness of Sinai across from the mountain (Ex 19,1–2); camp-space (Num 11,1–3) – their children must live as shepherds in the wilderness for forty years (Num 14,33) – wilderness as site of death as Yahweh's punishment for the Israelites' 'fornication' (disobedience, faithlessness) (Num 14,33–35)
3. Temporal	– Extended time period (Israel spends 40 years in the wilderness, Ex 16,35)[23] – Large scale temporal notices: In the third month after the departure from Egypt, they enter the wilderness of Sinai (Ex 19,1); on the first day of the second month of the second year after Israel left Egypt (Num 1,1)

As we see above, the texts do not offer many, detailed descriptions of the wilderness' materiality, climate and other physical conditions. The

23 A fantasmatic time period roughly corresponding to one generation. Human reproduction ceases while in the wilderness (no births take place there) (Cohn 1981,16); this lends the wilderness a temporal 'specialness'. The normal life cycle resumes upon entry into the land (Josh 5,5–7).

wilderness is not to any great extent described as 'nature'.[24] The representation of the desert space clearly retains enough geophysical description to appear as a realistic space. While there are many itinerary notices in the texts, and while they may communicate a pattern of significance (Davies 1983),[25] what is of importance here is that the journey is portrayed as taking place in a realistic space characterized by dryness, by its large size, by being uninhabited, unsown, and characterised by nomadic movement. Yet, as is apparent from the table above, what is in primary focus for the author/redactors is the desert wilderness' lack of food and water. The prime attention is focused on the alimentary area.

The word *midbār* used in these texts is also the most commonly used word for the desert wilderness in the Hebrew Bible overall (Talmon in TWAT s.v. 665).[26] S. Talmon has summarized the geophysical and material denotation of the word *midbār* in the Hebrew Bible (Talmon TWAT; see also Talmon 1966) as representing dry and semidry regions, which cannot be used for agricultural purposes, but which is sometimes used as grazing land for small flocks of goats and sheep (Talmon TWAT s.v. 664). As such, it could also be translated 'steppe' (Talmon 1966, 40). The majority of the occurrences of *midbār* in the Hebrew Bible designate this "drift" area, which is situated between the cultivated land and the desert proper; a space adjacent to the settlement or camp, but not a part of it (Talmon 1966, 41). *midbār* is, however, also used to designate the desert proper, the arid and desolate zones beyond the drift (Talmon 1966, 41). Often, when *midbār* and related words are used, it is the lack of water that is stressed (TWAT 668). However, it is important to also note that *midbār* also designated areas within "the land" proper. The boundaries between the agricultural land and the *midbār* were thus to a certain extent fluid and moveable (Talmon TWAT 671; Pastor 1997, 1–12). On the one hand, *midbār* areas could impinge on the areas of cultivated land in times of drought;

24 This is further highlighted by comparison with, for instance, the descriptions of the mountain forest wilderness of Sumerian religious texts (*Inana and Ebih*; *Lugalbanda in the Mountain Cave*, Black et al. eds. 1998).

25 According to Davies, the itinerary notices are redactional and function to tie together material as points on a route, they give the impression of a meaningful whole and a purpose to the diverse material, and thus they contribute to the task of restoring "the morale of the Judaean people" (Davies 1983, 8–13).

26 While *midbār* is used 271 times in the Hebrew Bible texts and *'arābāh* 59 (60) times and *šemāmāh* 57 times, the other related words like *y^ešîmôn* are used comparatively little (13 times; *ṣiyyāh* 15 times) (TWAT 667–8).

on the other *midbār* areas could, in some places of ancient Palestine, be transformed into cultivated land by means of irrigation (Pastor 1997, 10–12). Sheep and goat pastoralism and sedentary agriculture were always closely integrated practices in the Levant (Hiebert 1996,19–21; Bar-Yosef and Khazanov 1992). These concrete features of the landscape and climate of ancient Palestine, and the mixed ways of interacting with it, did of course not determine a specific wilderness conception, but provided a certain possibility of form, which must be taken into account (cf. Pedersen in this volume). The texts' alimentary perspective on the *midbār*, as well as the location of *midbār* as mobile, are important features to keep in mind. Now, I move to the actions and utterances of the personae.

3.b. Second- and Third-space aspects

I attend here to the events happening in the wilderness – and how the characters react to them – on the assumption that they are just as important as geographical description (cf. above). I will try to bring out how the narrative text functions as an arena for negotiation and contestation between spatial aspects as suggested above in the strategy of analysis.

A Entry into the wilderness – Exodus 15,22–16,36

In these chapters, Israel's first entry into the desert is narrated, and we are given the primary coordinates of the desert in the Hebrew Bible imaginary. When we examine the events, actions and the utterances of the characters represented as taking place in the wilderness, we see that the wilderness is, above all, the space of the deity's *marvellous events* and Israel's *reactions of rejection* (table 2). I take the characters' utterances and actions as indicative of thirdspace aspects, while I take the use of spatial paradigms and the utterances and actions of the deity and narrator as indicative of secondspace aspects.[27]

---

27 I have opted to include only the deity's and Israel's actions and utterances in table 2. The narrator is of course also relevant, but the narrator seems here primarily to offer support for the deity's voice. The other characters will be drawn in *ad hoc.* Moses occupies an interesting intermediary position, as he of course expresses textual ideals to some extent (note the frequent blurring between Yahweh and Moses), while also modelling recipient reactions in relation to the wil-

Table 2. Actions, reactions, utterances about the wilderness

Yahweh's actions and utterances	
1. Marvels/fantastic events:	*Alimentary marvels:* General: provides food and drink for a hyperbolic amount (cf. Ex 12,37)[28] of people and livestock (Ex 16,32), (Ex 17/) Bitter water to fresh water (Ex 15,22–25); Manna falls like dew (rain) in the morning in specific quantities (16,4.13–14); it rots and stinks and attracts maggots/ if stored (16,19); it comes in double amounts the day before Shabbat (16,22–24); it was like coriander seed, white, and it tasted like honey cakes (Ex 16,31). The sudden appearance of a flock of quails at the campsite (Ex 16,2–13); Provides water from a rock (17,5–6) *'Natural phenomenon'/'climate' marvels:* the pillar of cloud appears during the day to guide the Israelites (13,21–22); Yahweh communicates in the cloud pillar regularly (Ex 33,7–23)]; guidance (Ex 13,17; 16,32); the deity is accessible, available as/through cloud/fire pillars (13,22; 16,10) the pillar of fire appears during the night to guide the Israelites (Ex 13,21–22);

derness experiences. His persona is an interesting medium for negotiation of spatial aspects.

28 Ex 12,37 states that the number of Israelites involved in the Exodus was about 600,000 men. If they were married with families, then the total number would have been about two or three million people, plus cattle. Many commentators since Reimarus have noted the impossibility that numbers of this size could have been sustained with food and water for forty years in the desert (e.g., Johnstone 1990, 27 f). Num 1–4 confirm the number. The magnitude is conscious and purposeful. Grabbe calls attention to the normal size of armies in antiquity, which makes the numbers even more phantasmatic, and the lack of fear of Israel in other ethnic groups unbelievable (Grabbe 2000), and, I add, this also makes Israel's own doubt and lack of trust in the deity even more unbelievable. Meyers makes the good point that the great number points towards an inclusion of later groups of Israelites in the Exodus event (Meyers 2005,100).

Other actions/utterances:	*'Political' marvels:* Unspecified and ambiguous supernatural help in the war against Amalek (17,8–16)[29] Threatens Israel with diseases (Ex 15,26)[30] The wilderness trek is a deliberate detour (Ex 13,17); to make Israel wander aimlessly or confused (Ex 14,3); to test Israel (Ex 15,25–26) The purpose of the manna marvel is to test Israel, whether they will 'walk in my law/ instruction' or not (Ex 16,4) When the Israelites eat manna until they are full, then they will understand that Yahweh is their god (Ex 16,12) Reminders about Sabbath rest (Ex 16,28–29)
Israel's / the people's actions and utterances	Wandering (passim), successive camps / nomadic movement (passim), they break camp, wander, and rest when Yahweh's cloud communicates it[31] Complaint, doubt, apostasy (Ex 15,24; 16,2.8; 17,1–4; 17,7) Purpose of the wilderness trek according to Israel: to let us die from hunger (Ex 16,3) Hunger (Ex 16,3), nostalgia for the flesh pots of Egypt (Ex 16,3) Wonder at the marvel of manna (Ex 16,14–15); gathering manna (Ex 16,17–18) Disobedience/rule breaking (Ex 16,20.27–30)

As we see in table 2 above, the events that are characteristic of the wilderness are marvels, and the type of experience and reaction, for the human recipients, are alimentary deprivation and apostasy. The subject matter of the central marvels in the wilderness has been overlooked as communication about the wilderness. The primary subject matter of the marvels is alimentary – manna, water, and birds. The other marvels of the stories are also 'nature'-related marvels; e.g., the pillar of cloud and of fire communicate that Yahweh is present. Clearly, the nonhuman world is a primary arena of Yahweh's activity.

The alimentary type of fantastic event in Ex 16 represents a systematic inversion of what constitutes ordinary food according to the Hebrew Bible: manna is an imaginary, fantastic type of bread raining from the sky. It is a type of food which, contrary to ordinary types of food, does

29 For an analysis, see Feldt 2012.
30 There is a textual ambiguity as to whether this is uttered by Moses or Yahweh.
31 See also (Ex 17,1; Ex 40,34–38; Num 21,10–15), (Num 9,15–23).

not require work (cf. Gen. 3:17), it is free, it is collected, not harvested, and it cannot be stored[32] and it is a type of food which is in adequate supply.[33] In all of these respects, manna is oppositional to ordinary bread and it represents a systematic inversion of the normal in the Hebrew Bible: agriculture and subsequent bread-making (H.J.L. Jensen 2000, 39). In Ex 16, the manna is a curious, benign wonder, which attests to the deity's care for the people in the wilderness. The manna ceased upon Israel's entry into the promised land (Josh. 5:11–12; Ex 16:35) and this supernatural food characterises the wilderness wanderings more than anything. Here, in Ex 16, we get a predominantly positive story of the manna as a benign wilderness marvel. Ex 16:22–30 connects manna and the Sabbath and establishes the creation cycle of work and rest (Gen. 2:2–3) in the world (W.H.C. Propp 1999, 598),[34] as well as a weekly wilderness memorial practice.

The appearance of the flock of birds is equally oppositional to ordinary forms of food production in that it is also work-free and inverts not only cattle farming but also hunting, in that the prey here approaches the hunter, not the other way around. The water produced from the rock by striking it with the staff (Ex 17,1–7) also represents a magical-miraculous type of nourishment, which is work-free, instantly produced and which is oppositional to the ordinary means of procuring water. As such, all three types of alimentary marvels represent inversions of the ordinary, everyday world of agriculture. They communicate not only who is the lord of agriculture and 'natural' fertility: Yahweh, but also what Israel desires from its deity: 'natural' fertility and food. Ordinary, agricultural food production is here filmed over with the fantastic. The wilderness marvels point to Yahweh as the lord of that fertility of agriculture which ensures human alimentation and survival. The ideal landscape in the background of the texts is clearly an agricultural landscape (see also Talmon TWAT 685).

Further, the desert wilderness space comes to oscillate between a real and a phantasmatic presentation. For instance, the manna and the quails marvels have a supernatural cause, but the events are presented in a way that points towards a similar double readability between the natural and

32 It rots if stored, cf. Exod. 16:20.

33 It is not conspicuous consumption; everybody gets what s/he needs.

34 Manna is a heavenly, supernatural food, holy, and therefore fragile in the human world. The manna deposited at the (later) tabernacle (an instance of time warping, Exod. 16:32–34) does not decay, because here the nature of the food corresponds to the nature of the place.

supernatural,[35] as well as between the benign and the malign. The text's similes (bread, Ex 16,4, frost v. 14) and modes of interaction (gathering, decaying) contribute to the search for naturalistic explanations, while still retaining the marvel. Exegetes have pointed to migrating quails and 'honey-dew', sweet pellets secreted by plant-lice, as possible natural explanations of these fantastic events in the desert wilderness (Durham 1987, 224; W.H.C. Propp 1999, 599–600). While naturalistic paraphrase of the marvels is an unconvincing interpretational strategy, it is important to note that the texts do not present manna as a spiritual food; it is a very concrete food that can rot and melt and which the Israelites use and interact with in ordinary ways.[36] The double readability between the ordinary ways one can interact with these alimentary marvels and their 'supernaturalness' contributes to the desert's oscillation between the real and the fantasmatic.

The predominant reactions of Israel within the wilderness (cf. table 2 above) are doubt (e.g., 16,2–3)[37] and disobedience (e.g., 16,20, 27–28). The doubt, rejection and disobedience of Israel in the wilderness is so pervasive and so striking—so improbable, considering all the fantastic events—that it requires interpretation. Israel's later interaction with Yahweh in the wilderness is also a story of Israel's apostasy and lack of obedience. Israel's apostasy is, in the context of Ex 15,22-Ex 18, unbelievable,[38] considering the recent, marvellous events in Egypt and the crossing of the sea. The statements of the Israelites in 16:3 that they believe to have been led into the wilderness to die, and of Moses in 16:6 that Israel will now know that it was Yahweh who led them out of Egypt speak out the possibility that there was no supernatural presence (or another) behind the fantastic events. This is a significant statement, for if we take the fantastic events of the narrative at face value then how could Israel possibly doubt that Yahweh led them out of Egypt – considering the plagues of Egypt and the crossing of the sea, the guiding pillars of fire and cloud? This sentence confirms the ambiguity of the wilderness space as

35 Ex 15,22–26 and the bitter to sweet water marvel is a further example of a double readability, cf. Feldt 2012, ch.3.

36 Generally Israel's wilderness experience is presented as concrete and bodily. It thus differs from the later ascetic use of the wilderness.

37 Childs also finds that the people's complaint constitutes unbelief (Childs 1974, 285, 287, cf. also Durham 1987: 220, 230–231).

38 Yahweh does slay thousands for impiety in the desert (Exod. 32; Num. 11; 16–17; 21:4–9) and does not allow the rest to enter into the land (Num. 13–14; 20:1–13).

hovering between a realistic and a fantastic presentation. From this point on, when we hear of the disobedience, doubt and rebellion of the Israelites, we must remember that these reactions occur in a narrative context according to which Israel gathers its daily fantastic food, its daily miracle, every day (cf. above). This renders Israel's doubt, hesitation and disobedience even more extreme and noteworthy. I submit that these reactions become understandable if seen as reflecting an agricultural environment. I return to this in section 4.

The striking discrepancy between the fantastic events of Yahweh for the people and Israel's apostasy, disobedience and doubt characterises the wilderness more than anything. This discrepancy can be read as formulating a pressure towards self-transformation, towards obedience, belief, 'goodness', as communicating an ethos of religious transformation. I will return to this point in section 4.

B Instituting Wilderness: the mountain revelation, Sukkot, law, sanctuary

In the texts placed between Israel's entry into the wilderness in Exodus and the departure from Sinai and the further wilderness experiences in Numbers, decisive aspects of the religion and wilderness nexus are communicated. I do not offer full analyses of these texts, and many important issues cannot be treated here. However, I wish to point to a few aspects that I find important for the understanding of the wilderness and its religious function in the Pentateuch.

The importance of the dramatic episode in which Yahweh descends to the mountain in the wilderness to meet Israel in Ex 19 for the Pentateuch (and the Hebrew Bible) is difficult to overstate. The placement of the mountain of Yahweh's foundational revelation in the wilderness setting is emphasised in the first lines of Ex 19 with repetitive formulations that stop the action momentarily to emphasise the setting of 'the wilderness of Sinai' (*midbār Sînāy*). The ancient Near Eastern spatial paradigm of the cosmic mountain or mountain of god (Dozeman 1989,19–36) and the climactic moment of national formation and divine revelation is specifically placed in the "wilderness of Sinai" (Ex 19,1–3), demonstrating that the Hebrew Bible wilderness cannot be seen as solely negative in these texts. The text here draws on wider, ancient Near Eastern spatial paradigms of mountains as meeting places between humans and deities and as sites of divinely given fertility (Talmon TWAT 669–670 s.v. *midbār*; Simkins 1994,140), just as mountains are, in the Hebrew Bible generally, places of refuge and divine revelation (Cohn

1981,13). The spatial placement of the mountain of revelation in the wilderness is highly significant as a statement about the wilderness' positive qualities.[39]

Moreover, the discrepancy between Yahweh's perspective and that of Israel is resolved here in the 2nd space imagery offered by the deity, when Israel's confusion, thirst, hunger, and doubt on the trek from Egypt (cf. fig. 2 above), through the wilderness to the mountain, are re-described by the deity as "an exhilarating swift, smooth flight" (Pardes 2000, 66) on the wings of an eagle,[40] casting the deity in the image of a caring animal parent and the people as his fledgling (Pardes 2000,65–68). In this re-description, the wilderness chronotope is a space of nearness to the deity, a time of great intimacy (cf. also the special designations offered to Israel if they keep the covenant: *sᵉgullāʰ mamleket kōhanîm wᵉgôy qādoš*, 19,5–6), and of parental care (cf. Dt 8, Dt 32). This moment of intimacy between deity and people, the moment of unity at the mountain in the wilderness, where the people respond as one and accept the terms of engagement, is of pivotal, positive importance in the religious imaginary.

It is also important to observe the vital role that the nonhuman world plays in this text about the deity's decisive revelation. Yahweh's manifestation is verbalized in terms of aspects of the non-human world. It is through cloud, lightning, thunder, fire, vivid images of a mountain set ablaze by dramatic natural phenomena, that Yahweh's presence is imagined. The deity's appearance on the mountain, in and through which the relationship between Israel, deity, and land are verbalized, regulated, mediated, is constructed in 'nature' images. The image of the thundercloud, lightning, etc. (Ex 19:18–20) are basically thunder storm images. The thunderstorm on the mountaintop is relevant mainly for the farmers of the highlands in Palestine, and while it was the source of great destructive power, it was also beneficial in that it brought the seasonal rains necessary for agriculture (Lang 1998,146–158; Hiebert 1996; Simkins 1994, 146). This offers further support to the argument that the landscape which stands at the centre of the wilderness narratives and which

39 In the narrative of Elijah's encounter with Yahweh's fantastic events on Mt Sinai/Horeb, the wilderness and the mountain are also explicitly connected (1 Kings 19).

40 This image is also taken up in Dt 32:10–11.

is the focus of Israel's wishes is an agricultural landscape, and that the nonhuman world is of great importance in the Hebrew Bible.[41]

While the wilderness as a narrative frame space is of momentous importance in the Pentateuch, also for the law and cultic prescription materials, Sukkot is the only festal celebration explicitly prescribed as a ritual practice for remembrance of the wilderness events (Lev 23:39–43) within the Pentateuch/Torah wilderness framework. Sukkot is often seen as an originally agricultural festival connected to the (fruit) harvest during the dry season,[42] which was later ascribed a 'historical' background by being connected to the wilderness wanderings in Lev 23:40 (Spaulding 2009,41–42; 55–66). The commandment to 'dwell in booths' is connected to the exodus and wilderness wanderings in v. 42–43. Our understanding of the passage is not brought forward by observations that booths made of branches of green plants are impossible and anachronistic in the desert. Sukkot was prescribed as a time of joy, where the community would celebrate Yahweh's blessing, of which the fertility of the plants and the crops were seen as an expression. The present analyses of the wilderness stories show that the connection of Sukkot to the wilderness is not necessarily 'artificial' ascription[43] that marries a 'historical' concern on the one hand to a 'natural/agricultural' concern on the other, because the concern of the wilderness stories is also for agricultural fertility. The verbal forms involved stress Israel's dependence on Yahweh for survival (hipʿil of *yšb* and *yṣʾ*). In this way, a 'wilderness' perspective on agricultural fertility is offered. The connection between wilderness and Sukkot is intimate and meaningful in the religious imaginary: both the wilderness stories and Sukkot celebration concern human survival and the source of that fertility in 'agriculture-nature' that ensures human survival. To be sure, it is likely that the ritual prescriptions here marry a popular religious practice, an existing festal practice, to the wilderness stories, but

41 Often in biblical studies, nature/myth terminology is used of the other ancient Near Eastern cultures, while the Hebrew Bible gets the 'historical' predicate. But the image is much more complex – in the ancient Near East more broadly, deities were also seen as independent from the nature images with which they were characterised, and their sense of time was not necessarily 'cyclical' (Jacobsen 1987, Hendel 2005, 98–99; M.T. Larsen 1987, Albrektson 1967).

42 Jensen has pointed out how miraculous provisioning of food is related to the dry season, the season of abundance, of harvest, in the Hebrew Bible (Jensen 2000,514).

43 I do not rule out that the connection may be late; I just wish to point out that the connection is 'natural' in terms of the religious imaginary.

the relevant distinction is not that between history and nature. In Lev 23, the regulations for the celebrations are not especially intended for priests (as they are elsewhere), and they are not applicable to one place only (Ulfgard 1998,80–81) – a more likely distinction is popular vs. elite. The interpretative character of Lev 23:43 is unique and probably connected to an 'early Judaism' marked by exilic experience (Ulfgard 1998,216–218). The concern for agricultural fertility is, in this text, deterritorialized. The ideology of the ritual practices surrounding Shabbat and Sukkot concern agricultural fertility just as the Torah wilderness stories do. Together they communicate that these religious practices are not connected to any particular place.[44]

Lev 26 corroborates the peasant-agricultural view of what Yahweh-religion is about: If Israel follows the instructions, Yahweh will provide the rain at the right time, so that the land can yield its crops and the trees their fruits, the deity will ensure fertility among humans and animals and in the nonhuman world, and he will take up residence among the people of Israel (Lev 26:3–13). Yahweh clearly acts and communicates in and through the nonhuman world; he is the deity who blesses 'nature' and nurtures animals, humans, and plants alike and makes them thrive and multiply.[45] This image is just as outspoken in this chapter's curses: if the people do not follow Yahweh's commandments, he will send illness, turn the sky and the land into metal, so that the land cannot yield crops and the trees no fruit, wild animals will attack, human and animal fertility will decrease, and as a climax: the land will turn into wilderness and the cities to ruins (Lev 26:32–36).[46] Then the land will, then, finally get its Sabbath rest, which Israel has denied to it by not following Yahweh's commandments (Lev 26:14–45). We see here what is the lived space of the recipients of these texts: the agricultural environment of small-scale

44 The 'Holiness Code' generally communicates further ideals about Israel's purity and separation (Ulfgard 1998,89). In the prescriptions for Sukkot found in Numbers (Num 29,12–38), normally seen as P material, the amount of offerings are enormous and the festival is more temple-oriented, which is unique in the Pentateuchal & Torah tradition (Ulfgard 1998,91), and there is no explicit connecting of the celebrations to the wilderness stories, except for their overall setting within the Pentateuch. In the Dt Sukkot texts (Dt 16:13–16 and 31:10–11), there is no mention of booths, and we see the deuteronomistic ideas of concentrating the cult to one place and the obligation of pilgrimage.

45 Cf. what Bernhard Lang states for other Hebrew Bible texts (Lang 1998,183).

46 The word that I translate "wilderness" here is *šemamāh*, connoting desolation, wasteland, more so than *midbār*. Note that the theme of the attack of enemies is allotted a minor amount of attention in this text.

peasants who till the soil and tend their flocks, for whom the decisive function of the deity was the fertility (Lang 1998, 184) of the nonhuman world. This suggests that the peasant-pastoralist perspective plays a much larger role in the Pentateuch (cf. also Dt 7:12–14; Dt 11:14–15; Dt 26:15; 28:4–5) than is sometimes imagined.[47]

A final point is that the giving of the covenant and law and the ceremonious construction of the Tabernacle in the wilderness space are further reconciliatory aspects that complicate the negative image of the wilderness often given in biblical studies. In the wilderness space, Yahweh and Israel renew and redefine their relationship. We see this also in the texts about the fashioning of the wilderness sanctuary (*miškan*). As Ilana Pardes points out, the people here get that "tangible emblem of divine presence" that they craved in the Golden Calf episode, and the seriousness of the official discourse is balanced with the play of recipient voices, which are to some extent met (Pardes 2000, 85). From here on, then, Israel carries a mobile, material space of revelation with them on their trek, and the dialogue of Ex 19–20 continues in the space between the cherubim. Mark George's recent analysis of Israel's tabernacle as social space emphasises the importance of the portability and mobility of the tabernacle space. The social nature of the tabernacle space shows that Israel's deity is not only mobile, but also presented as the lord of any space, also the wilderness, and that Israel's social identity can be recreated anywhere (George 2009, 179, 188–194). This feature of the mobility of social space should not be overlooked as communication about the wilderness chronotope and its religious function. The wilderness placement of not only Israel's birth as a nation, but also its most intimate encounters with its deity, its reception of the law and the cultic instructions, all of which are evaluated positively in the Pentateuch, offers a message of deterritorialization and mobility as integral to Israel's relation to the land's fertility. To this message Israel's political history of exile and threats of dispossession from larger nations, as well as the shifting and mobile firstspatial aspects of the *midbār* presented earlier, provide a key support of form. The connection established, in the texts with a wilderness setting, of the Sabbath to the wilderness, as a memorial practice (Ex 16), as well as the

47 Where Lang sees conflicting ideological patterns in the Hexateuch between the warrior and wisdom ideologies, and sees the Dumézilian third function as applying primarily to Genesis and Job (Lang 1998, 184), I see more continuity across the Pentateuch and I suggest to grant his third function a larger role within the redacted Torah.

linkage between Sukkot celebration (Lev 23) and wilderness memory, support the communication that Yahweh is the lord of agricultural fertility even further.

C Wilderness vs. Land – Num 11–14

In Numbers 11–14, aspects of everyday life are introduced into the desert wilderness space that speak of how the distinction between the wilderness and everyday life in the land is constructed and negotiated.

Table 3. Actions, reactions, utterances Num 11–14

Yahweh's actions and utterances	
1. Marvels/fantastic events:	*Alimentary marvels:* Counterintuitive food work-free, sweet, falls from the sky with the dew, with regularity/tedium effect: Num 11:6–9 General: provides food and drink for a hyperbolic amount (cf. Ex 12,37)[48] of people and livestock (cf. Ex 16,32) Hyperbolic amount of meat: Num 11:19–20.22 Counterintuitive storm bringing quails in specific area: Num 11:31 Hyperbolic number of quails brought by storm: Num 11:31–32 Land gushing with milk and honey: Num 13:27; 14:8 Gigantic grapes in the land: Num 13:23–24; giants 13:22, 32–33 Land destroys/devours its inhabitants: Num 13:32[49] Land gushing with milk and honey: Num 14:8

48 Num 1–4 confirm the phantasmatic number of people in the wilderness.

49 A general question that can be raised here is that of metaphorical expression and 'how far' these images should be taken. I have discussed this question somewhat in Feldt 2011c in relation to the metaphors used of Ninurta in *Lugale.*

	Natural phenomenon'/'climate' marvels: Counterintuitive fire that consumes/kills (Num 11,1–3) Counterintuitive presence in cloud speaks to Moses: Num 11:25; Num 14,14 A speaking cloud pillar, pillar of fire: Num 12:5–10; Num 14,14 Counterintuitive cloud: Num 14:14 *The human body – fertility – death* Israel's exaggerated fertility/great number underlies: Num 11:10–15 and Num 11:21 Miriam from healthy skin to skin disease, semblance of death (*m*^{e}*ṣōrāʿat*), for seven days: Num 12:10, 12, 14–16 Plague/disease[50] caused by deity: Num 11:33 Counterintuitive killing of the spies: Num 14:36–28 *Other* Some of Yahweh's divine spirit transferred to seventy selected elders: 11:17, 25–26; when the deity's spirit dwells on them they prophesy (but do not continue): Num 11:25 Yahweh's *kbd* shows itself at the tent, superhuman voice speaks: Num 14:10
2. Other actions/utterances:	*Anger:* 11:1–3; 11,10; 11,19–20; 12:9.14; *Despair:* 14:11–12, 28–29, 33–35 *Detests* Israel: 14:27
Israel's / the people's actions and utterances	The Israelites break camp, wander, and rest when Yahweh's cloud communicates it[51] (passim; e.g., Num 9,15–23) *Complaint, rejection* Num 11:1–3; in Moses' words: 11,10–15 *Cry* for help to Moses against Yahweh's fire, rejection?: Num 11:2 *Distrust/rejection:* Num 11:4–5 *Rejection of ethnogenesis,* nostalgia for the food of Egypt, *doubt* of Yahweh's promises: Num 11:5; repeated in 11:18–20 *Rejection of Yahweh's gifts:* Num 11:6, 10, 13

50 For arguments that this means disease, see TWAT s.v. *nkh.*

51 See also (Ex 17,1; Ex 40,34–38; Num 21,10–15), (Num 9,15–23).

	Rejection of Moses: Num 13:30 *Doubt and fear:* Num 14:1–4 *Rejection* of Yahweh and fear: Num 14:7–9 *Rejection* of Moses, Aaron, Joshua, Kaleb: Num 14:10

In Num 11, the daily life in the wilderness, with its continual exposure to the fantastic food, makes it tedious, everyday-like, for the people and they crave variation (Num. 11:5–6).[52] Yahweh punishes Israel's doubt and rejection by means of an excessive number of quails, which here turns into a punishment on which to choke (Num. 11:20). Vomiting so violent that the food comes out the nose, that is, food transgressing the wrong bodily limit, is a horrific image likely to evoke disgust. The provisioning of quails, a fantastic event which was previously salvific (Ex 16), is now marked as its opposite, horrific. Manna is also now reframed: it is everyday-like, tedious.[53] These interactions between the fantastic food and the inversion of their previous (Exodus) evaluation draw attention to what is ordinary and what is extra-ordinary, to the differences between life in the land and life in the wilderness. Presenting the fantastic food of the desert through the lens of everyday life enables a re-description of everyday food through the lens of the fantastic wilderness food. In this way, the fantastic character of ordinary food (fish, cucumbers, watermelons, leeks, garlic, etc., Num 11:5) is communicated. Surrounded by fantastic events in the wilderness, in the presence of the deity, eating marvellous food every day (Num 11:6), what the people want is the food of everyday life. By presenting the people's nostalgia for everyday life and everyday food, the text communicates about the lived space of its recipients, and the major issues relevant to it: in everyday life in the land, the people easily forget who is the divine source of alimentation. Everyday life implies the two dangers of following other gods and following none,[54] heresy (polytheism) and apostasy (cf. Dt 8). By again presenting the Israelites as apostate and doubtful, regardless of Yahweh's wonders, regardless of

52 This theme re-occurs in Num. 20:3–5 and Num. 21:4–5.

53 In spite of the presentation of manna as tedious, the text proceeds to describe the curious wonder of the manna in detail—its appearance and treatment, and also the provisioning of meat for the people is presented as a marvel (Num. 11:18–23). Both types of fantastic food are at once curiosities the reader is invited to marvel at (Num. 11:7–9, 21–23: manna is attractive, versatile, convenient, tasty) and yet represented as punishment/tedium (manna Num. 11:6, 10; quails Num. 11:19–20).

54 E.g., Dt 8,17: *kōḥî wᵉʿōṣem yādî ʿāśāʰ lî ʾet-haḥayil hazzeʰ*.

the surface ideology of the deity and narrator and the gift of the laws and cult, the texts further emphasise the communication of Israel's identity as a discrepancy between what Israel is and what Israel ought to be, which amounts to a narrative attempt to install a desire to become "Israel" (as Israel ought to be).[55]

Num 13–14 concern the important transition to the new life with Israel poised on the threshold between the desert and the promised land. The coding of the two spaces of wilderness and land in the social imaginary is represented as a matter of controversy. Both areas are represented as oscillating between the real and the phantasmatic and both are represented by means of both positive and negative terminology. This is seen in the narrative of the first encounter with the land in Num 13. The list of names and lineage of the scouts in 13:5–15 contributes to the oscillation as do Moses' directions to the spies in 13:17–20. Considering Yahweh's repeated promise of a land flowing with milk and honey and of assistance in the conquest of the land, these directions are peculiar: they are to check whether the land is poor or rich (fertile), whether trees grow there, whether its inhabitants are strong or weak, few or many and so forth. This is not exactly a sign of a firm belief in Yahweh's promise of a land of abundance and help in conquest.[56] Divine help with the conquest is portrayed as equivocal,[57] Israel's relation to the land as insecure, and the land as hovering between a real and a fantastic presentation.

The land turns out to be peopled by giants, the descendants of Anak,[58] and the fruit that the scouts bring back is gigantic. At the entry of the land of promise, the land turns out to be too fantastic, it is so fertile that it is creepy, the homeland is *unheimlich*, and it scares Israel. What Israel yearns for is normal fertility, the fruits of normal everyday agricultural life, but what seems to be in store for them in the prom-

---

55 Therefore, the boundaries of Israel's identity are also porous: cf. that "recognition of Yahweh" is the basis for inclusion in Israel for P according to George (2010, 192).

56 The spies are sent on Yahweh's orders—does that mean that he is not sure himself? Mending this uncertainty, in Deuteronomy (Deut. 1:22–23, 37) the initiative comes from the people.

57 Milgrom mentions the possibility of the scouting as a test of faith because the men were chieftains, not trained scouts (*m*^{e}*ragg*^{e}*lim*; Milgrom 1990: 100).

58 See H. Rouillard in *Dictionary of Deities and Demons in the Bible* (Brill, Leiden 1995) s.v. REPHAIM) for details. The connection to the rephaim as deceased or denizens of the realm of the dead only adds to their eeriness.

ised land is fantastic-utopian, too bountiful for ordinary folk: it really (*gam*, 13:27) does gush with milk and honey, a phraseology which underscores the text's oscillation between the real and the fantastic. The utopian vision of the land of bounty is turned into its opposite, dystopia, in the words of the spies in 13:32, and the fundamental ambivalence towards the land in biblical religion (Pardes 2000,108), which is related to its phantasmatic-imaginary base, here comes out in the oscillation between a phantasmatic, motherly plenitude (milk and honey) and a cannibalistic appetite (Num. 13:32; cf. Pardes 2000, 114–115). The land is packed with others, and the old world of the patriarchs is not waiting there for Israel. The chapter ends as a threat fulfilled, when the Israelites, who in spite of Moses' warnings have attempted to conquer the land, are beaten (Num 14:45).

After Yahweh's punishment for the lack of trust in the wonders, the 'eyewitness' experience of the fantastic events of the Exodus and wilderness is sealed off within the desert space with the death of the Exodus generation (except for Joshua and Kaleb and the little ones, 14:30), and both the land and the wilderness are left in an oscillatory position between promise and threat, between the fantastic-utopian (milk and honey, giants, enormous grapes vs. a constant supply of manna, meat, and water, guidance) and the world of everyday experience (rich or meagre land, strong or weak inhabitants, tent camps or fortified cities, Num. 13:17–20 vs. thirst, alimentary deprivation, lack of orientation).

The wilderness of these stories sends signals in both directions: on the one hand, it is marked by the deity's guidance and marvels, but on the other these signs are revealed as ambiguous, repeatedly interpreted as caused by humans, rejected by the personae, and filmed over with elements from everyday life. While the texts clearly draw on existing, secondspace dichotomizing spatial paradigms of the wilderness and the land, with the land as the space of normalcy, home, plenitude, security, an accessible divine gift, and the wilderness as the region of lack and disturbance, of monsters, demons, and extraordinary events, of primordiality, these spatial paradigms are here reversed and presented for reflection; there is no simple transition from deprivation to fulfilment, from negative space to positive space, from wilderness to homeland. In the following chapters, the discord and disunity continues (15:32–36; 16:1–35;[59] 17:6–15) intermittently with the deity's orders (15:1–31, 38–41;

59 In Num. 16:13, the contingency of the fantastic constructions is blatant in the presentation of *Egypt* as the land of milk and honey.

17:1–5), the people's rejection and Yahweh's supernatural punishments and killings (as in, e.g., Num. 17:14–15), as well as reflections on the nature, knowability and veracity of Yahweh's marvels in the wilderness.

The wilderness narratives do thus not present the election and fantastic ethnogenesis in the wilderness as something that ensures Israel's possession of the promised land; Israel's relation to the land remains fragile (Pardes 2000: 15). The wilderness experience does not present the people, or individual Israelites, as moral exemplars or faithful worshippers, as they easily forget Yahweh and turn to 'secular' explanation, doubt, rejection and/or syncretistic religious practices. The idea of an inviolable relation between people and territory is called into question in these texts.[60] The wilderness is characterised more than anything by the relation between Yahweh's fantastic marvels and the reiterated emphasis on Israel's rejection, doubt, and apostasy. What we see, I suggest, amounts to the formulation of an ethos of religious identity transformation for recipient-Israel.[61] These texts suggest to their recipients a discrepancy in "Israel" between what Israel is and what Israel ought to be, which models religious identity formation as continual transformation.[62] The distinctions that ground identity for Israel (wilderness–land, wandering–settlement, alimentary lack-agricultural fulfilment, etc.), are thus not choices between a first part and a second, but instead an abiding, constitutive tension.

4. Wilderness and Hebrew Bible Religion

In this section, I call attention to four aspects of the religious construction and function of the wilderness space in the Torah/Pentateuch brought out here and how these aspects are relevant for broader discussions of the religion and wilderness nexus in Hebrew Bible religion.

---

60 This does not rule out that such a relation may be central in other strands of theology in the Hebrew Bible (H.J.L. Jensen 2003; Brueggemann 1977.

61 Cf. D. and J. Boyarin's essay on diasporism (Boyarin and Boyarin 1993).

62 Pardes says that "Israel's weaning is both terminable and interminable" (Pardes 2000, 62–63) and so, it seems to me, indicates the processual nature of Israel's identity (see also Pardes 2000, 79–80). W.H.C. Propp, similarly, notes that "the rebellion traditions... advise the Israelites of the dangers of disobedience and the rewards of fidelity (Propp1987, 125).

4.a. Ambiguity and transformation

As the above analyses show, the Hebrew Bible's wilderness idea is characterised by a continual slippage between the desert wilderness as natural-geophysical materiality and then the fantasy images of the desert in the social imaginary, between a realistic presentation and the fantastic marvels. The wilderness oscillates between a benign and a malign, a real and a phantasmatic, presentation, between terror and nostalgia, past and present, between Yahweh's bounty chamber and utter deprivation, between threat and promise, rescue and punishment. The desert wilderness ends up as an uncertain and ambiguous region, a hallucinatory wonder-object.[63] Contrary to much previous research,[64] which has focused on whether the Hebrew Bible wilderness was seen as primarily benign or malign, or on the wilderness as a real historical timespace and its chronology and concrete geography, the analyses here have shown that the wilderness is presented in an irreducibly ambiguous way. The wilderness' oscillatory position adds to its versatility as a space of experimentation, indeed, of testing, of the relationship between deity and people, of transformation and negotiation, of importance for religious identity formation. The wilderness of the Pentateuch is not presented as an area outside divine rule or the abode only of demons, or as an epitomical embodiment of 'nature'. The wilderness is a primary arena for intimate meetings with Yahweh,[65] it is a place of personal and communal religious transformation, where the relation and forms of communication between deity and humans are reflected upon, discussed, and negotiated. The wilderness is also a place of

63 The wilderness is thus not a wholly other world (contra Leach 1987, 586–587).

64 As mentioned, an exception is the ecotheological work of Leal, who stresses the extensive range of attitudes to wilderness encountered especially in the Hebrew Bible (Leal 2006, 223–25); Schofield also offers pointers towards a different view (2008).

65 While it is difficult for humans to survive there without divine help, the wilderness per se is not condemned or subjugated; it is a legitimate, necessary area. The presentation of wilderness differs from, e.g., the imagination of the wilderness as epitomized by an opponent monster, as in *Lugale* or *Inana and Ebih* (see Black et al. 2006). A comparative investigation with a Mesopotamian, and wider ancient Near Eastern "folkloristic" context is a desideratum, since in Mesopotamia at least it seems that the wild, open steppe, *ṣēru*, was the abode of wild animals, nomads, lawless people, enemies, and demons (cf. CAD s.v. *ṣēru*, Pongratz-Leisten 1994, 18–20). The idea of *ṣēru* as a social and cultural space in ancient Mesopotamia is, however, a topic that needs further investigation, and it is indeed possible that its positive qualities have been underestimated in previous research.

marvels that communicate Yahweh's 'natural' 'fertility power', to which I now turn.

4.b. Food, fertility and lived space

The analyses have also brought out how the subject matter of the marvels has been overlooked as communication about the wilderness and about the deity. Above all, this subject matter communicates that Yahweh is the lord and donor of agricultural fertility. The marvels revolve around securing food and fertility from the non-human world, specifically the agricultural field; the wilderness marvels are imagined up against a farmer's environment – the wilderness is where the crops cannot grow (*ʾereṣ nôšāvet*). This finding has consequences for traditional understandings of Hebrew Bible religion. Biblical scholarship has long argued, and in some quarters still does, that Hebrew Bible religion was, first and foremost, a historical religion, and the natural world unimportant to its concerns.[66] Based on the present analyses, it must contrariwise be noted that the ties of Hebrew Bible religion to the nonhuman world, as well as its ties to a particular human mode of engagement with the nonhuman world, namely agriculture, are strong.[67] This comes clearly to the fore in the wilderness stories, where deity and humans communicate through the medium of 'nature', where the aid Israel desires from its deity is expressed in 'natural' terms (fertility, food, crops, rain), where elements of the non-human world (fire, clouds, food, water) manifest the presence of the deity. It must thus be acknowledged that Yahweh as a deity is much more about "nature" than often thought in biblical studies. The narratives verbalise the interdependence of natural processes, deity-worship, and social identity; religious practice and the desired fertility in field, stable and bed as the deity's gifts are intimately linked. This study thus corroborates what scholars like Jensen, Lang, Propp, and Hie-

66 Hebrew Bible religion has been seen as very different from its ancient Near Eastern neighbours by not being bound to the cycles of nature or concerned with fertility. It seems that such a view of Hebrew Bible religion cannot be upheld (support for this based on other Hebrew Bible texts can be found in Hiebert 1996,143; Jensen 1998; Lang 1998).

67 In traditional biblical studies this is not a commonplace insight. However, in environmental studies, as Taylor's contribution to this volume shows, there is a strong tradition of connecting Judaism, Christianity and Islam to agriculture and imperialist expansion (Taylor this volume).

bert have argued for other Hebrew Bible texts (Jensen 1998, 2000; Lang 1998; Hiebert 1996, 142–143; W.H.C. Propp 1987,127–128).[68] 'Salvation' is not out of this world to a world beyond, but a peaceful life in the land with an adequate supply of food and fertility. Importantly, the wilderness stories analysed here bring out human survival as fragile and Israel's relation to the agricultural land as tenuous and contingent on religious identity transformation.

As for the wilderness texts' spatial aspects, the conclusion is that the relevant *lived space* is not the desert and the culture of nomadic pastoralism, but the culture of mixed agriculture and pastoralism and the agrarian environment of the arable highlands of Palestine, that of grain-based farming combined with fruit and vegetable production and animal husbandry. The religion reflected and idealized here relates not to the desert, but to small-scale agriculture-pastoralism.[69] In comparative perspective, several Mesopotamian texts verbalize the wilderness up against an urban environment, which is praised and celebrated (e.g., the Sumerian Lugalbanda, Gilgamesh and Huwawa stories, as well as the Babylonian, Ninevite Gilgamesh epic). In the Hebrew Bible narratives, the wilderness represents a disregard of, literally a journey away from, urban life (cf. Oehlschlaeger 1991,48–49). In the Pentateuchal wilderness stories, we are presented with the point of view of outsiders to urban culture (Mob-

68 And as W.H.C. Propp says of the wilderness journey: "… the historical tradition commemorated the alternation of sterility and fertility *in space*, in the journeys of Israel from Egypt, via the desert, to the mountain spring in the desert or to Canaan. The power of providing water in the desert demonstrated, *a fortiori*, Yahweh's power to irrigate his own land" (Propp 1987, 127–128. Emphases in original). J. Blenkinsopp criticizes the focus on salvation history in Christian exegesis (2011, 176–178).

69 Leal seems to distinguish between wilderness, nomadism and pastoralism on the one hand and urban civilization, sedentarism, and agriculture on the other (Leal 2006). Instead, a distinction between small-scale farming mixed with animal husbandry and then city life of the greater ancient Near Eastern civilizations is more likely, and this is corroborated by some studies (which show that the economic basis was mixed between agriculture and pastoralism, that it was not an either-or; Pastor 1997, 10). Considering the mixed economy, a split between small-scale farmers / animal herders and large city civilizations is more likely than a split between pastoralism and agriculture. I also disagree with the assumption that since Israel is encouraged not to have relations with "agriculturally oriented people" in the Hebrew Bible then Israel is not itself an agriculturally oriented people (Leal 2006, 145). What Israel wants from Yahweh is the same as the "agriculturally oriented" "Cana'anites" – Israel is also an "agriculturally oriented" people.

ley 1997,232; Gunkel 1913),[70] which may reflect the experiences of a people in the position of outsider in relation to more powerful urban cultures. Casting Israel's relationship with its god as forged in the wilderness, not in a city, and pointing clearly at small-scale farming and pastoralism as the desired way of life, these wilderness narratives may have been one way to actively discredit city life and large institutions in the exilic, post-exilic (cf. Lang 1998, 183; Dozeman 1996, 95–96) and Hellenistic periods, which was of continued relevance well into the Common Era. It is also worth pointing out that the wilderness is not to be domesticated and that conspicuous consumption is condemned;[71] land and humans alike are entitled to regular rest.

Israel's reactions to Yahweh's marvels in the wilderness are significant for understanding the religious function of the wilderness. In the wilderness, the forgetfulness and apostasy of the Israelites is unbelievable within the frame established by the narrative itself: if the marvels of the exodus, manna, water, the Sinai revelation, etc. are taken at "narrative face value", then the doubt and rejection of the Israelites is unbelievable. The agricultural everyday life of the text producers and recipients does go some way in explaining this: only in the framework of everyday, agricultural life, where Yahweh's presence is difficult to ascertain and easy to forget (cf. Dt. 8 and Dt. 32), is this doubt, forgetfulness and apostasy meaningful and predictable. Yet, I contend that this textual element has further consequences for the process of reception.

---

70 Gunkel already suggested that the Hebrew Bible adopts the point of view of Israel as a *Naturmensch* in their traditions about the premonarchical period (Gunkel 1913). He noted that Israel or Israel's heroic representative (in the Exodus and David and Goliath stories) defeats a technologically superior enemy through the agency of Nature itself, by means of a natural object at hand (a stone, the jawbone of an ass, a shepherd's staff). Samson triumphs because his wild strength is irrepressible. Also Dt 32,10 depicts the people of Israel in "wild man" terms as a foundling in the desert adopted by a divine godfather who nurses his son with honey from a rock (v13).

71 In Ex 16, the provisioning of food is dependent on the keeping of the law and the regulations. While the regulations are indeed presented as written, food and eating are presented as the elementary practice that leads to the understanding that "Yahweh is your god" (16,13). The model for food production is agricultural, while the model for food consumption is one of adequate supply – not conspicuous consumption. The rhetoric of violence of expansion and subjugation is aimed at the "Cana'anites".

4.c. An ethos of religious identity transformation

Above, the importance of the embedded character reactions was noted. Israel's dominant reactions of doubt, rejection and apostasy in the face of the many marvels verge on the incredible within the frame set by the narrative itself. Not only does this point to the texts' lived space, but the reactions of the people in the wilderness also construct a discrepancy between an ideal – what Israel ought to be (devout, law-abiding, remembering Yahweh) – and a 'reality' – what Israel 'is' (apostate, disobedient, forgetful); a discrepancy which works as a call to transformation. The strong reactions of apostasy formulate the necessity of transformation towards an ideal, the aim of which is to secure Israel's continued access to the fertility of the agricultural land that is vital to survival. The wilderness narratives analysed here communicate that Israel's relation to the land is tenuous. The land is often placed in situations of dislocation and peril, or represented as endangered in the Hebrew Bible (cf. Num 13–14, Judg 6–9, et al.) The shifting and mobile location of 'wilderness' and 'land' areas refers to spatial practices inside the land (cf. section 3a of this paper, e.g., that out of the six regions of the land, one is designated *midbār* (Talmon TWAT). The wilderness is thus an integral and mobile part of the geography of the land, just as it is an abiding tension in the religious identity of "Israel". The reactions of the text-internal personae not only reflect recipient responses but also suggest an itinerary of identity formation for the recipients – an ethos of religious transformation. The wilderness narratives communicate an invitation to regard religious identity as unfinished. The religious function of the wilderness lies in its ability to maintain contradictory aspects in a productive tension, thus enabling, modelling, a continual work of self-creation, in the recipient, towards a more ideal "Israelite". Through the discrepancy between ideal and reality in Israel's responses to the marvels, in the depiction of 'Israel' as internally split, an ethos of religious identity transformation is verbalised.

4.d. Wilderness Memorial Practice?

In his book *Not Bread Alone* (2009), Nathan MacDonald analyses food as a vehicle for memory in Deuteronomy. In the Exodus and Numbers wilderness narratives, I suggest, food is also set forth as a memorial practice; food production / consumption and Sabbath observance are incorporated forms of memory that further the inscribed memory (for these distinctions, see Connerton 2003, 76–79). In both Ex 16 and Num 11 narratives, everyday, agricultural food production is filmed over by the memorial narratives of the wilderness, because manna, as shown above, functioned as an inversion of normal, agricultural practice. The texts – the inscribed memory – attempt to evoke an incorporated memory in this way. The Sabbath, according to Ex 16, provides a temporal abstention from the gathering of manna. The connection of the Sabbath to the manna in the wilderness thus supports the communication that Yahweh is the provider of the fertility of the agricultural land, that humans are not the real owners or producers of food and land. Through the everyday, agricultural production of food and the weekly Sabbath observance, recipient 'Israel' is joined to the 'Israel' of the wilderness narratives. Agricultural food production and Sabbath observance are incorporated memorial practices of a type that can be repeated anywhere; they are very mobile anchors for religious memory. Their wilderness origin adds another layer to this mobility, as it offers a deterritorialized perspective on the land area and a 'theodicy' for Israel's life outside of the land.[72]

The specific literary form used here also presents both 'wilderness' and 'land' spaces as ambiguous and subject to multiple interpretations. Interestingly, while writing and textuality are not the subject of explicit reflection and stipulated memory practice in these texts, as for instance in Deuteronomy, the only access to the wilderness is here through text and narration. As we can see, there was a preference for the preservation and transmission of a multivocal wilderness tradition. The textual practices to which the wilderness narratives bear witness clearly recognize the multiple possibilities of interpretation. The multivocality of the desert tradition may indeed be an expression of cultural and ideological ambivalence,[73] which is what meets the recipients of these texts. It is fair to assume that the desert's ambiguity is socially formative in specific ways. What S. Fraade argues for the later rabbinic textual practices (Fraade 2007,12–14), I suggest is equally pertinent for the preservation and transmission of the Torah itself, namely that the texts depict interpretation of the marvels of the wilderness and the wilderness space itself as originary to it (Fraade 2007,26), and as multiple. This encourages the recipients to recognize, apprehend, and produce multiple understandings of

72 Brueggemann says that the land is a gift from Yahweh, that its fertility is not inherent and its possession not inalienable (1977, xii, 2, 54).

73 This does not rule out that it is also an expression of sedimentary source layers.

spatiality as a response to varying situations. It is at the redacted textual stage that the texts are performatively and dialogically engaged by their recipients. At this stage, the wilderness events are represented as ambiguous, and access to the fertility of the agricultural land as contingent on religious transformation. This transformation is based on legal, ritual, narrative practices based in and on *text.* The reuse and rewriting of, and intertextual references to, the wilderness narratives and events within the Hebrew Bible itself (e.g. between Exodus and Numbers's versions of the manna marvel) further speak of a writing-oriented culture in which the ongoing development of the tradition was a literary-textual activity. The ambiguous narrative representations of Israel's wilderness origins were hardly motivated by an interest in history and nor do the interest in and valuation of the nonhuman world and fertility support the designation of 'historical' religion. Rather, the desire was to anchor, model, and authorize certain religious identity formation practices. In these wilderness texts, everyday food production, Sabbath observance, and Sukkot celebration are the means used to anchor the wilderness memory both in everyday life and in the yearly communal celebrations, as an incorporated-inscribed memory that presents the religious identity of the social group Israel as unfinished and access to the desired fertility of the land as contingent on religious transformation. As such, it is a 'theodicy' for Israel's existence away from the land, for this group's mobility.

5. Conclusion

The wilderness is, in the Hebrew Bible, a crucial domain. It plays an important role in and for Hebrew Bible religion and it functions as an arena for reflection on, and negotiation of, central issues and tensions in religious identity. The analyses have shown that the wilderness of these Hebrew Bible texts is presented as irreducibly ambiguous, that it oscillates between a malign and a benign, a realistic and a fantasmatic presentation, and that this enhances its versatility for processes of transformation. The subject matter of the wilderness marvels is used to communicate that Yahweh is the donor of the fertility of the agricultural land, which among other things shows that the lived space of the texts is the mixed culture of agriculture and pastoralism, not desert nomadism. The fruits of the material, agricultural world are celebrated and longed for, the deity's manifestations come in natural terms, and "salvation" is not out of this world but into an unscathed everyday life. The reactions of rejection of the Israelites in the wilderness were interpreted here as reflecting and modelling lived-space recipient responses[74] and as reflecting an agricultural everyday life, as well as playing an important role in verbalizing an ethos of transformation. By bringing out the tension between ideal and reality in the social group of "Israel" (the unbelievable apostasy and forgetfulness vs. the marvels and the clearly formulated ideals, rules, regulations), the wilderness narratives achieve their pedagogical aim: to communicate the idealized response of religious transformation to the recipients.

This analysis thus offers cause to question and nuance traditional, dichotomous cosmological grids in favour of a sharper focus on the concrete production of space, on the specific literary form and literary strategies of religious texts with regard to space, and, not least, on the effects of the literary form on the process of reception. The religious texts that may embed, produce, sanction and /or negotiate specific spaces and spatial practices may affect the process of reception in variable ways.

Based on these analyses, Hebrew Bible religion as embedded and espoused in these texts is a religion which focuses on this world in the specific sense that it is not about anticipating an afterlife in a transcendent, supernatural, other-worldly realm, but a religion that values "nature", the materiality and fertility of the non-human world (Taylor 2005, x).[75] Consequently, it seems fair to say that the traditional distinction in exegesis between "nature" and "history" does not serve these texts well. The wilderness narratives of Exodus and Numbers reflect that the deity steers the order of nature and, in particular, the processes which produce crops and offspring. What interests the producers and recipients of these texts is what Yahweh does, and they communicate clearly about this: he is the

74 Of course the wilderness of these texts could function as a second-space paradigm for its recipients. But, as shown, a spatial paradigm can reflect, incorporate, and use thirdspace experiences and emotions.

75 It is often helpful to distinguish between nature-as-sacred-religion (in which nature is worthy of reverent care) and the "natural dimension" of religion, which encompasses the entire "religion-and-nature/religion-and-ecology"-field (Taylor 2005,x, cf. Albanese 1990), but it seems to me that neither term is apt for religion in Mesopotamia or for Hebrew Bible religion. These are religions in which the desired exchange goods, between transempirical beings and humans, are "natural" and immanent; in which salvation is not about rescue from this world, but where the emphasis is on securing survival and an unscathed life in the "here and now"; where humans need to entertain relations with the deities / deity imagined to control "nature", or specific domains in "nature", in order to survive (to get food, fertility, health).

donor of agricultural, "natural" fertility,[76] but he is also a mobile deity who may be worshipped elsewhere than in the land. In this perspective, possession of the land, and the access to agricultural fertility, is made contingent on religious transformation. The wilderness stories respond to the problems of the contingency of Israel's' possession of the land, and the fragility of the social group of "Israel". The answer they present is a religious ethos of transformation and a deterritorialized perspective on the land.

6. References

Addinall, Peter. "The Wilderness in Pedersen's *Israel*". In *Journal for the Study of the Old Testament* (1981) 20, 75–83.

Albanese, Catherine. *Nature Religion in America.* Chicago: Chicago University Press 1990.

Albrektsson, Benny. *History and the Gods. An Essay on the Idea of Historical Events as Divine Manifestations in the Ancient Near East and in Israel.* C.W.K. Gleerup, Lund 1967.

Bal, Mieke. *Narratology.* University of Toronto, Press, Toronto 1997.

Bar-Yosef, O. and A.M. Khazanov, eds., *Pastoralism in the Levant: Archaeological Materials in Anthropological Perspective.* Madison, WI: Prehistory.

Berquist, Jon and Camp, Claudia. eds. *Constructions of Space II. The Biblical City and Other Imagined Spaces.* T&T Clark, London 2008.

Blenkinsopp, Joseph. *Creation, Un-Creation, Re-Creation. A Discursive Commentary on Gen 1–11.* London: T&T Clark 2011.

Bingham, Nick and Thrift, Nigel. "Some New Instructions for Travellers. The geography of Bruno Latour and Michel Serres". In M. Crang and N. Thrift, *Thinking Space.* Routledge, London 2004, 281–301.

Black, Jeremy A., Robson, Eleanor, et al. eds. *The Electronic Text Corpus of Sumerian Literature.* Online publication of Sumerian literature fully edited: http://etcsl.orinst.ox.ac.uk The University of Oxford, Oxford 2006.

Boyarin, Jonathan and Boyarin, Daniel. "Generation: Diaspora and the Ground of Jewish Identity". In *Critical Inquiry* 19,4 (1993), 693–725.

76 As Dozeman points out for Deuteronomy, the focus on the mountain of god, Sinai/Horeb, in the wilderness amounts to a criticism of the urban Zion theology (Dozeman 1989). This wilderness conception thus speaks also of the social space of the authors, who were, likely, part of a cultural elite removed from the land during the exile and some of whom returned to Palestine in post-exilic, Persian times (cf. Dozeman 1989, 1998). Blenkinsopp (2011,180) notes that the political developments in Judah of the 6th century BCE made it easier to release the deity from territorial and ethnic attachments. The deterritorialized perspective was clearly of continued relevance into the Common Era and beyond.

Brueggemann, Walter. *The Land. Place as gift, promise, challenge in biblical faith.* London 1978.

Camp, Claudia. "Storied Space or Ben Sira Tells a Temple". In Gunn, David M., McNutt, Paula and Flanagan, Jim W., eds., *"Imagining" Biblical Worlds: Studies in Spatial, Social and Historical Constructs in Honor of James W. Flanagan*, JSOTSup 353. Sheffield Academic Press, Sheffield 2002, 64–80.

Carr, David. "Torah on the Heart: Literary Jewish Textuality Within Its Ancient Near Eastern Context". In *Oral Tradition* 25/1 (2010): 17–40.

Certeau, Michel de. *The Practice of Everyday Life.* University of California Press, Berkeley 1984.

Childs, Bernard S. *The Book of Exodus.* Old Testament Library. The Westminster Press, Philadelphia 1974.

Coats, George W. *Rebellion in the Wilderness. The murmuring motif in the wilderness traditions of the Old Testament.* Abdingdon Press, Nashville and New York 1968.

Cohen, Jeremy J. *Of Giants: Sex, Monsters and the Middle Ages.* University of Minnesota Press, Minnesota 1999.

Cohn, Robert L. *The Shape of Sacred Space. Four Biblical Studies.* AAR Studies in Religion. Scholars Press, Chico, CF 1981.

Connerton, Paul. *How Societies Remember.* Cambridge: Cambridge University Press 2003 [1989].

Corrigan, John. "Spatiality and religion". In Warf, B. and Arias, S. *The Spatial Turn. Interdisciplinary perspectives.* Routledge, London 2009,157–172.

Davies, G.I. "The wilderness itineraries and the composition of the Pentateuch". In *Vetus Testamentum* 33 (1983), Jan., 1–13.

Dozeman, Thomas. *God on the Mountain. A Study of Redaction, Theology and Canon in Exodus 19–24.* Atlanta: Scholars Press 1989.

Dozeman, Thomas. *God at War: Power in the Exodus Tradition.* Oxford University Press, Oxford 1996.

Durham, J.I. *Exodus.* Word Biblical Commentary. Word Books: Waco, Texas 1987.

Feldt, Laura. "Monstrøsitet som kulturel og religiøs diskurs". *Religionsvidenskabeligt Tidsskrift* 42 (2003), 43–64.

Feldt, Laura. "Reading the Monstrous" in Eksell, K. and Feldt, L. eds. *Readings in Eastern Mediterranean Literatures*, Ergon Verlag, Würtzburg 2006.

Feldt, Laura. "Religious Narrative and the Literary Fantastic". In *Religion* vol 41 (2011 (a)), nr. 2, s. 251–283.

Feldt, Laura. "Fantastic Recollection – cultural vs. autobiographical memory in the Exodus narrative". In A.W. Geertz and J.S. Jensen, eds. *Religious Narrative, Cognition and Culture.* London: Equinox 2011 (b), 191–217.

Feldt, Laura. "Monstrous Identities – narrative strategies in Lugale". In F. Hagen et al. eds. *Narratives of Egypt and the Near East. Literary and Linguistic Approaches.* Orientalia Lovanensia Analecta 189. Peeters, Leuven 2011 (c), 123–164.

Feldt, Laura. *The Fantastic in Religious Narrative from Exodus to Elisha.* London: Equinox 2012.

Fraade, Stephen D. "Rabbinic Polysemy and Pluralism Revisited". In *AJS Review* 31,1 (2007),1–40.
Frankfort, Henri. *Before Philosophy. The Intellectual Adventure of Ancient Man.* Harmondsworth 1949.
Garrard, Greg. *Ecocriticism.* Routledge, London 2004.
George, Mark. *Israel's Tabernacle as Social Space.* Society of Biblical Literature, Atlanta 2009.
Grabbe, Lester L. "Adde Praeputium Praeputio Magnus Acervus Erit: If the Exodus and Conquest had really happened" in J.C. Exum, ed., *Virtual History and the Bible.* Leiden: Brill 2000, 23–32.
Gunkel, Herman. 'Simson'. In *Reden und Aufsätze.* Göttingen: Vandenhoeck und Ruprecht 1913, 38–64.
Hendel, Ron. *Remembering Abraham. Culture, Memory and History in the Hebrew Bible.* Oxford University Press, Oxford 2005.
Hendel, Ron. "The Poetics of Myth in Genesis". In S. Daniel Breslauer, ed., *The Seductiveness of Jewish Myth.* State University of New York Press, 1997, 157–170.
Hiebert, Theodore. *The Yahwist's Landscape. Nature and Religion in Ancient Palestine.* Minneapolis: Fortress Press 1996.
Ingold, Tim. *The Perception of the Environment. Essays in livelihood, dwelling and skill,* London: Routledge, 2000.
Jacobsen, Thorkild. "The Graven Image," in P. D. Miller, P.D. Hanson, and S.D. McBride, eds., *Ancient Israelite Religion: Essays in Honor of Frank Moore Cross.* Philadelphia, 1987, 23–29.
Jasanoff, Sheila. "A New Climate for Society" in *Theory, Culture & Society* 27 (2010), 233–253.
Jensen, Hans Jørgen Lundager. *Den fortærende ild. Strukturelle analyser af narrative og rituelle tekster i Det gamle Testamente.[The Fire that Devours. Structuralist analyses of narrative and ritual texts in the Old Testament].* Aarhus: Aarhus University Press, 2000.
Jensen, Hans Jørgen Lundager. "Hvad drejer profetismen sig egentlig om?" [What is Prophetism Really About?] In E. Holt, ed., *Alle der ånder skal lovprise Herren! Det gamle Testamente i tempel, synagoge og kirke.* Copenhagen: Anis 1998, 91–125.
Jensen, Hans Jørgen Lundager. *Gammeltestamentlig religion.* Copenhagen: Anis, 1998(a).
Johnstone, William. *Exodus.* Old Testament Guides. JSOT Press, Sheffield 1990.
Kawashima, Robert S. *Biblical Narrative and the Death of the Rhapsode.* Bloomington and Indianapolis: Indiana University Press, 2004.
Kirby, Peter W. *Boundless Worlds. An Anthropological Approach to Movement.* New York: Berghahn Books, 2010.
Knott, Kim. *The Location of Religion. A Spatial Analysis.* London: Equinox, 2005.
Lang, Bernhard. *The Hebrew God. Portrait of an Ancient Deity.* New Haven: Yale University Press, 1998.
Larsen, Mogens Trolle, "The Mesopotamian Luke-Warm Mind. Reflections on Science, Divination and Literacy" in F. Rochberg-Halton, ed., *Language, Lit-*

erature, and History: Philological and Historical Studies Presented to Erica Reiner, New Haven: American Oriental Society 67, 1987: 203–225.

Leach, Edmund. "Fishing for Men". In R. Alter and F. Kermode, *Literary Guide to the Bible.* Cambridge, MA.: Harvard University Press, 1987, 579–599.

Leal, Robert B. *Wilderness in the Bible. Toward a Theology of Wilderness.* New York: Peter Lang, 2006.

Lefebvre, Henri. *The Production of Space.* Oxford: Basil Blackwell 1991. [Translated from *La production de l'espace*, 1974].

Lee, Won."The Concept of the Wilderness in the Pentateuch" in Pomykala, K. ed. *Israel in the Wilderness.* Leiden: Brill 2008, 1–16.

Lied, Liv I. *The Other Lands of Israel – imaginations of the land in 2nd Baruch.* Leiden/Boston: Brill 2008.

Lied, Liv I. "Another Look at the Land of Damascus. The Spaces of the Damascus Document in the Light of Edward W. Soja's Thirdspace Approach". In Campbell, J.G., Lyons, W.J., and Pietersen, Lloyd K., eds., *New Directions in Qumran Studies. Proceedings of the Bristol Colloquium on the Dead Sea Scrolls 8–10 September 2003*, T&T Clark London 2005, 101–125.

MacDonald, Nathan. *Not Bread Alone. The Uses of Food in the Old Testament.* Oxford: Oxford University Press, 2009.

Meyers, Carol. *Exodus.* The New Cambridge Bible Dictionary. Cambridge: Cambridge University Press, 2006.

Milgrom, Jacob. *Numbers.* New York: Jewish Publication Society, 1990.

Mobley, Gregory. "The Wild Man in the Bible and in the Ancient Near East". In *Journal of Biblical Literature* 116, no. 2 (1997), 217–233.

Morton, Timothy. *Ecology without Nature. Rethinking Environmental Aesthetics.* Cambridge: Harvard University Press, 2007.

Nash, Roderick. *Wilderness and the American Mind* (4th edition). New Haven: Yale University Press, 2001 [1967].

Oelschlager, Max. "Wilderness Religion" in Taylor, B., ed. *The Encyclopedia of Religion and Nature*, Continuum 2005, s.v.

Oelschlager, Max. *The Idea of Wilderness. From Prehistory to the Age of Ecology.* New Haven: Yale University Press, 1991.

Pardes, Ilana. *The Biography of Ancient Israel.* University of California Press, Berkeley 2000.

Partridge, Christopher. *The Re-Enchantment of the West*, 2 vols. T&T Clark, London 2006/2004.

Pastor, Jack. *Land and Economy in Ancient Palestine.* London: Routledge 1997.

Pedersen, Johannes. *Israel I-IV.* Copenhagen: Branner og Korch 1958–1960.

Pedersen, Poul. 1995. "Nature, Religion and Cultural Identity: The Religious Environmentalist Paradigm". In *Asian Perceptions of Nature: a critical approach*, edited by O. Bruun and A. Kalland: Curzon Press, 258–276.

Pongratz-Leisten, Beate. *Ina šulmi īrub. Die kulttopographische und ideologische Programmatik der akitu-Prozession in Babylonien und Assyrien im I. Jahrtausend v. Chr.* Mainz: Baghdader Forschungen 16, 1994.

Propp, William H.C. *Water in the Wilderness. A Biblical Motif and Its Mythological Background.* Atlanta: Scholars Press, 1987.

Propp, William H.C. *Exodus 1–18. A New Translation with Introduction and Commentary.* Vol. I. Anchor Bible Commentaries. New York: Doubleday, 1999.

Schaper, Joachim. *Die Textualisierung der Religion.* Tübingen: Mohr Siebeck, 2009.

Schofield, Alison. "The Wilderness Motif in the Dead Sea Scrolls". In Pomykala, K. *Israel in the Wilderness.* Leiden: Brill 2008, 37–54.

Simkins, Ronald A. *Creator and Creation. Nature in the Worldview of Ancient Israel.* Peabody: Hendrickson Publishers, 1994.

Soja, Edward W. *Thirdspace. Journeys to Los Angeles and Other Real-and-Imagined Places.* Cambridge Mss: Blackwell 1996.

Soja, Edward W. 1989. *Postmodern Geographies.* New York: Verso.

Spaulding, Mary B. *Commemorative Identities, Jewish Social Memory and the Johannine Feast of Booths.* London: T&T Clark 2009.

Talmon, Shemaryahu. 'The Desert Motif' in the Bible and in Qumran Literature", in Altmann, A., ed., *Biblical and Other Studies.* Cambridge: Harvard University Press, 1966, 31–63.

Talmon, Shemaryahu. *midbār* in G.J. Botterweck and H. Ringgren, eds., *Theologisches Wörterbuch zum Alten Testament* [TWAT], Stuttgart, s.v.

Taylor, Bron. *The Encyclopedia of Religion and Nature.* Continuum, London 2005.

Thrift, Nigel and Whatmore, Susan. *Cultural Geography. Critical Concepts in the Social Sciences.* London: Routledge, 2004.

van der Toorn, Karel. *Scribal Culture and the Making of the Hebrew Bible.* Cambridge, MA.:Harvard University Press, 2007.

Ulfgard, Håkan. *The Story of Sukkot. The Setting, Shaping and Sequel of the Biblical Feast of Tabernacles.* Tübingen: Mohr Siebeck 1998.

Veldhuis, Niek. "Mesopotamian Canons." In *Homer, the Bible and Beyond: Literary and Religious Canons in the Ancient World.* Ed. by M. Finkelberg and G. G. Stroumsa. Leiden: Brill 2003, 9–28.

White, Lynn Jr. "The Historical Roots of Our Ecologic Crisis". In *Science*, March 10 (1967), 1203–1207.

4. "The mountain, a desert place": Spatial categories and mythical landscapes in the *Secret Book of John*

Ingvild Sælid Gilhus

1. Wilderness and the *Secret Book of John*

The *Secret Book of John* (*Ap.John*) is a learned treatise, which was an object of processes of re-interpretation and re-writing in antiquity. Most likely the treatise was originally written in Greek, but it exists now in four manuscripts in Coptic, three of them found at Nag Hammadi in Upper Egypt in 1945 (Codex II, III and IV). The fourth manuscript is part of Codex Papyrus Berolinensis 8502 (BG), probably from Achmim, also in Upper Egypt. The Nag Hammadi versions are usually dated to the second part of the fourth century, and BG to the beginning of the fifth century. Codex II and IV contain a long version of the text, while Codex III and BG feature a short version.[1] Only the version in BG preserves the lines that are our main target in this article, "the mountain, a desert place" (*eptooy euma nsjaeie*).[2]

An underlying presupposition is that the space referred to in the frame story of *Ap.John*, "the mountain, a desert place", can be seen as wilderness. According to a modern definition wilderness is "those parts of the world that are perceived to be untouched by humanity" (Bauman,

---

1 *Ap.John* is the first text in Nag Hammadi codices II, III and IV and the second text in BG. The main differences between the four versions are that the long version (II and IV) includes an enumeration of the demons connected to the different body-parts of humans and ends with a hymn (the so-called Pronoia hymn). The text edition and translation that are used in this article is that of Michael Waldstein and Frederik Wisse (Waldstein and Wisse 1995).

2 In Codex III the end of the expression is missing, while Codex II and IV have lacunae at this point. The various translators translate the lines of BG 20: 5–6 in a roughly similar way. M. Waldstein and F. Wisse: "the mountain, a desert place; B. Barc: "vers la Montagne, vers un lieu désert" (Barc 2007: 218); K. L. King: "to the mountain, which was a place of desert" (King 2006: 26); B. Layton: "the barren mountain" or "desert" (Layton 1987: 28, note g).

Bohannan, and O'Brien 2011: 235). The scholarly term "wilderness" is an etic concept, which is based on and is in dialogue with emic concepts. The scholarly concept of wilderness presents a new type of conceived space on a higher level of abstraction than the types of space that are constructed in a text or lived by a group. Connecting the scholarly concept of wilderness to spatial theories as presented by Edward Soja and Henri Lefebvre, and applying them to *Ap.John* bring out the special way in which the text uses biblical conceptions of wilderness while at the same time reworking these conceptions. How the wilderness is seen in *Ap.John* is further dependent on whose point of view it is seen from – the author or the text's various readers and users.

In this article I will discuss how "the mountain, a desert place" was conceived in *Ap.John* and what its uses and functions were. Drawing on critical spatiality theory, I examine the spatial categories of the text, including its elusive lived space, which is the type of space from which the text was produced and kept in circulation.

2. The frame story of *Ap.John*

According to the frame story, which covers little more than three of the fifty-seven pages of the text in BG, John, one of the sons of Zebedee, goes up to the temple after the resurrection of Jesus. A Pharisee named Arimanias approaches him and asks where John's master is. John says that he has returned "to the place from which he came" (BG 19:15–16). The Pharisee retorts, "With deception did this Nazarene deceive you (*planan*), and he filled your ears with [lies] and closed [your hearts], and turned (*afkte*) you [from] (*ebol*) the traditions of your fathers." (BG 19:17–20:3). The text then switches from a third-person to a first-person narrative: "When I, (John) heard these things, I turned away (*aikot ebol*) from the temple and to the mountain, a desert place *(eptooy euma nsjaeie)*" (BG 20:3–6). Here John asks doubtful and agonizing questions about the status of the Saviour, his father and the imperishable aeon (BG 20:6–19). Then John receives a vision.

In the vision the heavens open, the whole creation below is illuminated, and the world is shaken. John is afraid: "and behold, a child appeared to me, and [it changed] (its) likeness into an old man [while] the light [existed] in him. [As I looked] at it, I did not [understand this] wonder, whether there was a [likeness] with multiple forms [because of the] light –

since its forms appeared through each [other – or] whether it was one [likeness because] it had three faces" (BG 21:3–13).

Two of the "faces" are identified as a child and an old man. The third "face" is not described in BG.[3] The vision presents itself: "John [why do you] wonder and (why are you) [afraid]? Surely you are [not] unfamiliar with [this likeness]. Do not be [faint-hearted – I] am the one who is with you (pl.) always. I am [the Father], I am the Mother, I [am the Son]. I am the eternal One, the undefiled [and] the uncontaminated [One]." (BG 21:14–22:2). Then this person, who is identified as Jesus, offers to teach John.

The main part of *Ap.John* is presented as the teaching of Jesus and consists in a theogony, a cosmogony and an anthropogony, followed by John asking questions to Jesus, about the fate of the souls after death. The text ends with an admonition to write down the things that have been said and keep them secure. Jesus curses "everyone who will exchange these things for a gift, whether for food or for drink or for clothing or for another such thing." (BG 76:10–15). Then Jesus disappeared and John went to his fellow disciples and began to tell them what Jesus had told him (BG 76:17–77:5).

3. Turning away

When the frame story presents a contrast between the temple with Arimanias and "the mountain, a desert place" including the vision of the multiple forms of divinity, it anticipates the contrast between the world-creator and a superior godhead, which is a main theme in *Ap.John.* The text uses several pages to describe the superior godhead, while repeatedly stating that this godhead is indescribable. In addition to the supreme godhead, *Ap.John* promotes a biblical-demiurgical myth, which says that the material world is created by an evil/stupid world-creator. The world-creator is

3 According to Codex II, the vision changes likeness again, becoming like a servant (II 1:30–2:5). The Coptic word for servant is *hal* which means "slave, servant", probably from Greek *pais.* Guy Stroumsa points out a possible parallel to Paul's letter to the Philippians: "let the same mind be in you that was in Christ Jesus, who, though he was in the form of God, did not regard equality with God as something to be exploited, but emptied himself, taking the form of a slave, being born in human likeness. " (Phil 2: 5–7) (Stroumsa 1981: 412–34).

called Ialdabaoth and is a caricature of Yahweh.[4] The temple, which is referred to in the frame is probably the temple in Jerusalem where Yahweh was worshipped.[5] Similar to other temples in the ancient Near East the temple in Jerusalem was the centre of the earth, sacred and set apart, the place of daily rituals and sacrifices and an economical centre, in short a place of religious and political power.

Connected to this temple is the Pharisee Arimanias, who is unknown from other sources. Bernard Barc sees the frame narrative in relation to Mat 17: 1–9 and to the transfiguration on the mountain, and Arimanias as referring to Joseph of Arimathea (who witnessed the burial of Jesus, but not the resurrection). In Barc's interpretation the temple and Arimanias are symbols of pharisaic Judaism (Barc 2007: 318). Michael Waldstein presents a different explanation of the Pharisee. He points out that Arimanias is a Greek form of the Persian Zoroastrian name Ahriman ("the angry spirit") (Waldstein 2001: 101).[6] Waldstein's interpretation of Arimanias is convincing because Arimanias in *Ap.John* functions as one who raises doubt and thus acts like a traditional demon. But Barc is obviously right in seeing an opposition to pharisaic Judaism in this text. Pharisaic Judaism is most likely used to define types of Christianity that *Ap.John* opposes, which implies those who saw the God of Genesis as the father of Jesus – in other words the text includes a strong opposition to what became the Christian canonical interpretation of the Hebrew Bible. If John's opponent is associated with the Persian demon, in line with Waldstein's suggestion, the antagonism towards Judaism could still be present in the name "Arimanias", because a tradition existed according to which the Jews were descendants from the Magi.[7]

The important point, however, is that John's turning away from the temple has a parallel in his turning away from the tradition of his fathers, and that "the mountain, a desert place" presents an alternative to the temple where Yahweh was worshipped. "The mountain, a desert place" is cast as a superior sacred space.

4 This creature is known from several texts, somehow related to *Ap.John*, but the origin and the meaning of the name Ialdabaoth are uncertain. Because *Ap.John* casts Jesus as its narrator, the text also makes him the guarantor of the validity of the idea of Yahweh as an evil demiurge.

5 *Ap.John* says that John goes *up* to the temple, which probably reflects topographical knowledge about the temple in Jerusalem, which was situated on a hill.

6 Cf. also Turner 2001: 69, note 14.

7 Diogenes Laertius, *Prologue* 1.8–9.

4. The vision

In the frame narrative there are several movements, horizontal as well as vertical. These movements cover three spaces, one connected to the temple, one to "the mountain, a desert place" and one to the heavens. In the frame-narrative John goes up to the temple (where Arimanias approaches him), turns away from the temple to "the mountain, a desert place", and finally, at the end of the text, goes to his fellow disciples. The text does not mention where the disciples are. Jesus is, according to John's answer to Arimanias, gone to the place from which he came, which is the world above. Jesus was, according to John, sent into the world by his Father. When the heavens open, the illuminated likeness with multiple forms, which is Jesus, shows itself. At the end of the text Jesus disappears.

The locative centre of the narrative is "the mountain, a desert place" situated between the temple and the heavens, where the movement from the temple and the movement from the heavens – one horizontal and one vertical movement – converge in the asymmetrical encounter between John and the multiform version of Jesus. The vision and the encounter mark "the mountain, a desert place" as a specific type of place, betwixt and between the superhuman world above and the material world below.

The polymorphic Christ is known from 2nd and 3rd century texts, and "Christ's polymorphic epiphanies cannot be reduced to a single motif." (Thomassen 2009: 37–8).[8] The multiform version of Christ deviates from the natural form of a human being. He retains his human form, but is illuminated, in flux and reflects several age stages at the same time (cf. King 2009). The polymorphic Christ could be characterized as monstrous because his fluidity is strange and unnatural compared to the normally fixed human form. The vision connects very well to the barren landscape of "the mountain, a desert place", because this landscape seems to exist outside ordinary geography and to be closely connected to the spiritual world above. Since the vision of Jesus in "the mountain, a desert place" seems to happen after the resurrection, Jesus has transcended life and death and takes part in the spiritual world above and

8 According to Paul Foster, the polymorphic Christs are in the second and third centuries used in vastly divergent Christologies. Some of these Christologies reflect a docetic conception of Christ, other reflect proto-orthodox Christologies. In the first case Christ is "used to denote transcendence over the material world", while the proto-orthodox Christologies illustrate "that Jesus is not constrained by the forces of mortality, but rather that he has entered a higher state of physical existence." (Foster 2007: 66; cf. also Lalleman 1975).

in a supra-biological life, not limited by space and time. We also note in passing that the contrast, but also connection, between life and death on the one hand and the supra-biological life on the other is a prominent theme in the text, as seen when its last part consists of questions about what happens to different groups after death.

There is further a conflict in the text between the fluidity of the multiform Jesus and the theological explanations that exist as a potential in the text and which work in the direction of fixity. Different layers of interpretation have probably been at work in this text: When the vision with multiple forms and three faces presents itself as Father, Mother and Son, probably based on Plato's triad in Timaeus this nuclear family can also be seen as a version of a divine triad, which appears in a wider group of texts called Sethian (Turner 2001: 250–53; Waldstein 1997: 178–87). It is further likely that in the last phase of the text's life in Egypt, the fourth and fifth century, the triad was interpreted in line with Trinitarian theology (Stroumsa 1981: 426).[9]

That the different versions of Jesus were seen simultaneously is an expression of a spiritual and superior way of seeing, made possible in the special space of "the mountain, a desert place". The spiritual way of seeing stands in contrast to "normal" sight and the eyes of the flesh.[10] Characteristic of the vision in *Ap.John* is its fleeting and dynamic character, before it is defined as three forms and stabilized in the figure of Jesus.

9 In the third century Origen had created a subordinationist schema with three levels: The Father operates with the whole cosmos, the Son with rational beings and the Spirit with sanctified beings. At the Council of Nicea the Holy Spirit was touched upon in the formula, "We believe in the Holy Spirit", and further elaborated on at the Council of Chalcedon: "And in the Holy Spirit, the Lord, the Life Giver, who proceeds from the father, who with the Father and the Son is worshiped and glorified, who spoke by the prophets." In the fourth and fifth centuries the status, dignity and working of the Holy Spirit were debated, especially in the Eastern Church, even if no treatise on the person of the Spirit seems to have been written before 350 (Wright 1997: 1142–47).

10 It is worth dwelling on the sensual character of the vision, which starts with a special and superhuman kind of sight and continues as an audition and dialogue. The Alexandrian Jewish philosopher Philo (ca. 20 BC-50 AD) connects taste, smell and touch to the brute beasts and to slaves. Hearing and sight had, according to him, a link with philosophy and are more preeminent, "but the ears are in a way more slow and effeminate than the eyes". God made sight "a sense of queen of the rest, placing it above them all". This sense is in the closest connection with the soul" (*On Abraham*, 149–50, cf. Chidester 1992: 30–43; Frank 2000: 123). Christian theologians in the main agreed with Philo and regarded sight as the privileged sense (Frank 2000: 118 ff.).

The place of the vision is not described with specific mountains, caves, stones or living beings (Kelsey 1992: 143). There is no geophysical information about this area except for being described as "the mountain, a desert place". These characteristics open up two main interpretations; either an identification with a specific or non-specific mountain (see below), or to understand the characteristics as no more than the most elementary descriptions of a desert. In short, one may discuss to which degree this is a mountain or a mountainous area like one would find in the desert. Both interpretations have probably been at work during the text's life. The last interpretation – "the mountain, a desert place", conceived of as a desert – means that the landscape is stripped bare, only its label is left, which suggests that its space is like a clean slate to be written upon. It is an intermediate space between the world above and the world below, which is the world of humans. "The mountain, a desert place" functions as "transitional" space, a concept originally coined by D.W. Winnicott, which is a space that opens up to creativity, innovation and play.[11] In *Ap.John*, it is a space where the psychological tension and confusion of John finds its outlet and where it blends with external space.

Space and time are both in flux in "the mountain, a desert place", and therefore in consonance with each other. This type of fluidity is also in line with the possibilities for change and the special mutability that are opened up. The immobility of the temple is challenged by the capacity of "the mountain, a desert place" to create change and renewal. The possibility of change is exactly what John is offered when Jesus, in the main part of *Ap.John* offers a re-interpretation of Genesis and teaches him the difference between the spiritual, supra-biological life and the biological life of procreation and death. In the words of Karen L. King, John is taught "how to distinguish what is pure, undefiled, and eternal, from what is impure, mixed, and subject to becoming and destruction." (King 2006: 66). Laura Feldt characterizes the biblical landscapes of the desert as *locus transformationis:* "The desert is where you can be lead astray, into idolotary or death, or where you can discover a new subjectivity. The ambivalence of the desert holds out the possibility of trans-

11 "Transitional space" combines D.W. Winnicott's ideas of potential space and transitional phenomena. Potential space is the overlapping space between mother and child, seen in the interaction between them and where the subject-object relation is blurred. In this space we find transitional objects and transitional phenomena. (Winnicott 2005 [1971]). I am grateful to Robert Segal for drawing my attention to D. W. Winnicott.

formation." (Feldt 2009: 139). The narrative of *Ap.John* is inscribed in this type of space and unfolds itself from within it.

5. Perceived space and conceived space

Space is a cultural and social construct that humans create and shape by means of their practices. The philosopher Henri Lefebvre investigated the interplay between mental space and real space and inspired the work of the geographer Edward W. Soja (Lefebvre 1991; Soja 1996). According to their thinking space is an active force and the production of space serves to support certain interests. Lefebvre distinguishes between perceived, conceived and lived space, which are, respectively, the physical or geographical space; thoughts about space; and the social relations that produce thoughts about space. Soja takes the three spaces from their Lefebvrian origin in Marxist thinking into post-modernity and relabels them Firstspace, Secondspace and Thirdspace (Soja 1996: 10–12). In Lefebvre and Soja's thinking a trialectic of spatiality is grafted onto the dualisms of objective/subjective, material/mental and real/imagined because the lived, practiced and inhabited space is built on and communicates with Firstspace and Secondspace and modifies them. Accordingly the three spaces should be seen as three dimensions that exist at the same time and always interact with each other (Lied 2008: 16–17). For analytical purposes, however, they can fruitfully be distinguished.

In *Ap.John* we can distinguish between things in "the mountain, a desert place"; the mental maps or imagery applied to describe this place; and the lived experiences of the place. In line with this, *Ap.John* and "the mountain, a desert place" speak to the conceived space of the imagination and is, in the words of Soja, "entirely ideational, made up of projections into the empirical world from conceived or imagined geographies." (Soja 1996: 78–79).

The conceived space in the frame story of *Ap.John* draws its sap from various types of biblical contextualizations and inter-textual references that must have been at work for those who read and used the text.[12] "The mountain, a desert place" could, as already mentioned, be seen as referring to the mountain of transfiguration (Mt 17:1–9) (Barc 2007:

12 The frame of *Ap.John* draws the text into the Johannine tradition (Turner 2001: 69–70, note 14, cf. King 2006). For the use of allusions and intertextuality in Nag Hammadi texts more generally, see Lundhaug 2010: 2–4.

218), which means that the expression refers to a biblical landscape. Bentley Layton points out that the "Judean Desert begins at the back of the Mount of Olives, which on the other side overlooks the site of the temple" (Layton 1987: 28). As mentioned, the expression "the mountain, a desert place" is ambiguous because it can both be seen as referring to a geographical place in the form of a specific mountain (Mountain of Transfiguration or Mount of Olives including the Judean desert), at the same time as it more vaguely refers to the desert. In both cases it can be seen as wilderness and as characterized as that which is outside the city (Endsjø 2008).

The pentateuchal tradition presents the wilderness as a place of divine access, both to God and the law, for in the wilderness period God was unusually accessible (Schofield 2008: 44; cf. also Schäder 2010). The wilderness or desert is "a recurring backdrop" in "the stories of the patriarchs and the early Israelites to the degree that one may even speak of them as a wilderness people" (Schofield 2008: 37). There "can be no doubt that a crucial strand of Israelite self-understanding is tied to the desert" (Feldt 2009: 138, note 300). Accordingly, the wilderness in the Hebrew Bible is connected to a place, a period and a group of people.

The New Testament conceptions of wilderness are built on the Old Testament, and three motives can be identified: The idea that a true prophet of God receives divine revelation within a wilderness context; the idea of God's self-revelation to the chosen people in the context of the wilderness journey; and the idea of wilderness as a threatening presence (Jefford 2008: 152–72). In line with the conception of wilderness in OT and NT, "the mountain, a desert place" is in *Ap.John* a place where access to God is easy and immediate and, not least, it is a space of teaching where revelation and true knowledge are offered to them that look for it. *Ap.John* also references the second motif; this happens when John is admonished to teach the message that he has received to the disciples, in other words to those that have been chosen. The text ends with John going to the disciples to tell them what he had been told. This means that John is on the verge of sharing the message of the Saviour with the disciples, but *de facto* he has already shared it with the readers of the text. This supports the inclusion of the reader as an implicit partner in the dialogue between John and Jesus. Finally, it is said that John was afraid when he saw that the heavens opened, so the idea of the frightening quality of the wilderness, or better of what might happen there, is present as well.[13] Similarly to how the wilderness is cast as a space of doubt and apostacy in the Hebrew Bible, John's initial conflict with Arimanias over the status of Jesus and his doubtful questions signal apostacy from the biblical conception of Yahweh.

The description in *Ap.John* of "the mountain, a desert place" is further in accordance with how Athanasius in his orthodox and model narrative of the Christian ascetic life, describes the landscape where he places his hero, Antony. The terms that Athanasius most frequently uses are *oros*, "mountain" and *eremos*, "desert", which are equivalents to the Coptic *tooy* and *ma nsjaeie*, which are used in *Ap.John*.

In his book about desert asceticism, especially related to Athanasius' *Vita Antonii,* Dag Øistein Endsjø, establishes a strong connection between Greek geography and Athanasius' construction of the desert (cf. Endsjø 2008). According to Endsjø, Athanasius describes the desert as a traditional Greek wilderness, which is dependent on the existence of an ordered centre, and "initially defined by its opposite, the Greek city or *polis*" (Endsjø 2008: 31) Several of Endsjø's observations about the wilderness/desert in *Vita Antonii* and what happens there fits *Ap.John* very well, even if the contrast between the temple and the desert in the case of *Ap.John* seems to be more dependent on biblical models than on Greek geography.

Athanasius describes the desert as "a landscape apparently devoid of life" (Endsjø 2008: 17) and Endsjø characterizes this landscape as "coun-

13 In addition to intertextual interplays between the wilderness in *Ap.John* and wilderness-motives in the bible, one can also add the introduction to the vision in Ezekiel where "the heavens were opened, and I saw visions of God" (Ezekiel 1.1) as a possible intertext to the vision in *Ap.John.* The common structure between the two narratives consists in an initial illumination, a fixed number (four in Ezekiel, three in Ap.John) and the vision of God/Jesus. A general lesson in Ezekiel, as interpreted by Christian theologians/church-fathers, and a theme in *Ap.John* as well, is the incomprehensibility of God (cf. Christman 2005: 63–98). In Ezekiel's vision there is a combination of different creatures, and what "is kept separate in this world is combined in this vision of the heavenly world" (Nielsen 2008: 101). In *Ap.John* there is a combination of different human forms and ages, and similar to Ezekiel, the vision combines and transcends mundane categories. Ezekiel also presents a contrast between the earthly temple and the heavenly temple (Ezekiel 40–48), which could be one of the inputs for the contrast between the temple and the vision. Different to *Ap.John*, though, the vision of Ezekiel does not take place in the desert, Ezekiel sits by the river Chebor (cf. Turner 2001: 244). I am grateful to Jan Bremmer who pointed out a possible intertextual connection to Ezekiel.

ter-space" and uses the Foucauldian concept of *heterotopia* to describe it. Endsjø makes the apt observation that "whatever wonders there were in this landscape did not refer to the physical geography itself but to the *ideas* brought to it" (Endsjø 2008: 18). This insight can fruitfully be applied to "the mountain, a desert place" of *Ap.John* as well, which arguably functions as an intermediate zone between the world above and the material world and is staged as a supernatural place of superhuman revelation.

"The mountain, a desert place" in *Ap.John* is, to an even higher degree than the desert of Antony, devoid of characteristics. There are no animals or demons that wrestle with John, as they did with Antony, though there is an acute conflict between John's own experience of Jesus and Arimanias' interpretation of Jesus. The outstanding event in the desert is John's encounter with the vision of the polymorphic Christ, which is a fluid sort of reality that is made possible exactly because of the empty space of the wilderness.

6. Lived space as ascetic withdrawal

New social movements tend to create new types of spaces. Christian asceticism created the desert. One important dimension of early Christianity was the experience of the desert as the praxis aspect of Christian ascetic life (cf. Guillaumont 1975). James E. Goehring writes about the ideological power and function of the desert and its associations to ascetic withdrawal:

> The popularity and power of the desert myth over time and across geographical boundaries depended finally on its successful equation of the Near Eastern desert landscape with the spiritual interpretation of Christian withdrawal from the world. As the Coptic word for mountain (*toou*) came to be associated with monastery and desert, so the meaning of "desert" itself came to equate less with the nature of the land than with the general concept of ascetic withdrawal.
>
> (Goehring 2003: 446)

Philip Rousseau has recently made a pertinent point about ascetic texts. First, and in line with much research on asceticism, he points out that the ascetic texts "exalt a 'desert' in the eyes of those who live in the 'city'" (Rousseau 2010: xxv). According to him the ascetic texts "were designed, often by urban authors, to evoke and reinforce implications, not least about spiritual authority, that were to be realized within the urban milieu, no matter how withdrawn, how anchoretic in the literal sense, the exemplars might seem to be. The texts made, to that extent, a 'political' point – a point, as I say, about the Episcopal milieu, about the milieu of church's pastoral endeavour" (Rousseau 2010: xxv). But secondly and equally important, Rousseau also stresses that there were people who did not live urban lives and did not primarily rely on the authority of the bishops: "many a monastery and many a solitary cell, across the changing Roman landscape, bore witness to a different sense of authority, a different understanding of legacy, a different deployment of power" (Rouseau 2010: xxvi).

What was the lived space of *Ap.John?* What sort of social relations have produced "the mountain, a desert space" in the vision of the frame narrative? What kind of lived experience generated this text? To answer such questions it is necessary to make the concept of space more complex. Gert T. M. Prinsloo has written about different worlds in biblical texts: "Three worlds interact in the interpretation of biblical stories: the socio-historical milieu from which the work arose, the world of the reader or hearer, and the world created in the work itself" (Prinsloo 2005: 458–9, note 9). The three spaces of Lefebvre and Soja (the perceived, conceived and lived space) are dependent on which position they are seen from, either from the narrator or from the audience.

In the case of *Ap.John* one can imagine people who saw "the mountain, a desert place" as a literary *topos*, like its author(s) might have done. It is not far-fetched to imagine the authorship of *Ap.John* to have belonged to an urban centre, with Alexandria as a likely candidate – Karen L. King mentions an urban school setting in Alexandria (King 2006: 9–17). For the author(s) the desert might have been a distant place best savoured within the walls of the city and the confines of a text. The text criticises the authority of the church and its leaders by means of the idea of Yahweh as a subordinate god. The original milieu of the text might very well have been in line with what Rousseau has suggested, people who lived in cities and exalted the desert. At the same time it is also easy to imagine – still in line with Rousseau's ideas – subsequent readers, transmitters and re-interpreters of *Ap.John* to have been ascetics, ascetic groups and monasteries that were much closer to or even lived in the desert (Rousseau 2010: 18–21).

Soja makes the important point that this "process of producing spatiality or 'making geographies' begins with the body, with the construction and performance of the self, the human subject, as a distinctively spatial entity involved in a complex relation with its surroundings" (Soja 2000: 6). The people who used *Ap.John* were people whose ideas about desert as living reality were made to interact with *Ap.John*, for instance in the vicinity of Nag Hammadi where three of the versions of this text was found. In this area, Pachomian monasteries were located, and a Pachomian ownership has been suggested for the codices (Wisse 1978, cf. also King 2006: 18–21), but it has also been pointed out that other types of religious communities were present in the area. A close reading of *Ap.John* reveals that the treatise seems to be negative to sexuality and procreation (Gilhus 2008), and that it ends on a note of asceticism: "Cursed be everyone who will exchange these things for a gift, whether for food or for drink or for clothing or for another such thing" (BG 76: 10–15).[14]

A hermeneutical process always consists in complex interplays between exegesis and "*eisegesis*", reading things out of texts and reading things into them. The readers and users of *Ap.John* most likely saw "the mountain, a desert place" in dialogue and interplay with biblical texts. But they would also have seen it in a dialogue with a natural landscape, which for them was the Egyptian desert area with rock hills and mountains, a landscape characterized by a sharp border to vegetation and to the cultivated area. The narrated space of the text was mixed and blended with the geography of the area where the text's interpreters and readers lived and most likely intimately intertwined with their ascetic lives.

7. Cosmology and geography

The emptiness of the desert makes it into a type of space where things are changed and at the same time illuminated. As already mentioned, fundamental in this change is that *Ap.John* rewrites the script of Genesis by introducing the evil demiurge Ialdabaoth in place of Yahweh. The contrast between the temple and the desert and the contrast between the two gods correspond with each other, because the desert is a place that vouches for the truth of God versus the demiurge Ialdabaoth. Another observation is pertinent here. The references to the temple and "the mountain, a desert

14 The other short version of the text in codex III is even more elaborate, here the saviour curses "everyone who will exchange these things for a gift, whether of silver or gold, whether for drink or for food or for clothing or for another such thing" (III 39:24–40:4).

place" are to actual places, but with strong mythological and ideological groundings, probably the Temple Mount and the Mount of Transfiguration or the Mount of Olives with adjacent areas of desert. These are real places, but in the text they appear as conceived places. With time these specific geographical places may have receded into the background as the Egyptian desert landscape gradually came more into focus. Except for "the mountain, a desert area", there are, arguably, no other references to geographical places in the entire text.[15] The text points to cosmology, not to geography. When it elsewhere refers to places (Coptic *ma* and Greek *topos*), these places are situated in the world above and are organized hierarchically in relation to each other, or they are in the underworld, which is Hades (BG 57:7). In opposition to "the mountain, a desert place," these other places are not part of the geography of this world.

Ap.John describes a cosmological system and its mythological entities. It refers to Timaeus as well as Genesis (King 2006) and works with a complete apocalyptic scenario including what belongs to the world above and what belongs to the world below – the world of spiritual beings and the world of humans. Thus *Ap.John* offers a mythological map of the world, which is, like most mythological maps, ideological and strategic. The cosmology presents the underlying patterns and forces that determine life in this world and puts human life into perspective. The cosmological narrative has its only possible meeting point with an earthly landscape in the empty space of "the mountain, a desert place", and because this space is stripped, naked and devoid of all life, it is thus opened up for divine revelation. It is a superhuman place of transformation and revelation that has been almost completely taken over by the illuminated vision and the divine voice. The text is written on this space in a way that erases nearly all traces of real geography. In this way the empty space of the frame story becomes a receptacle for the detailed cosmology with all its spiritual beings, and the space of the desert is lifted out of its earthly context and made to reflect the world above.

15 *Ap.John* refers to Paradise, where Adam and Eve are placed. But Paradise seems not to be a place in a similar way as "the mountain, a desert place". Sometimes in antiquity Paradise was thought of as a real place, but not always: "Who is stupid enough to think that like a farmer God planted a garden in Eden to the east and in the garden made a tree of life, visible and perceptible to the senses, of such a kind that he who ate its fruits with his physical teeth would receive life." (Origen, *Princ.* 4.3.1.).

8. Conclusion

The concepts of wilderness and space have helped to focus on the frame narrative of *Ap.John*, which in previous research on this text has, so far, not received the attention that it deserves. Since *Ap.John* is the first text in three of the four codices in which it is included, the frame story serves also as the introduction to these codices. John is the mediator between Jesus and the disciples/potential readers, but the fact that the disciples remain, throughout, on the margins of the text probably also implies that the impact of the text on its readers becomes more forceful because they identify with the intended recipients of the message, the absent disciples. When *Ap.John* ends with John who "went to his fellow disciples, and began to tell them what had been told to him by the saviour" (BG 76:18–77:5), the readers/listeners have already received the message. As the immediate recipients of the message of *Ap.John*, the readers are clearly invited to identify with John as well as with his "fellow-disciples". In this way, the space where John speaks with the polymorphic Christ – "the mountain, a desert place" – becomes the textual space where the readers receive the message of the text as well.

The use of critical space theory has been instrumental in teasing out the complexities of the wilderness in *Ap.John* and its various references and contexts: The perceived wilderness of biblical mountains and/or the Egyptian desert; the conceived wilderness of biblical mythology; as well as the lived wilderness of ascetic life. These three types of spaces are not static. One must distinguish between the author and the readers of *Ap.John* on the one hand and between different groups of readers on the other who always create the three spaces in different ways, some of them within the walls of cities, some living in closer vicinity to the desert, and some living in the desert. The complexities of the wilderness in *Ap.-John* are not captured by the simpler dichotomies of wilderness/civilization, cosmos/chaos and nature/culture, which is in line with one of the main hypotheses of this volume (cf. the Introduction, p. 9).

The analysis of *Ap.John* has further shown that the wilderness it contains – "the mountain, a desert place" – is not only "those parts of the world that are perceived to be untouched by humanity", which was the initial definition of wilderness in this article. On the contrary, this study also shows that there is every reason to question the conventional dualism that envisages wilderness as "a domain void of humans, opposed to civilization", as Laura Feldt does in the introduction to this volume (cf. Introduction p. 18). "The mountain, a desert place" becomes a wilderness exactly because humans touch it, not only by their conceptualizations, but also by their specific ways of living. In this way there is an ambiguity in the concept of wilderness: The wilderness of *Ap.John* is attractive because it is perceived to be untouched by humans at the same time as it is touched by humans by being constructed as a space for divine revelation and mythology and progressively also as a habitat for the angel-like lives of ascetics.

9. References

The Apocryphon of John. Synopsis of Nag Hammadi Codices II,1; III,1; and IV,1 with BG 8502,2. Edited by Michael Waldstein and Frederik Wisse. Leiden/New York/Köln: Brill, 1995.

Bauman, Whitney A., Richard R. Bohannan II, and Kevin J. O'Brien, eds., *Grounding Religion. A Field Guide to the Study of Religion and Ecology.* London/New York: Routledge, 2011.

Barc, Bernard. "Livre des secrets de Jean", in *Écrite gnostique. La bibliothèque de Nag Hammadi*, edition publiée sous la direction de Jean-Pierre Mahé et de Paul-Hubert Poirier. Paris: Gallimard, 2007, 217–95.

Chidester, David. *Word and Light: Seeing, Hearing, and Religious Discourse.* Urbana: University of Illinois Press, 1992.

Christman, Angela Russell. *"What did Ezekiel see?" Christian Exegesis of Ezekiel's vision of the chariot from Irenaeus to Gregory the Great.* Leiden/Boston: Brill, 2005.

Endsjø, Dag Øistein. *Primordial Landscapes, Incorruptible Bodies: Desert Asceticism and the Christian Appropriation of Greek Ideas on Geography, Bodies, and Immortality.* New York: Peter Lang, 2008.

Feldt, Laura. *Signs of Wonder – Traces of Doubt: the fantastic in religious narrative from Exodus to Elisha.* Ph.D. Dissertation, Section of the Study of Religion, Faculty of Theology. Aarhus University: 2009.

Foster, Paul. "Polymorphic Christology: Its Origins and Development in Early Christianity", *Journal of Theological Studies*, 58, 1, (2007), 66–99.

Frank, Georgia. *The Memory of the Eyes. Pilgrims to Living Saints in Christian Late Antiquity.* Berkeley: University of California Press, 2000.

Gilhus, Ingvild Sælid. "Sexuality, Marriage and Knowledge in the *Apocryphon of John*". In: Martti Nissinen and Risto Uro, eds., *Sacred Marriages: The Divine-Human Metaphor from Sumer to Early Christianity.* Winona Lake, Eisenbrauns, 2008, 487–510.

Goehring, James E. "The Dark Side of Landscape: Ideology and Power in the Christian Myth of the Desert", *Journal of Medieval and Early Modern Studies*, 33, 3, (2003), 437–51.

Guillaumont, Antoine. "La conception du desert chez les moines d'Egypte", *Révue de l'histoire des religions*, 188, (1975), 3–21.

Jefford, Clayton N. "The Wilderness Narrative in the Apostolic Fathers". In: Kenneth E. Pomykala (ed.), *Israel in the Wilderness. Interpretations of the Bib-*

lical Narrative in Jewish and Christian Traditions. Leiden: Brill, 2008, pp 157–72.

Kelsey, Neal. "The Body as Desert in the *Life of St. Anthony*". In: Vincent L. Winbush (ed.), *Discursive Formations. Ascetic Piety and the Interpretation of Early Christian Literature. Part 1* (Semeia, 57), 1992, 131–51.

King, Karen L. *The Secret Revelation of John.* Cambridge Mass./London, England: Harvard University Press, 2006.

King, Karen L. ""In your midst as a child" – "In the form of an old man". Images of Aging and Immortality in Ancient Christianity". In: T.K. Seim and J. Økland (eds), *Metamorphoses. Resurrection, Body and Transformative Practices in Early Christianity.* Berlin/New York: Walter de Gruyter, 2009, 59–82.

Lalleman, Pieter J. "Polymorphy of Christ". In: Jan Bremmer, ed., *The Apocryphal Acts of John.* Kampen: Pharos, 1975, 97–118.

Layton, Bentley. *The Gnostic Scriptures. A new translation with annotations and introduction by B. Layton.* USA: SCM Press LTD, 1987.

Lefebvre, Henri. *The Production of Space.* Oxford: Blackwell, 1991 [1974].

Lied, Liv Ingeborg. *The Other Lands of Israel. Imaginations of the Land in 2 Baruch.* Leiden: Brill, 2008.

Lundhaug, Hugo. *Images of Rebirth. Cognitive Poetics and Transformational Soteriology in the* Gospel of Philip *and the* Exegesis to the Soul. Leiden: Brill 2010.

Nielsen, Kirsten. "Ezekiel's Visionary Call as Prologue: From Complexitiy and Changeability to Order and Stability?" *Journal for the Study of the Old Testament* 33.1 (2008), 99–114.

Philo. *The Works of Philo. Completed and Unabridged.* Tranlated by C.D. Yonge. USA: Hendrickson Publishers, 2002 [1993].

Pomykala, Kenneth E. ed., *Israel in the Wilderness. Interpretations of the Biblical Narrative in Jewish and Christian Traditions.* Leiden: Brill, 2008.

Prinsloo, Gert T. M. "The Role of Space in the šîrē hammalôt" (Psalms 120–34), *Biblica* 86/4 (2005), 457–77.

Rousseau, Philip. *Ascetics, Authority, and the Church. In the Age of Jerome and Cassian.* Notre Dame, Indiana: University of Notre Dame Press, 2010 [1978].

Schäder, Jo-Marí. "Understanding (the Lack of) Space in Psalm 47:6 in Light of its Neighbouring Psalms: A Spatial Reading of Psalms 46–48", *Old Testament Essays,* 23, 1, (2010), 139–60.

Schofield, Alison. "The Wilderness Motif in the Dead Sea Scrolls", In: Kenneth E. Pomykala, ed., *Israel in the Wilderness. Interpretations of the Biblical Narratives in Jewish and Christian Traditions.* Leiden: Brill, 2008, 37–53.

Soja, Edward W. *Thirdspace. Journeys to Los Angeles and Other Real-and-Imagined Places.* Cambridge. Mass.: Blackwell, 1996.

Soja, Edward W. *Postmetropolis. Critical Studies of Cities and Regions.* Oxford: Blackwell, 2000.

Stroumsa, Gedaliahu G. "Polymorphie divine et transformations d'un mythologème: L'"Apocryphon de Jean" et ses sources", *Vigiliae Christianae* 35 (1981), 412–34.

Thomassen, Einar. "Jesus in the New Testament apocrypha." In: Olav Hammer, ed., *Alternative Christs.* Cambridge: Cambridge University Press, 2009, 33–50.

Turner, John D. *Sethian Gnosticism and the Platonic Tradition.* Québec/Paris: Les Presses de l'Université Laval/ Éditions Peeters, 2001.

Waldstein, Michael. "The Primal Triad in the *Apocryphon of John.*" In: John D. Turner and Anne McGuire, eds., *The Nag Hammadi Library after Fifty Years: Proceedings of the 1995 Society of Biblical Literature Commemoration.* Leiden: Brill, 1997.

Waldstein, Michael. "Das Apocryphon des Johannes (NHC I,1; II,1; III,1; and BG2)". In: Hans-Martin Schenke, Hans-Gebhard Bethge and Ursula Ulrike Kaiser, eds., *Nag Hammadi Deutsch.* 1. Band: NHC I,1-V,1 (Koptisch-Gnostische Schriften II). Berlin: Walter de Gruyter, 2001, 95–150.

Waldstein, Michael and Frederik Wisse (eds). *The Apocryphon of John. Synopsis of Nag Hammadi Codices II,1; III,1; and IV,1 with BG 8502,2.* Leiden/New York/Köln: Brill, 1995.

Winnicott, Donald W. *Playing and Reality.* London/New York: Routledge, 2005 [1971].

Wisse, Frederik. "Gnosticism and Early Monasticism in Egypt". In: Barbara Aland, ed., *Gnosis. Festschrift für Hans Jonas.* Göttingen: Vandenhoek & Ruprecht, 1978, 431–40.

Wright, David F. "The Trinity." In: Everett Ferguson, ed., *Encyclopedia of Early Christianity.* 2 vols. New York and London: Garland Publishing, 1997, 1142–47. 1990.

5. "The truth is out there": Primordial lore and ignorance in the wilderness of Athanasius' *Vita Antonii*

Dag Øistein Endsjø

1. Introduction

"The truth is out there" was a well known phrase in popular culture in the 1990s, in connection with the successful mystery series *The X Files* (1993–2002), where the leading couple of federal agents would repeatedly encounter various aliens and other mythical creatures and phenomena. Regardless of how much knowledge we have accumulated within our modern civilization, as the rationale of this claim goes, we will never be able to grasp the essence of it all if we do not leave behind the fundamental structures of our society, the logical limitations of how we are taught to experience reality, and, ultimately, also the physical limits of our civilization. Here, we enter the geographical wilderness.

This claim does, of course, not originate with this popular television series. Indeed, these beliefs about truth itself somehow being situated in the wilderness have been widespread for, literally, millennia. As such there is reason to suspect that one of the explanations of the success of *The X Files* lies exactly in the way the apt scriptwriters reformulated these ancient beliefs within a modern framework.

The very belief that certain forms of knowledge, or even "truth" itself, may have a special relation to particular geographical areas connects to the most basic ideas about how space is never a neutral entity, but always imbued with various cultural values. As the cultural geographer Henri Lefebvre points out: "The concept of space is not in space … Like anything else, space is an historical product in the classical sense of the term" (Lefebvre 1991 [1974],299; 1976 [1970],31). As long as we do not lie on the ground sniffing the earth, and perhaps not even then, is it impossible to get a sense of space *per se.* Space can only be understood through the filters of tradition, cultural and religious ideas, artistic presentation, and language. Or one may phrase it as Edward Soja, another prominent geographer: "Space in itself may be primordially given, but the organization and meaning of space is a product of social translation, transformation, and experience" (Soja 1989,79–80). In this way we realize that the whole concept of wilderness, regardless of whether it is seen as connected to any profound knowledge or not, also is a cultural construct which will have different meanings depending on which culture we examine. The wilderness will nevertheless have a more universal aspect by the way that this essentially refers to areas where the culture that operates with the concept is not in control. Any notion of the wilderness, regardless of how it is understood or limited, is generally a reflection of the "other", and will thus tell just as much about the space that which it is not and about the culture which is defining it from a distance. Somehow the wilderness will always be something akin to what Michel Foucault defines as *heterotopia*, the places which "are absolutely different from all the sites that they reflect and speak about" (Foucault 1986 [1967], 24).

2. The lore of the periphery, and its apparent denial

Although the notion of the wilderness as the eternal spatial other may be a concept common to most cultures, the conviction that the truth is "out there" is not universally held. This is just one of countless beliefs that sometimes are, and sometimes are not, seen as connected to the uncultivated geography. We can just look at our own Western clichés of secluded wise men and ignorant country bumpkins, noble savages and primitive natives. But the very idea that there is a special relation between knowledge and the wilderness, regardless whether this area is considered to enhance or diminish one's wisdom, remains a good example of the conviction that space really may influence our very human nature, and, moreover, how these beliefs actually form our sense of reality. In this context late antiquity, the time when Christianity had its breakthrough in the Greco-Roman world, presents itself as a good arena for examining how various ideas about knowledge and the wilderness interacted, and how the encounter between the two contributed to the formation of new ideas.

Turning to the *Vita Antonii*, Athanasius of Alexandria's biography of Anthony, the first known monk who made the wilderness his home (written briefly after the hermit's death in 355), we find a classic presentation of the belief that true wisdom is found in the wilderness, and what at a first glance seems to be a direct criticism of this belief. One day, when speaking to a number of visiting monks who had gathered around him in the desert east of the Nile, Anthony had this to say:

> The Greeks live abroad and cross the sea to learn letters, but we have no need to live abroad for the kingdom of heaven or to cross the sea for excellence. For the Lord has said: "The kingdom of heaven is within you" (Athanasius *Vita Antonii* [henceforth just *VA*] 20.4, paraphrasing *Luke* 17:21).

What Athanasius made Anthony refer to here, and apparently also deny, was the particular traditional Greek belief that in the wilderness, in the area that may be referred to with the traditional term *eschatia*, wisdom was considered to be found free from the restrictions of human community; that travelling in the uncultivated geography could somehow make you wise. This was an ancient idea within the worldview of the Greeks. Celebrated sages, like Orpheus, Anacharsis, Melampus, Musaeus, Daedalus, Homer, Lycurgus, Solon, Pythagoras, and Plato, were all considered to have obtained much of their wisdom from various parts of the *eschatia*, while the wisest of peoples were frequently held to live far off. But when David Frankfurter argues that "one of the hallmarks of Greco-Roman culture was the allure of the peripheral, where might be found the exotic, the true, the most ancient, and the surest route to the spiritual truths" (Frankfurter 1998,19), one must, at the same time be aware of that the Greek wilderness did not just pertain to the most distant periphery. The essence of the *eschatia* was, indeed, its uncultivated state as contrary to the cultivated space of the *polis*. Thus, the *eschatia* in the Greek worldview sums up the wilderness, the uncultivated geography, from the areas immediately outside the city walls to the very ends of the earth (cf. Hartog 1988 [1980],13; Scully 1990,10; Endsjø 2000,359–60, 366–67). From the most distant and mythical era of the Greeks, all the way up to the Christian era, people who operated within the Greek worldview considered the wilderness not only a place filled with sagacious figures but also a region into which one could venture in search of wisdom: We have the most ancient stories about figures like Dionysus, Jason, Achilles, and Odysseus receiving their education in the most remote and in the not-at-all-so-distant wilderness, sometimes from the semi-human centaurs; we have the reports of wise men travelling to the uttermost periphery of the world and of Hesiod being instructed by the Muses on the wild slopes of Mount Helicon just a few kilometres from Thebes (Hesiod *Theogony* 22–23); we have the continuous use all through this era of the wilderness immediately outside the city-walls in coming-of-age rituals and the great appeal of the secret lore of exotic peoples from far away used in a number of popular cults. Sometimes barbarian words were said to represent central components of the secret sayings uttered during the rites (Iamblichus *On the Mysteries* 7.4–5), non-Greek musical instruments played an important part in the Dionysian mysteries, while the participants would don the guise of satyrs, nymphs and other prototypal characters of the periphery (Plato *Laws* 815c; cf. Burkert 1985 [1977],259). In the first century B.C., Diodorus of Sicily placed the very origins of Greek mysteries in either Egypt, Phrygia, or in the blurry Pre-Greek past of Crete (Diodorus of Sicily *Library* 1.29.2–4, 5.64.4, 4.77.3). Consequently Heracles, after travelling to the ends of the earth and beyond, could claim that he no longer had any need for the Eleusinian rites: "I have been initiated into truer mysteries" (cf. Endsjø 2000,351).[1]

As Anthony himself was living in the wilderness, this apparent disparagement of obtaining knowledge in distant places is all the more remarkable because it may even seem like he was made to belittle his own status. As if Athanasius wants to be sure that we get the point, the hermit is even depicted as taking this to a more personal level, when two Greek philosophers visiting Anthony in the wilderness see him in exactly this traditional Greek manner of the wise men belonging to the uncultivated geography: Insisting that they consider Anthony "very wise", the hermit himself asks of them: "Why did you trouble yourselves so much to see a *foolish* man?" (*VA* 72.3–4, my emphasis).

This alleged self-representation of Anthony as "a foolish man" may seem puzzling as this man of the wilderness, of course, is both the hero and ultimate Christian role model in this biography penned by the mighty patriarch of Alexandria. But these two episodes really are not in any way exceptional. Indeed, a distinct attempt to present Anthony as completely uneducated is evident in the composition. As a child of lower middle class ethnic Egyptian parents, Anthony would not be someone who the contemporary reader would expect to have received much classical education. But Athanasius did not leave it at that:

> As a small child he was brought up by his parents, knowing nothing besides them and his home. Later when he had become an older boy ... he refused to learn letters, not wanting to be in the company with other children. All his

---

1 *mustêria [pollô a]lêthestera amemuêmai.* The claim of Heracles on a second century A.D. papyrus found in Egyptian Tebtynis is published in Vogliano 1937,177.

> desire was, as it is written, to live unaffected (*aplastos*) in his home (*VA* 1.2–3).

According to the *Vita Antonii*, it was completely out of the question for Anthony to have had any formal schooling. Not only his childhood was lacking in formal training, but Athanasius also made it explicitly clear that Anthony *never* learned to read or write (*VA* 1.2, 72.1, 73.1, 78.1). He also had, still according to Athanasius, no knowledge of the Greek language, which in the still predominantly Hellenistic Roman province of Egypt was generally the same as having received the most basic instruction. The celebrated desert father was instead described as someone depending on interpreters (*VA* 72.3, 74.2, 77.1).

Anthony's illiteracy and general ignorance were of importance to Athanasius. This is further demonstrated by certain facts indicating that the real life Anthony probably was not at all completely uneducated. The letters of Anthony, which, as it has convincingly been argued, must be authentic, "reveal that he must not only have been literate but also possessed of some education" (Rubenson 1990,141). Athanasius even slips a couple of times himself in his attempt to present Anthony as in all manners uneducated, having him, apparently unassisted, writing letters both to the emperor Constantine and to a Roman general (*VA* 81.5, 86.2).

As such, Athanasius took great care to depict the celebrated desert father as an ignorant yokel, even though other evidence indicates that he must have had at least some basic education. But if we really are to accept this utterly simplistic presentation, why did anyone bother about this monk at all? Why did he not only become an instant hero to so many Christians in his contemporary Egypt, but the subject of a biography of the extremely well educated patriarch Athanasius, and even a great star in Christianity in general as his story became more widely known? This dilemma does not turn any less strange when we take into account how Anthony completely transformed the very essence of monasticism at large, as the very geographical locus of this phenomenon was moved from cities and villages to the wilderness, through emulation of this pioneering desert father.

Anyone may be shielded from proper education, and in fourth century Egypt it was often not even a question of shielding: many, perhaps most people, did not have a chance at all of receiving any schooling. In this way, the uneducated Anthony was not at all exceptional. His living in the wilderness, on the other hand, was quite another case. His very claim to late antique super-stardom is instead intimately connected to his being the first Christian monk who made the desert his home. For that reason it is relevant to examine whether Anthony's allegedly complete ignorance has a special meaning when connected to his exceptional geographical locus.

3. The essence of the Greek wilderness

Anthony's wilderness as presented in the *Vita Antonii* of course in many ways reflects the geographical notions of the Hebrew Bible, which repeatedly depicts the desert as a place of divine intervention, not least seen in the connection with various prophets like Moses and Elijah. This originally Judaic worldview is also indirectly reflected in the way the biography of Anthony presents the wilderness as a place of divine trial as is the case with John the Baptist and Jesus in the New Testament. At the same time, the geography of Athanasius' *Vita Antonii* is just as much a Greek landscape. Schooled in the Classical tradition himself and a Grecophone native of the Hellenistic metropolis of Alexandria, Athanasius operates clearly within a traditional Greek spatial framework. Not only is Anthony's wilderness a place of similar wonders, riches, challenges, and confusion, as the traditional Greek *eschatia* (cf. Endsjø 2008,65–69; 83–85, 101–17), but it also starts immediately where the cultivated lands end (Endsjø 2008,33–34), thus accurately reflecting the traditional Greek spatial dichotomy of intimately juxtaposed *polis* and *apolis:* The marvels Anthony encountered did not emerge with his moving to the most distant parts of the desert, or even to the desert itself; they appeared as soon as Anthony departed for the cemetery which was situated only "a distance from the village" (*VA* 8.1): they take place immediately as Anthony leaves the cultivated space of his homestead for the uncultivated geography; where the dead were buried.

The *Vita Antonii* also exhibits a direct continuity with such undisputedly Greek traditional *religious* beliefs that may seem surprising in a book written by a leading Christian theologian of his time: Just the way the *eschatia* traditionally was the place where Greeks could encounter their gods (Endsjø 2008,51), Athanasius filled Anthony's wilderness with Greek deities, although certainly not acknowledging their divine nature, instead defining them as "demons whom the Greeks themselves consider gods" (*VA* 94.2, cf. 22.1, 37.1–3). The hermit is even made to encounter a figure resembling a Greek satyr (*VA* 53.2). It would nevertheless be a mistake to assume that Athanasius made a direct attempt to present the Egyptian wilderness in Greek terms. As a Grecophone native of Egypt himself, Athanasius would quite naturally consider the uncultivated landscapes of his own country in the same category as any uncultivated area in Greece proper. He simply used the geographical language, the terms, notions, and beliefs, that he would find everywhere in the Hellenistic culture he belonged to. Even more importantly, this is how a Grecophone audience would understand his text.

The Egyptian wilderness generally equalled a waterless and treeless region with scorching sun, or what most people today would consider a prototypal desert, but this does not alter the understanding of the major dichotomy, and it really does not seem to have affected how Anthony's *eschatia* was viewed. The Greek word most usually translated into desert, *erêmos*, actually denoted just a desolate landscape, an uncultivated wasteland, or an untamed wilderness. As such the *erêmos* is really synonymous with the *eschatia*. In Greece proper most of the land that was considered uncultivated wasteland consisted of rugged mountains. It is therefore interesting to notice how Athanasius at times used the two terms – *erêmos* and *oros* – interchangeably on the Egyptian landscapes without distinguishing the one from the other (*VA* 11.1–3, 14.7). Accordingly, he also named Anthony's famous "inner mountain" (*VA* 51.1, 82.1, 91.1) the "inner desert" (*VA* 49.4).

In the basic idea of what constitutes the desert or wilderness, there really is, to a large degree, a parallel between the traditional Greek view and the biblical understanding. The Hebrew words usually rendered as "desert", *midbar*, *arabah*, *horbah*, or *yeshimon*, are more correctly translated with "wilderness". In both cases, what really made an area understood as opposite to the cultivated lands was not the level of fecundity, but the level of proper human activity. In Egypt where the border between desert and cultivated land was so clear and distinct, large parts of the *erêmos* populated by the monks were, as James E. Goehring points out, "within a stone's throw" of settled land (Goehring 1993,298). As demonstrated in the fourth century biography of Pachomius, a somewhat later desert ascetic, the Egyptian *erêmos* could designate arable land that had only recently been abandoned (*Vita Pachomii* 12).

Any part of the *eschatia* could easily be turned into a cultivated landscape by proper agricultural practice. At this moment the wilderness ceased to be wilderness. But the *eschatia* could yield even extensive crops and still remain *eschatia* as long it never was the subject of the *properly* performed agricultural activity. Athanasius, accordingly, also stressed the essential difference between Anthony's little oasis by the "inner mountain" and the cultivated *chôra.* Athanasius consequently described the palm trees found where the hermit finally settled as "unkept" or "wild-growing" (*amelêthentes*, 49.7). This understanding of the ascetic wilderness was preserved as the use of the *eschatia* was imitated in other parts of Christendom. In early fifth century Syria, Theodoret of Cyrrhus wrote of his own ascetic "*erêmos*" as a place with "many high mountains, … some covered with *unproductive* vegetation" (Theodoret *Epistle* 42; my emphasis) – in other words a desert that was not at all a barren landscape, just not *properly* cultivated. In Gaul, the ascetic "desert" was identical with the rather lush islands of the Lerins right outside modern Cannes, as can be seen in Eucherius of Lyon's fifth century text *In Praise of the Desert.*

4. Folly and wisdom, side by side

As the wilderness of *Vita Antonii* is a landscape that so thoroughly reflects traditional Greek ideas about the uncultivated geography, Anthony is really presented as a fool placed right in a land of wisdom. And more than that: He is a fool criticizing the very belief that the landscape around him is associated with profound knowledge. But for all the wisdom traditionally connected with the *eschatia*, this insistence that the wilderness could be a place of folly and ignorance really would not surprise anyone operating within a Greek worldview. Just as the uncultivated geography was where one could find the most profound wisdom, it was at the same time a place considered fundamentally barbaric, uncivilized, and illogical. Athanasius himself referred to the denizens of the wilderness as "having wildness innate in their behaviour" (Athanasius *De Incarnatione* 52.9–10).

But the one did not exclude the other: Barbaric folly was repeatedly found juxtaposed to barbaric wisdom. If we, then, look again at the traditional Greek beliefs about knowledge and geography, we find that the *Vita Antonii*'s *critique* of people travelling to the *eschatia* for wisdom is as essentially Greek as the *belief* which is criticized. But when we look even closer, we find that the *Vita Antonii* does much more than just present and criticize the traditional belief that wisdom was to be found in the *eschatia.* Indeed, the text just as much promotes the very idea itself: Also the wilderness of Anthony was a place that really connected to the ultimate truth.

The two philosophers expecting that Anthony, as the ultimate denizen of the *eschatia,* would be "very wise" are countered, in the text, by a second group of philosophers thinking they are about to meet a Christian moron. This group is just as surprised as the group entertaining high expectations greeted by a hermit insisting on being "a foolish man". The philosophers with low expectations are "amazed at seeing so much understanding in" what they still defined as an "*idiôtês* – an ignorant man": "For his manner were not rough as though he had bad been reared in the mountain and grown old there, but graceful and civilized, and his speech was seasoned with divine salt" (*VA* 73.3–4). Again and again, Athanasius' repeated emphasis on the ignorance of Anthony is countered by an equally repeated stress on the boundless wisdom of the hermit. He is generally depicted as an outstanding theological prodigy, always in perfect harmony with the biblical truth. The hermit is, moreover, described with a profound prophetical talent which he frequently demonstrated. For example, "in regard to those who came to him, he often foretold days or sometimes a month in advance what was the cause of their coming" (*VA* 62.1). He also prophesied correctly about Arian attacks on the Athanasian church (*VA* 82.4–13), about the demise of a Roman general persecuting Christians (*VA* 86.2), and, finally, also about his own death (*VA* 89.2).

But if the *eschatia*, which was what made Anthony stand out so uniquely among his many contemporary monks, really connected to *both* the most utmost ignorance and to the most boundless wisdom, why did not Athanasius restrict himself to present the hermit as the ultimate sage? Why would he go to such great length at simultaneously presenting Anthony as both the wise man and the fool?

5. The primordial landscape

To understand this dilemma, how both knowledge and the complete lack of it are connected to the wilderness, we must look at the very essence of how the Greeks traditionally understood their *eschatia*. What was it about the spatial understanding of the uncultivated geography that made it an area that either opened or closed the ability to attain the truth?

In his seminal work on Western orientalism, Edward Said concludes, "all kinds of suppositions, associations, and fictions appear to crowd the unfamiliar space outside one's own" (Said 1978,54). Good or bad, whatever is not ours may apparently be found out there. Said traces the historical connections of these ideas all the way back to ancient Greece (Said 1978,55–58), and as such this is a claim that also refers directly to the essentially Greek geography of Athanasius. When Said argues that "to a certain extent modern and primitive societies seem thus to derive a sense of their identities negatively" (Said 1978,54), we find that the Saidian model at a first glance seems to fit well with the traditional Greek worldview. But Said's explanation is too simplistic. The otherness of the *eschatia* was no haphazard combination of just anything that was not found in the civilized geography. There really was a pattern.

According to the Greeks, the construction of the *polis*, the Greek city, was what terminated the primordial state, regardless whether this original state was considered a golden age or a miserable time when humans could not be distinguished from the wild beasts. As James S. Romm points out: "Perhaps the most fundamental act by which the Archaic Greeks defined their world was to give it boundaries" and the most fundamental spatial border was that of the *polis* (Romm 1992,10). But when the *polis* was distinguished from the greater wilderness, the wilderness itself was left as boundless as ever. Since the border of the historically constructed entity of the *polis* represented the most important distinction in the Greek worldview, geography was characterized by temporal differences. The Greek city was an area of the present tense, whereas the *eschatia* forever reflected the primordial past, a place where the proper *polis* had never been established and the land thus never been properly cultivated. This was how any part of the Greek periphery could be considered to reflect something like an eternal continuation of either a golden age or a beastly dystopia. This was an area where history never had started but which, at the same time, was the past. As an area identical with the primeval chaos that once was everything, the *eschatia* was somewhere time never had been properly distinguished (cf. Endsjø 1997,382–83). As such, the past was still out there.

The spatial dichotomy between wilderness and cultivated geography was thus considered the result of a historical event when the *polis*, as the only proper space for humans, had been distinguished from a generally undifferentiated realm. The essential trait of the diverse landscapes that was considered to belong to the *eschatia* was the lack of proper cultivation – these areas had not gone through the humanizing process that the *polis* had gone through. The spatial separation that had created the *polis* was something that had not affected the basic nature of the wilderness: Fundamentally unaltered, the wilderness was seen as still reflecting the primordial landscape before humans had established their communities and made the space itself human. This general idea of the *eschatia* as reflecting the primordial past is not only found in the *Vita Antonii* in the way the wilderness is presented as the wild and boundless area contrary to the cultivated geography, but it is also translated into an understanding which agrees with Christian ideas of history: Anthony's intimate relation with God and his peaceful coexistence with the wild beasts resemble humankind's original existence in Eden (Endsjø 2008,42–49). As his story was presented within the frames of a traditional Greek worldview, Athanasius was able to convincingly depict Anthony's wilderness as a place of a paradisical potential, at times practically the equivalent of Eden, without having to force his way past the guardian cherubs.

The Greek spatial dichotomy of the cultivated geography and the wilderness did not, however, reflect an absolute difference between the human and the inhuman: it equalled instead man's present and properly balanced condition. This was found in the polis as opposed to the primordial potential of everything that was found in the *eschatia.* As a place where human identity never had been properly defined, where humans often were not distinguishable from either beasts or gods, the wilderness of the Greeks forever reflected the primordial past. Going out into the *eschatia,* one would consequently find the remnants of what had once been the universal state, regardless whether this was considered a golden age or a more dystopic past: Here one found both lands inhabited by the wisest and overly cultivated peoples, and areas full of the fiercest and most uncivilized savages. There were the fabulous lands of the Ethiopians and the Hyperboreans, and the depressing realms of the Cyclopes and various anthropophagous peoples, side by side. Indeed, one could, in theory, encounter this boundless existence, where there was nothing to distinguish humans from nonhuman beings, by just leaving the city walls behind. But whether one would encounter the superhuman or the subhuman aspects of the most distant past in the unregulated geography was never certain. This is most clearly demonstrated in Euripides' *Bacchae.* Here, the wilderness right outside of Thebes was, first, miraculously turned into a wondrous setting where milk, wine and honey appeared out of the ground (Euripides *Bacchae* 704–11). Suddenly this idyllic image was transformed into a gruesome scene of boundless carnage (*ibid.* 731–64). Then, this depressing scene was just as abruptly made idyllic again (*ibid.* 765–68), only to be turned once again into a nightmarish drama (*ibid.* 1058–1152). These contradictory scenes mirrored each other as they both, both separately and together, connected to the primeval chaos that still constituted the wilderness. At the same time they functioned as a reminder of where humans belonged. Those who were tempted to venture into the *eschatia* by its boundless *positive* potential immediately risk encountering the boundless *negative* potential of this realm the moment they went out to get this.

The Greek wilderness, the vast spatial margins surrounding the polis, was thus the very image of the primordial landscape where no proper order had been established. But it was more than just a confusion of the proper elements of the cosmos. As a state where no proper borders had been established, everything could be found in its original, boundless, primordial potential. Establishing the borders of the human *polis*, men did not only distinguish themselves from the spatial chaos of the wilderness: they simultaneously established all the limitations that pertained to humanity. Although this distinction was necessary in order to distinguish men from the boundless brutality, misery, and the generally beastly existence of the wilderness, the spatial limitation of the *polis* simultaneously separated humans from all the boundless *positive* aspects also found in the wilderness: the endless riches, the boundless fecundity, and even the possibility of eternal life (Endsjø 2009,78–82). Most importantly, in this context, is how the wilderness, as the space in which everything may be found in its limitless primordial state, was the place of both boundless original ignorance and boundless primordial wisdom.

6. Primordial lore and original ignorance

As men limited themselves from their primordial past, and thus became human, they simultaneously limited their wisdom – as well as their ignorance – from the boundless potential of the wilderness. When defining humanity from that which it was not, securing it from the boundless challenges and dangers of the *eschatia* men had, at the same time, severed the connection to the original truth. In the traditional Greek worldview, all true wisdom was therefore considered essentially primordial. As seen already in Homer and Hesiod, and continuously emphasized in sources all through the Classical and Hellenistic period, man's evolution was frequently seen as one of progressive degradation (Lovejoy and Boas 1997 [1935],23–101). This was an historical understanding of knowledge that was easily harmonized with the Christian idea of how everything had its beginnings with God. Only in the state *before* everything was divided into various separate categories – only before man turned away from God with the Fall – was it possible to reach that absolute and undivided truth. Or as Athanasius would put it: "In the beginning" there was "no obstacle to the knowledge of the divine" (Athanasius *Contra Gentes* 2.1–2).

Here we touch upon the reason why all through antiquity, people were either looking back in time for answers, or, as Athanasius pointed out, journeyed "across the sea" (*VA* 20.4). While primordial time was understood to have been terminated by the establishment of the *polis*, the process could be reversed by means of movements in space. By deserting the city for the wilderness, one not only left civilization behind, one turned back time and entered man's own primordial beginnings. Take away the cultivated geography and one removed not only the essence of humanity, but even the aspect of present time, and entered the past. But this is still no absolute dichotomy of culture versus nature: all the aspects of human culture were considered to exist in the wilderness, but it was only in the *polis* that one could find these aspects properly limited and differentiated in way that constitutes culture. As the wilderness reflected the same undifferentiated state as the beginning of time, one could therefore find here the primordial wisdom because it had never been limited (as in the cultivated geography). In the *De Incarnatione*, Athanasius supported these ideas of the *eschatia* even more directly:

> So nothing in creation had been lead astray in its conceptions of God, only man. By all means, neither the sun, nor the moon, nor the sky, nor the stars, nor the water nor the air have changed their course, but knowing their creator and king the Word they remained the way they had been created. Only men turned away from the good (Athanasius *De Incarnatione* 43.10–15).

This statement of Athanasius' would be a very accurate description of the *eschatia* according to the Greek traditional worldview. As untouched by human hands, nature was considered to reflect the primordial truth. It was the same basic understanding Jerome advocated when, just a few years after the *Vita Antonii,* he let "beasts speak of Christ" in the very same landscape that Anthony made his home (Jerome *Vita Pauli* 8).

As we look at the nature of Anthony, at Athanasius' presentation of him, a pivotal term here seems to be *aplastos.* This term is used to describe Anthony as a young boy (*VA* 1.3). The term, meaning unaffected or unformed, refers to the essential aspects of Anthony's alleged ignorance. As he was sheltered from any outside influence, he was a man as original as they came – or, more precisely, as original it was possible to be in the postlapsarian geography of the *polis.* When Anthony was depicted as perfectly uneducated, this meant that his mind was close to how the Lord had created it. But this, as already pointed out, would be true of any uneducated village urchin. What made Anthony stand out was, of course, that he went into the wilderness. It was, indeed, the primordial landscape that made it possible for Anthony to reach man's ultimately original state of primordial wisdom. As a connection always remained between space and human identity, it was impossible to reach this absolute primordial state in the cultivated geography of the *polis.*

Anthony in the wilderness is the very image of the primeval man exhibiting man's primordial wisdom. He is "completely in balance as he was being guided by reason and remaining in a natural state" (*VA* 14.4). Consequently one does not need anything except what one will find in one's natural and original state. As Anthony asked the second group of Greek philosophers:

> "What do you say? Which is first, mind or letters? And which is the cause of which – mind of letters or letters of mind?" And when they answered mind is first and what made up letters, Anthony said: "Anyone who has a healthy mind has therefore no need of letters" (*VA* 73.2–3).

The folly of Anthony was thus not just the ordinary ignorance traditionally seen as connected to the wilderness, but of a certain kind: It was the ignorance of mere human knowledge, without the complications, misunderstandings, corruptions, and outright untruths, which had evolved after the Fall, when the soul "swerves and turns away from its natural state", when it did not have "its intellectual part in its natural state as it was created" (*VA* 20.7). As Bernard McGinn argues, "*The Life of Anthony* presents the hermit as uneducated and illiterate in order to stress the divine source of his philosophy" (McGinn 2006,157), but it was the wilderness that made Anthony's connection to primordial lore possible. As the traditional scholarly and essentially non-Christian wisdom of the *polis* was considered a corruption of original wisdom, the direct access to the undiluted divine truth was to be found in the uncultivated geography. *Both* Anthony's lack of knowledge and his deep wisdom were thus inseparably connected to the wilderness and its primordial nature. Anthony's absolute ignorance equalled the original ignorance of man, the innocent unawareness of all things corrupted and untrue; Anthony's absolute wisdom was identical to that divine original truth from the beginning of time, as this had been revealed through Scripture. The *Vita Antonii* turned the fool's ignorance and the boundless wisdom, both traditionally found in the wilderness, into one. Where previously the two had been confused, the ignorance of the various complicated matters of human civilization was now presented as just another side of the primordial truth forever found in the *eschatia.*

Although the presentation of the folly and primordial lore of Anthony thus is intimately tied to traditional Greek beliefs about space, primordiality, and knowledge, this was an understanding which was easily translated into Christianity. This is exactly what Athanasius did. As only God has existed forever, God's words were identical with that original truth from the beginning of time, the same natural wisdom still found in the uncultivated wilderness. Athanasius also made Anthony draw a direct connection between this primordial wisdom and Christian beliefs: "Without having learned letters, we believe in God, recognizing through His works His providence over all things" (*VA* 78.1). The revealed scripture was, of course, identical with the primordial wisdom still found in the wilderness as both reflected original truth. And in presenting Anthony's primordial beliefs as always identical with what he himself held to be true, Athanasius, the mighty patriarch of Alexandria, thus made a solid case for his particular theology being in complete harmony with how the Divine truth was found in nature itself.

7. Conclusion

Anthony and the other monks who later made the wilderness their home represented a major challenge to the formal hierarchy of the dominant Church of Athanasius. Although the many monks in the Egyptian cities and villages proved at times an unruly bunch, they were still within easy reach of the Church and when assessing their contemporary esteem we should remember that they, as individuals, all became completely forgotten as they were totally eclipsed by the later desert fathers. The monks of the wilderness were not only literally far removed but became, as amply demonstrated in the case with Anthony as the very first one, instant superstars the very moment they proved their ability to remain in the *eschatia.* The difference of geography was the only thing that made the desert fathers stand out from their numerous ascetic predecessors, and so it was the wilderness as such that enhanced the whole monastic experience. The perceived primordial state of the *eschatia* explained why someone looking at this realm within a traditional Greek worldview could not only find the paradisical elements in Anthony's desert plausible, but recognize the desert father as someone closer to man's original state, how man originally had been created by God. As such, the desert fathers, with what was considered a direct connection to primordial truths, represented a rival source of authority to the traditional Church hierarchy with its command of formal theology. By almost immediately after Anthony's death publishing a biography where there was complete agreement between the theology of the desert father and the Church in Alexandria, Athanasius managed to take control over this potentially dangerous source of truth and use it in order to forward his own theological views.

When looking at the connection between knowledge and the wilderness in the *Vita Antonii*, we find that space plays a part on many levels. There is, of course, first of all the most basic understanding that space itself is never an objective entity, but always something that is presented, viewed, and understood through various cultural ideas. As such, we find here a basis of traditional Greek worldview where the wilderness and the cultivated geography represent fundamentally different realms. Then there is, both in the general contemporary understanding of the desert father and in Athanasius' presentation of him, the question of how this traditional Greek worldview is reformulated by means of Christian beliefs. In addition to this, one must recognize that Athanasius must have made a number of deliberate choices in order to present the spiritual truths reflected in the desert father and his geography in a way that supported his own position as the head of the Church.

In this way we find that we are operating with various different spaces when analysing this text. There is the very physical landscape of fourth century Egypt which lies as the basis of this depiction; there is the contemporary cultural view on geography based on Greek and Christian ideas which at the time would be hard to separate from the physical realm itself; there is the perceived primordial essence of this landscape which provides this area a strong and unique essence which again influences everything and everyone connected to it.

Returning to the very role of the wilderness as presented in the *Vita Antonii*, we find that this represents not just a separate aspect in this text, but really the key for its understanding. As it was the spatial dimension that originally made Anthony what he was, all other aspects in his biography connect to this. Not least how the desert father was a figure combining boundless wisdom with primordial folly in a way that made him an ultimate Christian role model.

8. References

Burkert, Walter. *Greek Religion.* Cambridge, Mass.: Harvard University Press, 1985 [1977].

Endsjø, Dag Øistein. *Greek Resurrection Beliefs and the Success of Christianity.* New York: Palgrave Macmillan, 2009.

Endsjø, Dag Øistein. *Primordial Landscapes, Incorruptible Bodies. Desert Asceticism and the Christian Appropriation of Greek Ideas on Geography, Bodies, and Immortality.* New York: Peter Lang, 2008.

Endsjø, Dag Øistein. "To Lock Up Eleusis: A Question of Liminal Space," in *Numen* 47 (2000), 351–86.

Endsjø, Dag Øistein. "Placing the Unplaceable. The Making of Apollonius's Argonautic Geography," in *Greek, Roman and Byzantine Studies* 38 (1997), 373–85.

Foucault, Michel. "Of Other Spaces," in *Diacritics* 16 (1986), 22–27 [1967].

Frankfurter, David. "Introduction: Approaches to Coptic Pilgrimage" in David Frankfurter (ed.) *Pilgrimage and Holy Space in Late Antique Egypt.* Leiden, Boston & Cologne: Brill, 1998.

Goehring, James E. "The Encroaching Desert: Literary Production and Ascetic Space in Early Christian Egypt" in *Journal of Early Christian Studies* 1 (1993): 281–96.

Hartog, François. *The Mirror of Herodotus: An Essay on the Representation of the Other,* trans. Janet Lloyd. Berkeley & Los Angeles: University of California Press, 1988 [1980].

Lefebvre, Henri. *The Production of Space,* transl. Donald Nicholson-Smith. Oxford & Cambridge, Mass.: Blackwell, 1991 [1974].

Lefebvre, Henri. "Reflections on the Politics of Space" in *Antipode: A Radical Journal of Geography* 8 (1976), 30–37 [1970].

Lovejoy, Arthur A. & George Boas. *Primitivism and Related Ideas in Antiquity.* Baltimore & London: Johns Hopkins University Press, 1997 [1935].

McGinn, Bernard. "Withdrawal and Return: Reflections on Monastic Retreat from the World" in *Spiritus* 6 (2006), 149–157.

Romm, James S. *The Edges of the Earth in Ancient Thought.* Princeton: Princeton University Press, 1992.

Rubenson, Samuel. *The Letters of St. Antony: Origenist Theology, Monastic Tradition and the Making of a Saint.* Lund: Lund University Press, 1990.

Said, Edward. *Orientalism.* New York: Pantheon Books, 1978.

Scully, Stephen. *Homer and the Sacred City.* Ithaca & London: Cornell University Press, 1990.

Soja, Edward W. *Postmodern Geographies: The Reassertion of Space in Critical Social Theory.* London & New York: Verso, 1989.

Vogliano, Achille. *Papiri della R. Università di Milano: Volume Primo.* Milan: Ulrico Hoepli, 1937.

6. Wilderness as a Necessary Feature in Hindu Religion

Marianne C. Qvortrup Fibiger

1. Introduction

Hinduism is full of paradoxes. It can even be argued that this paradoxicality is the core feature of Hinduism – a religion or a religious complex so difficult to define because it does not share core features with the other so-called world religions as Buddhism, Christianity, Islam and Judaism. This can lead to the conclusion that Hinduism as an -ism and as a single belief system is a construction made by the English colonists and, after them, Western scholars or theologians with their predilection for neat categories.[1] Yet, I will argue that it can also lead to the conclusion that Hinduism as an -ism simply has other core features and other ways to deal with issues normally related to religion, and that these may challenge standard definitions of religion both as a notion and as a phenomenon. The paradoxicality implicit in Hinduism is one of the interesting features that can be used not only in a comparative study as such, but also in relation to the theme of this book, namely wilderness.[2] This article offers two examples of analyses of the idea of wilderness from the Hindu tradition. I argue that the wild or wilderness should, in relation to religion,

1 See the many books on Orientalism, e.g. Inden (1986 and 1990), Nandy (1983), King (1999), or Bloch et al. (2010), just to mention a few.

2 The term paradoxicality is not as far as I know a normal used term in introductory books on Hinduism, but it is always indirectly touched upon. Gavin Flood writes, in his introduction to his book on Hinduism, how he wants to explain its internal coherence as well as its apparent inconsistencies (1996, 3). Wendy Doniger finds the term "Cross-fertilization" between different layers of the tradition" a useful term (2009, 6), Axel Michaels mentions these references to others: "a potpourri of religions, an all-absorbent sponge, a net ensnaring everything, an upside down banyan tree with countless roots" or with reference to Goethe, who writes in a letter in 1824: "I am by no means averse to what is Hinduism, but I am afraid of it, because it draws my power of imagination into formlessness and deformation." (1998, 17).

not always be seen in contradiction to the tamed or ordered world or as something to avoided, because the wild beasts, or the situation related to the wild, brings disorder. It can also function as a necessary feature of the ordered world and as essential for the upkeep of order. In Hindu religion, wilderness is something that has to be encompassed, embraced and lived through, not something that has to be shunned. The function and use of ideas of wilderness offer a pregnant example of the paradoxicality inhering in Hinduism, and one for which the use of simple dualistic analytical models fall short.

The idea of finding a path to the spiritual world in the wilderness is something we see in nearly all religions (Christian monks and ascetics, sufis, shamans etc.), and of course this invariably involves interrelations between those who seek and those who stay behind in society. But in Hinduism the two spheres are interwoven in a very complex inter-relation.[3] In the context of Hinduism, the two spheres of wilderness and ordered society can be understood as counterweights in a system according to which both are necessary for attaining the state of equilibrium that is the goal at all levels – both in relation to the individual person, who lives in the wilderness for a while or for good, and in relation to the world as part of society as well. In relation to an analysis of the Vedic sacrifice, Heesterman (1993, 1–44) points to the fact that the Vedic texts on the one hand stress the opposition between the domestic world (grāma) and the wild or jungle world (āraṇya) and on the other see those two worlds as interwoven. While the sacrifice is placed in the domestic and ordered world, it does not mean that the jungle is left out in the sacrificial order; it is seen as a necessary complement to the domesticated world. This is seen in relation to the textual description of the horse sacrifice, which should include both the domestic horse and a

3 Geoffrey Samuel (2008) suggests that this form for dialectic is already present in the Buddhist and other early ascetic orders (śramaṇa-traditions) in India. He writes p. 239: "…that renunciate ascetics placed themselves outside ordinary society in formal terms, however closely they might in fact interact with it subsequently. In practice, there was an ongoing dialectic between forms of renunciate practice that became more or less assimilated to the needs of the society, and others that positioned themselves at the edges or outside society." While the groups who placed themselves outside society did it firmly by having other ideals for example by having a special death cult as some of the śaiva ascetics as the Pāśupatas and the kāpālikas and others by doing magic of dubious character; others saw it as their most prominent role to keep a relationship to society. It is the latter group that is of interest, when it comes to the interwoven relationship between the ascetics living in the wild as a wild and the tamed society.

wild animal. But while the horse is to be sacrificed, the wild animal must be released after consecration. "In this way, they are sacrificed and at the same time not sacrificed" (Heesterman 1993, 30). This has to do with the understanding of the two worlds; one is understood as the world of man (the domestic world), and the other as the wild world – the world of Gods and ascetics who strive for living in opposition to the domestic world. Wilderness in Hindu religion should, therefore, not only be understood as a chaotic place; while it may be "out of place" from one perspective, it is "in place" from another. It is out of place in relation to persons living in the world, who follow norms related to society and who strive to maintain it by producing goods as well as having children. It is "in place" in relation to persons who live as ascetics and who are guided by norms that place them on the outside of society or a "life-in-the-world" for a while or for good. We therefore fail to grasp the religious function of wilderness, if we see it only as an opposition to the ordered world to be avoided or abhorred, or as a temporary scene for a *rite de passage.*

In Hinduism, wilderness is characterised by opposition to the tamed, profane world. Therefore, wilderness is geophysically or naturally placed in the mountains, in the forests or in the desert, where isolation from the profane cultivated world is possible. But wilderness can also be placed in a human being, who by his behaviour and look lives as a wild person and who by different practices – for example meditation or other ritualistic behaviour – isolates himself from the profane world while still being a visible part of it.

What is of interest in relation to the broader field of religious studies is therefore that wilderness within Hinduism can, on the one hand, be understood as a chaotic place, where demons or other counterproductive beings live and where disorder rules, and, on the other hand, as a place that leads to an order beyond the order determined by upholding society, namely to the ultimate goal of the liberation of rebirths or mokṣa. I will give two examples of how wilderness is both related to the profane life and to an order beyond it.

The first example touches upon Hindu ideals of life. In the Dharma Śāstra literature, the ideals involve on the one hand a householder (gṛhastha), who secures the socially based life by having children and producing goods, and on the other an ascetic (sādhu or saṃnyāsa), who turns his back on life in the world by living in the wild or as a wild man. The householder's most prominent role is to maintain and uphold society, and for that reason he has to orient himself towards the norms and morals related to a life in this world. On the contrary, the ascetic has to orient himself to a life "out of this world". Therefore, he follows norms that are oriented towards the transcendental world, and by so doing he can loosen his ties to life in this world. One of the necessary outcomes is that he, in many ways, lives a life in opposition to the cultured world, with other ideals to live up to and so he, from a profane viewpoint, lives in the wild or as a wild man.

To live in opposition to the ideals of the socially based world is not only necessary for his individual pursuit of attaining mokṣa or mukti (liberation from the cycle of rebirths), it also secures equilibrium in the world as such because he represents the necessary counterweight to the profane world.[4] The Dharma Śāstra literature presents the two ideals as mutually dependent – one cannot exist without the other. If there were no householder to secure and maintain the profane world, there would be no room for releasing the ascetic from his profane duties. If there are no ascetics to represent the path of liberation, the man-in-the-world cannot accomplish a better karma by offering food to the begging ascetics. Equilibrium in the world as such cannot be maintained if the one side is missing.

The second example is related to the understanding and the worship of the Great Goddess (Mahādevī). She is a figure who is both represented as the bearer of the ordered world, a benign goddess and the creator of life (epitomizing an ideal of the tamed world) and as the wild, untamed and horrific destructor of life (epitomizing an ideal of the wild, unordered world). She encompasses both aspects and is worshipped accordingly. One day she is worshipped as a Mother, the idealised wife, and as the Great Creator; the next day she can be worshipped as an independent Virgin and as the Great Destructor. This form of worship is found all over India and it can be seen either as an –ism in itself, where the goddess is understood as an independent figure, or as a part of a divine couple.[5] Goddess worship shows not only how paradoxicality is embraced in

4 The understanding of the equilibrium discourse is found in different philosophical schools in a micro-macro-cosmos system of correspondence, according to which it is crucial to find equilibrium between seeming oppositions in the individual as well as in the universe.

5 To call goddess worship for Śaktism, making it an independent tradition in itself is an ongoing debate among indologists. Gavin Flood underlines, that the Śaktā tradition is less clearly defined than Śaivism and Vaiṣṇavism (1996, 175). Wendy Doniger emphasised the independence of the Goddess, but does not call it an –ism (2009, 387–392), and Geoffrey Samuels debates this issue in his book on the origin of yoga and tantra (2008)

Hindu religion, but also how wilderness is assigned crucial functions. In Hinduism, wilderness is not to be defeated, domesticated, but the wild is, instead, assigned crucial functions. The implication is that the "order" of Hinduism entails an embrace of paradoxicality. First, however, I will discuss where the paradoxicality implicit in Hinduism seems to be in play in order to move towards the specific understanding of wilderness as a necessary part of order.

2. Paradoxicality in Hinduism

It is important to stress that paradoxicality can be found in many different aspects of Hinduism and in many different contexts. Still, its most explicit expression is in the Upaniṣads (from 600–300 BCE), where the understanding of the embedded paradoxicality of life, the nature of God, the world, and even the rituals, is represented as the key to salvation or mokṣa. A crucial example is the description of brahman[6] in the Kena Upaniṣad II, 3, where it becomes obvious how the seeker of truth only comprehends brahman if he or she realises that it is a paradox. The seeker of truth must recognise that one cannot understand brahman by means of deductive, intellectual exercises; the understanding of brahman goes beyond that. The text goes as follows: "He who thinks it not thinks it, he who thinks it knows it not; it is not understood by those who know and known by those who do not know." (Brockington 1981, 48).

This implicit paradoxicality can be found already in the Ṛgvedic texts, especially in the many cosmologies presented in Ṛgveda 10, with the hymn Nāsadīya Sūkta, X. 129, 1–7 as a very good example. Paradoxicality is here the core feature, not only in the comprehension of the creation of the world, but also in the understanding of the reasons behind everything in the created world. The myth begins in śloka 1 with a statement about what can be said about the stadium before creation and is

6 Brahman is a neuter noun and understood as the essence of the universe or the totality of the universe. In the earlier texts, the Brāhmaṇas, from around 800 BCE, brahman was related to the power within the ritual. In the Upaniṣads it became a principle both referring to the power or essence in the ritual and to the essence of the universe, as well as the essence within all living appearances. Salvation from the cycle of rebirths is reached when the seeker of truth understands that the inner self (ātman) is identical with Brahman, described in famous dictum from the Chandogya Upaniṣad: Tat tvam asi "You are that, or That you are" (Brockington 1981, 41–50; Flood 1996, 47–49).

followed by questions that try to pinpoint what made creation happen. It goes as follows:

> There was neither non-existence nor existence then; there was neither the realm of space nor the sky which is beyond. What stirred? Where? In whose protection? Was there water, bottomlessly deep? (1)
>
> There was neither death nor immortality then. There was no distinguishing sign of night nor day. That one breathed, windless, by its own impulse. Other than that there was nothing beyond. (2)
>
> Darkness was hidden by darkness in the beginning; with no distinguishing sign, all this was water. The life force that was covered with emptiness, that one arose through the power of heat. (3)
>
> Desire came upon that one in the beginning; that was the first seed of mind. Poets seeking in their heart with wisdom found the bond of existence in non-existence. (4)
>
> Their cord was extended across. Was there below? Was there above? There were seed-placers; there were powers. There was impulse beneath; there was giving-forth above. 5)
>
> Who really knows? Who will here proclaim it? Whence was it produced? Whence is this creation? The gods came afterwards, with the creation of the universe. Who then knows whence it has arisen? (6)
>
> Whence this creation has arisen – perhaps it formed itself, or perhaps it did not – the one who looks down on it, in the highest heaven, only he knows – or perhaps he does not know. (7) (O'Flaherty 1981, 25–26).

At least two things are made clear for the reader or listener.[7] First of all, the text asserts the limitations of the human mind in comprehending creation. This is done by having the text end where it began, by arguing in a circle, asking nearly the same question at the beginning and at the end. Secondly, it concludes that creation is a never solved paradox, because it is stressed at the end that no one really knows how existence came to be. In my reading of the ślokas, this can lead to the conclusion that paradoxicality is not only something you have to live with, but also something that you have to live by, as an unavoidable and indispensable part of life and the basic feature of existence.

The texts quoted above are of course developed in a special period of time, namely in the later Vedic period from around 1000–300 BCE. Still, they are crucial for understanding the later developments of different soteriological trends in Hinduism, most explicitly found in the different philosophical schools or Darśanas from around the beginning of the

7 The vedic texts were, over a long period of time, orally transmitted.

Common Era. However, I will argue that this basic feature can also be traced or pinpointed in the way the ordinary Hindu copes with his or her life.

2.a. Compartmentalisation strategies

If Hinduism, as I have suggested above, has paradoxicality as one of its core features, it means that Hindus have to cope with this interwoven paradoxicality in different ways depending on social stratification. One approach is to ponder philosophically what this paradoxicality stands for in philosophical investigations. This approach is only a possibility for the intellectual elite. The other main approach would be to apply a compartmentalisation strategy[8] that would situate this paradoxicality in a special space and thereby enable individuals to cope with it in everyday life. In one way, this paradoxicality is a part of life that may explain why life does not always turn out as planned; in another it is mostly related to the transcendental world. As for the latter, paradoxicality can be coped with by means of a compartmentalisation strategy where profane life, "life-in-the-world", is placed in one category. Here, paradoxicality is not lived by as an interwoven part of life but as an explanation of life. "Life-out-of-the world" can both categorise the transcendental world, or the life of the ascetics, who have chosen to live in opposition to "life-in-the-world." Here paradoxicality is lived by as a core feature, which has to be encompassed in life as such.

We also find this compartmentalisation strategy at work in Hindu religion in a micro- and macro-cosmos corresponding system, which shows

8 The analytical use of the idea of "compartmentalisation" which I here apply to the distinction between everyday life and the soteriological quest is inspired by Milton Singer (1971 &1972) and his elaboration on how Madrasi Hindus categorized the industrial public sphere and the private domestic sphere as two distinct areas, not only socially and culturally but also in relation to behaviour, norms and conduct precepts. Singer concludes: "In this compartmentalization, the work sphere is "ritually neutralized", that is, relatively freed from customary norms and ritual restrictions." (1972, 349). I think this distinction is fruitful to use in relation to understand how Hindus cope with the implicit paradoxicality within Hinduism. The compartmentalisation makes it possible to differentiate between everyday life and the quest for doing well in the world, and the soteriological path. A key point is that, in spite of this mental and spatial compartmentalisation, the two spheres are in permanent interrelated communication.

how the two above mentioned categories of the profane and the transcendental are interwoven. Such a system views the individual that is part of society as a micro-cosmos in relation to the universe, the macro-cosmos. The individual and the universe are mirror images and therefore closely related. Disorder – understood as anti-dharma or anti-dharmic powers – at one level causes disorder at the other and vice versa. For this reason, it is very important for every individual to live up to his or her specific dharma (moral, ethic, rules), outlined according to sex, life stage and caste, and in relation to a life either as part of a "life-in-the-world" or as part of "life-out-of-the-world". At the same time, this compartmentalisation strategy can be used to outline how moral, ethical and behavioural rules can change in relation to setting or context. In one context, for example in the temple or at religious ceremonies, one set of rules should be followed. In another, for example when doing business, another set of rules should be followed. This also means that the relation between the micro- and macro-cosmos corresponding system changes according to context. For a man "living out of the world", the ultimate goal is first to understand the interwoven paradoxicality in life, so that he is later able to remove it. The compartmentalisation strategy in play between the man "living in the world" and the man "living out of the world" is to realise that different rules have to be followed according to which sphere you form part of. In some ways, the two life-spheres are in opposition; in others they are closely related and depending on each other. In the overall scheme, both spheres are necessary in the overall pursuit of equilibrium in the world, as well as out of the world, which leads to a dissolution of all oppositions.

As a crucial example of how "the master plan" is not thought of as establishing oppositions for good, but that the oppositions are, rather, temporary and necessary oppositions that exist in order that they may be overcome or integrated later, I turn to a myth from the late Vedic period, namely Śatapatha Brāhmaṇa from around 1000–800 BCE. In this text both the gods and the demons are created as the progeny of Prajāpati (The Lord of creation). The demons are not, as in other religions, created due to some kind of mistake, or as fallen gods, or as the necessary opposition to the Gods or the positive powers. Rather, they play a crucial role in making the world an ordered place, in the creation of the world as we know it, with sky and earth and with day and night (O'Flaherty, 1988,

29–30).[9] Here the oppositions are established for two main reasons: Firstly, to make the world an ordered place for humans and other creatures to live in; secondly, to give the seeker of truth the possibility of integrating them in himself again or to make a de-differentiation possible between these two seeming oppositions.

The crucial point in the myth is, therefore, not the creation of the gods and the demons in themselves, but their simultaneous birth from the same breath of Prajāpati while he is doing yoga (see below). Both gods and demons are found in the outer world (macro-cosmos) as well as in the individual (micro-cosmos). What may be even more important is the idea that because the demons are created and because they are created as evil, they can also be overcome. In a soteriological perspective, this means that the presence of evil is necessary as an opposition to the presence of good, in the outer world as well as in the individual, that it may be overcome. Or, as Heestermann points out with regard to Prajāpati's relation to the sacrifice: Prajāpati "conquers" both worlds by integrating them in the sacrifice. This is done by his sacrifice of both a wild and a domestic animal (Heestermann 1993, 30). This same pattern can be found in this myth. It is only through the process of overcoming evil that equilibrium, understood as a de-differentiation of opposites, can be attained. Only through the process of overcoming evil can equilibrium, understood as a de-differentiation of opposites, be attained. Therefore this opposition, as all kinds of oppositions in the world, has to be integrated in the life of the individual in order for him/her to overcome it or to dedifferentiate it:

> Prajapati was born to live for a thousand years. Just as one might see in the distance the far shore of a river, so he saw the far shore of his own life. He desired progeny, and so he sang hymns and exhausted himself, and he placed the power to produce progeny in himself. From his mouth he created the gods, and when the gods were created they entered the sky (*divam*); and this is why the gods are gods (*devas*), because when they were created they entered the sky. And there was daylight (*diva*) for him when he had created them, and this is why the gods are gods, because there was daylight for him when he had created them.
>
> Then with his downward breath he created the demons; when they were created they entered this earth, and there was darkness for him when he had created them. Then he knew that he had created evil, since darkness appeared to him when he had created them. Then he pierced them with evil, and it was because of this that they were overcome.

9 Here it is important to note that an "ordered" does not mean the ideal world or state of being, but it explains how the world is as it is.

> Therefore it is said, "The battle between gods and demons did not happen as it is told in the narratives and histories, for Prajapati pierced them with evil and it was because of this that they were overcome". And so the sage has said, "You have not fought with anyone for a single day, nor do you have any enemy, O bountiful Indra. Your battles which they tell about are all magic illusion; you fought no enemy today, nor in the past" (*Rig Veda* 10.54.2).
>
> The daylight which had appeared for him when he had created the gods he made into day; and the darkness which had appeared for him when he had created the demons he made into night. And they are day and night. (*Shatapata Brahmana* 11.1.6.6–11; O'Flaherty 1988,30).

The overall goal in most of the philosophical schools within Hinduism is to integrate seeming oppositions, which are seen as merely human constructions. The best way of realising this is to try to integrate those oppositions within yourself or in your life. Similarly, wilderness is both seen as an opposition to the ordered world or to culture, *and* as an integrated part of order or culture. In other words, the wild or wilderness in itself is not something to be avoided, but to be coped with as an integrated part of the ordered world and of life. This is also the way to reach mokṣa, which can be understood as a stage of equilibrium where there are no differences in all kinds of matter.

This soteriology is explained clearly in the philosophy of Saṃkhya, where equilibrium is the stage where matter or prakṛti and the guṇas (the three strands of all life) are undifferentiated (Brockington 1981, 99–101). To make this possible, the seeker of mokṣa must work with his or her body as a micro-cosmos by integrating all kinds of oppositions in it. But it can also be noticed in orthodox Hindu observance of moral standards in relation to varṇāśramadharma (different moral/ethics in relation to social class and stage of life),[10] where a person lives through different oppositions during his or her life and in that way integrates them through the course of life. First as a brahmacārya, then as a householder – both aśramas where the person is committed to a life in the world or so-

10 The four aśramas or life stages are as following: 1) Brahmacārya, the celibate living student, who after the upanayana initiation ritual, where he acquires the holy thread and becomes dwija or double born, is now born or bound to the tradition and shall from now on study the Vedic scriptures, 2) Ḉhastha, the householder, who after the marriage ceremony as the initiation ritual to the second aśrama is now bound to the life in this world and is therefore from now on dedicated to family life, getting children and doing well in society, 3) Vanaprastha, the hermit or forest dweller, and finally 4) Saṃnyāsa/sạ̄mnyāsin, ascetic (Michaels 1998, 108; Flood 1986, 61–65).

ciety, and later as a renouncer – where one is now obliged to turn one's back on one's former life and live on the outskirts of society (Flood 1986, 61–65).

This distinction between the householder and the renouncer is a focus point in Louis Dumont's book *Homo Hierarchicus* (1980 [1966]). Here he argues that Hinduism can be seen in terms of a dialogue or interrelationship between those two groups. The householder or "man-in-the-world" is defined by his or her social existence and therefore lives within the rules of society – which, according to Louis Dumont, is the caste system. Because of the social restrictions of the caste system, the person "living in the world" can only be defined in relation to others and therefore has no individuality or existence in himself. On the contrary, the renouncer has established individuality by stepping outside the society and its social network (Dumont 1980 [1966], 272). Yet, according to the Law of Manu, one of the most famous of the Dharma Śāstras, the two groups cannot exist without each other. While the householder is the fundament for upholding society, the renouncer is the fundament for creating soteriological values, which then again enter into society. Hinduism cannot function without the dialectical framework between these interrelated oppositions.

The renouncer, alias the ascetic, both relates himself to wilderness and he is related to wilderness by other people in two ways: Firstly, in relation to wilderness as a place seen from the perspective of the profane world, and secondly in relation to wilderness as space, the soteriologically oriented path or world. Wilderness as a place means living on the outskirts of society, in the mountains or in the jungle, where the renouncer isolates himself and lives a simple life following simple principles, removing him from all materialistic yearnings. In that way, he can follow wilderness as space understood as the opposition to the tamed or profane world. In wilderness as space, he can find tranquillity as well as equilibrium within himself. In the end, this should lead him closer to mokṣa, where the relationship to the materialistic or empirical world is dissolved. By following the aśrama-system by first living by the ideals that oblige him to uphold society and later by renouncing those same obligations, he takes in two opposite, but according to the Dharma Śāstras, equally important poles. In that way he can attain equilibrium within himself. While staying in the wilderness, he attains equilibrium in society too by representing the opposite pole to "The Man-Living-in-the-World", whose most prominent role is to represent the opposite pole to the renouncer, for the purpose of securing equilibrium in the world.

The same way of uniting oppositions can be found in Tantric rituals, where for example the consumption of alcohol, the eating of meat and fish, and sexual intercourse with low-caste women – things that are expressly forbidden for twice-born men in the Laws of Manu – create the necessary pollution in an individual, so that he, by himself, can overcome or transcend these differences and thereby attain equilibrium. Or as Gavin Flood writes: "...the tantric Brahman can pursue his soteriological quest for power and liberation through transcending his social inhibition in a controlled ritual context. It is one thing to perform erotic worship with a low-caste woman in a ritual setting, but quite another to interact with her outside that context" (Flood 1996,192).

2.b. Upholding equilibrium in society

The same idea of upholding equilibrium goes for society as such and leads to the above-mentioned all-encompassing compartmentalisation strategy, which again forms the background for a thorough division of labour and the clear-cut understanding of what or who is clean/unpolluted and what or who is unclean/polluted. The same is the case in relation to dharma where the dharma for a "man-in-the-world" and the dharma for the ascetic differ (Flood 1996, 12–13).

When it comes to compartmentalisation strategies in classical Hindu society, the Brahman-priest, the king and the ascetic (saṃnyāsin or sādhu) or renouncer (vanaprastha) are seen as superior in different domains or spheres. The Brahman-priest is superior in the ritualistic public or religious institutionalised sphere and he ensures good relations to the transcendental powers. This secures equilibrium in the transcendental macro-cosmos as well as in the collective micro-cosmos (society). The king is superior in the profane sphere and accordingly the superior figure in profane matters. In other words, the king secures equilibrium in society – a necessity for the work of the Brahman. His work forms a necessary opposition to the life of the ascetic, who is superior in the individual and soteriologically oriented micro-cosmos. Renunciation is, in this way, not only seen as a necessary stage on the individual path to liberation (mokṣa) from the eternal cycle of rebirth (samsāra), but also as a necessary counterpart to the reign of the king and the reign of the Brahman.

While the first two figures, the Brahman-priest and the king, uphold and relate to moral and ethical virtues in culture, the tamed or ordered world, the ascetic relates to wild and untamed nature. In relation to the social world, he is therefore set in opposition to the other two. He wanders around in the wild, or at the outskirts of the civilized world as a wild man. His appearance communicates it: he is mostly half-naked, has uncut matted hair, his beard is long and his body is dirty and often smeared with ashes. But without his presence, the work of the other two figures cannot be put in perspective. The opposition is needed. The ascetic has to seek the wilderness. He has to be part of the wilderness to represent, bodily and spatially, the counterpart of the reign of the king of culture. In this way, wilderness both functions as a necessary counterpart to the socially based culture in the binary or multilayered world (the collective level) and, on the individual level, as a station on the way to mokṣa in relation to the four life stages (aśramas) an orthodox Hindu male has to live through during his life (Michaels 1998, 108; Flood 1986, 61–65).

In other words, the way of achieving mokṣa on a personal level and of securing equilibrium in the macro-cosmos is to bring together, and live through, the polarities or dualities between the tamed and the wild, culture and nature, clean and unclean, male and female, and the different dharmas.

The ascetic or renouncer is only one of the figures related to the wild or the wilderness in classical Hinduism. Hindu mythology is full of demons and other monstrous figures which, especially in the Upaniṣads, are explicitly synonymous with elements that uphold samsāra and, at the same time, are seen as necessary key figures in the attempt to attain mokṣa. In other words, it is not possible to achieve mokṣa if it is only understood as in opposition to samsāra and if the figures that uphold samsāra are not there. The same seems to be the case in the great Hindu epics like the Mahābhārata and the Rāmāyana, where the idea of wilderness plays an important role as the representation of the anti-dharmic world that not only has to be overcome but also lived by and experienced before the ordered dharmic world can come into being (Brockington 1981, 62–63).[11]

Having offered an image of the function of wilderness in Hinduism more broadly, I will now concentrate on a more detailed analysis of the

11 It is obvious in the Rāmāyana, where Rāma, the eponymous hero of the epic, has to live through fourteen years of exile in the wilderness to show not only that he can live up to his dharma as a warrior (the relativistic ethic) because he protects the ascetics living there, but also that it is a necessary stepping stone towards his deification (the universal ethic). (Brockington,1981, 62–63).

role of ascetics in Hindu society. Axel Michael rightly calls ascetics key figures in understanding Hindu religion both in history and in understanding living Hinduism of today (Michael 1998, 347). Then, I turn to Goddess worship (Śaktism) especially in relation to the worship of the Great Goddess (Mahādevī). The Goddess is a figure of worship that epitomises paradoxicality, a figure embracing tameness and wilderness together and symbolizing the necessary coexistence of opposites. I regard Mahādevī as the most crucial example of how tameness and wilderness are integrated in one and the same figure, and of how both perspectives depend on each other in a multilayered understanding of life and death, sickness and health, dharma and antidharma. These two examples touch upon the same relation between wilderness and tameness, but with two important differences: 1) While the Goddess is related to the transcendental world but with obligations to both the transcendental world and the profane world (this can be called a top-top-down relation), the ascetic is related to the profane world and with obligations both related to the profane and the transcendental world (this can be called a down-down-top relation). 2) While the ascetic represents the wild or wilderness in relation to the man in the world, the Goddess encompasses both the wild and the tamed. The one and the same Goddess can both be described as wild and furious and as tamed and domesticated. While the ascetic is either or, the Goddess is both and. Therefore, in Hindu religion, the Goddess and the ascetic are, in a sense, mirror images with inverse profane/transcendental indicators, and wilderness is both idealized and sanctioned in ascetic practice as well as in Goddess worship, and assigned a necessary function for the maintenance of order both within the world and in the individual.

3. Case 1: Ascetics

The ascetic lives in fundamental opposition to the householder's domestic life and therefore also to the domestic fire rituals. For that reason, he has to give up all the utensils used in the domestic rituals and instead, in his ascetic practice, attempt to heat the inner fire (tapas) in the body or micro-cosmos. In other words, the Vedic fire ritual is internalised. The sacrifice becomes a self-sacrifice. The ascetic is socially dead; he lives in opposition to the socially engaged life. This makes him pure because he does not touch or engage with anything that can make him impure. He no longer performs the fire ritual to communicate with the gods;

he is the sacrificial fire himself. For that reason, many ascetics sit in the middle point between four fires, looking at the fifth – the sun – or they sit with a bowl of hot coals on their head. The ascetic is smeared with ashes (vibbhūti) for similar reasons (Flood 1996, 90–91; Michaels 1998, 347–348).

There are two forms of ascetics in living Hinduism of today. First, the vanaprastha and the saṃnyāsin, who follow the pattern of what I call the relativistic dharma of varṇāśramadharma – meaning that a person cannot renounce the world before he has lived through the life stage as a householder. Second, the very diverse group of sādhus, who choose the ascetic path as their overall life style as a kind of a soteriological call and renounce the world no matter their position in life, and who withdraw from following the path of varṇāśramadharma.

3.a. Vanaprastha – the third aśrama or stage of life

A vanaprastha (from Sanskrit vana: "forest" and either prastha: "gone to" or prus: "dwelling") denotes the third aśrama or stage of life out of four in the Vedic aśrama system prescribed in the Manusmṛti. The aśrama begins when a person is between the age of 50 and 75, when his skin has become wrinkled and his hair grey, and when he has seen the son of his son, which means that his son is now able to take over the commitments in society as the head of the family (Flood, 1996, 63). The vanaprastha disengages himself from all family ties except that of his wife, who may accompany him to the wilderness if she chooses to do so. The vanaprastha stage of life, which should be without any form of material desires, is as already mentioned in opposition to the former stage of life as a gṛhastha. One's position in this stage of life is shown by, for one thing, one's clothing – usually, the vanaprastha wears only a single piece of clothing, and he does not cut his hair, beard and nails. Further, he is required to live not only in the wilderness but also *by* the wilderness, subsisting exclusively on food growing wild in the forest. He eats raw or uncultivated food, such as roots, green herbs, wild rice and grain. The only thing he brings along to the wilderness from his former life is the sacred fire and the implements required for the daily and periodical offerings. He is not allowed to accept gifts from anyone, except for what may be absolutely necessary to sustain life. Yet, he must honour, to the best of his ability, those who visit his hermitage. This shows how he still has important obligations toward society.

He lives as a hermit for the sake of himself, but he still has to engage in society.

The hermit's time in the wilderness must be spent reading the Veda, making oblations and undergoing various kinds of austerities. This helps him to subdue worldly passions and gives him the tools to manipulate his mind, so that he slowly feels indifferent to all kinds of worldly objects (Flood, 1996, 63). This stage does not only denote a beginning transition from a material to a spiritual life but also a beginning transition of the mind, where everything becomes undifferentiated or of equal importance. The vanaprastha stage is therefore a stage in between that of a full born member of society with the obligations related to society and that of an ascetic. Wilderness, mostly related to the woods or the jungle, is understood as the place where this transition can happen. By living in, and by, the opposition to culture and the materialistic life in the wilderness, the vanaprastha is on his way to a stage of equilibrium within himself (the micro-cosmos).

The vanaprastha stage shows how wilderness, in a Hinduistic context and in relation to the aśrama system, can be understood in at least three ways. Firstly, as a concrete geographical place a person needs to go to find inner peace and to forget all material desires; secondly as a way of living, where the person lives as a wild with no materialistic goods and relates himself to the wild by looking like a wild and eating food mostly found in the wild; and thirdly as a state of mind where the wild is no longer understood as oppositional to the tamed, but as something that has to be encompassed in a unified system where a clear-cut differentiation between oppositions known from the profane world disappears. The fourth aśrama, living as a saṃnyāsin, shows that if the process of transition in the third aśrama is succeesful, the transition of the mind should be completed.

3.b. The fourth stage of life as a saṃnyāsin

As a significant difference between a vanaprastha and a saṃnyāsin, as a total renouncer, is the use of fire. The saṃnyāsin shows a total disengagement in life in the world: he no longer uses or maintains his sacred fire. He lives only to beg or eat raw food because he is no longer able to cook his own food. This way of living fits perfectly into the structural system of Lévi-Strauss (Levi-Strauss 1963, 221–230), as also suggested by Gavin Flood (1996, 63). It also fits the system of oppositions that I have outlined here: If fire and cooked food are the symbols of culture or the tamed world, and raw food is the symbol of nature, the untamed world, of wilderness, then the saṃnyāsin, by renouncing the sacred fire, also has renounced culture. The only relation he has to society lies in his begging of food, which is done for the sake of the people living in the world, who by giving food to a saṃnyāsin can attain good karma.

3.c. The dedication of life to asceticism

The Sanskrit terms *sādhu* (good man) and *sādhvī* (good woman) refer to renouncers who have chosen to live a life apart from or on the edges of society in order to focus on their own spiritual development. They either live alone or together with a sectarian group of equals in the mountains, at the sacred riverbanks or on cremation grounds, all places which can be categorised as wilderness in the sense of a place in opposition to ordinary life in society. They differ from the group of vanaprasthas and saṃnyāsin because they do not necessarily follow the aśrama scheme but choose, whenever they feel ready to do so, to renounce "the life in the world", dedicating themselves to the soteriological path of achieving mokṣa while still living (jīvanmukti). Conversations with sādhus, though, indicate that a personal crisis in relation to family problems, exclusion from the extended family, crime, and avoidance of an arranged marriage are often mediating causes for choosing to become a sādhu (Michaels1998, 349).

No matter the motive for becoming a sādhu, their religious goal is to live in a state of transcendence with no individual attachments to the profane world. This does not mean that they are not engaged in society at large or that they perform no religious functions. Theologically, they are seen as the counterweight to people engaged in the profane world, adhering to the relativistic dharma according to varnāśrama (social class and life stage) as śvadharma (individual dharma). And the life of the sādhu and the sādhu himself is seen as one big sacrifice or sacrificial offering (Michaels 1998, 354).

Around 4 or 5 million sādhus live in India today, and they are widely respected because of their ability to live without any attachments to ordinary material life, their indifference to all forms of physical pain and food, and their complete abstinence from sex (Samuel 2008, 181). However, they are also feared because, by living in a state of transcendence, they are believed to be able to manipulate things in the world, to do magic, or to curse people. Sādhus occupy a unique and important place in Hindu society, particularly in villages and small towns, where they are considered to be the living embodiments of transcendence and the living images of religious illumination and liberation from the cycle of birth and death (Michaels 1998, 348).

In the widest sense, the ambiguity towards the sādhus, who embody wilderness both by living in concrete "wilderness" areas and by living as wild persons in opposition to the tamed ordered world, is a manifestation of the ambiguity Hindus have towards wilderness. It is threatening, because it is out of order. It cannot be controlled and therefore one never knows what the outcome will be if one relates to it in some way. But wilderness is also attractive, because it is a place where transitions can happen. The profane world or "the-life-in- the-world" can be seen from without, and consequently coped with as something that should not be striven for. In that way a person engaged in the wilderness can release himself from the bonds that bind him to the empirical karmic world, and, in the end, he can thereby transcend it.

There are three primary sectarian divisions within the sādhu community, which follow the overall secterian division of the Hindu community as such: Śaiva sādhus, Vaiṣṇava sādhus and Śākta sādhus. They devote themselves to different gods in the Hindu pantheon in their endeavour to transcend themselves, and, in the state of transcendence, also transcend the gods. Their lifestyles seem to differ; especially the Śaiva sādhus seem to follow a more austere and radical path than that of the Vaiṣṇavas, and in relation to the social world, they often emphasise the need for a radical separation more and a complete commitment to a life lived in contrast to the social world. The Aghorīs are a crucial example. They live on the cremation grounds, eating and drinking out of human skulls. Conversely, most of the Vaiṣṇava sādhus openly emphasise an obliged engagement in the social world, and by begging and chanting religious songs while wandering around in the villages, they are believed to bestow good karma on the villagers (Michaels, 1998, 348–354).

Within these overall and general sectarian divisions, numerous subsects may be identified, each of which reflects different lineages and philosophical schools and traditions. What binds the group of sādhus together is that they live on the edge in the untamed and wild world and in contrast to the relativistic dharma, the rules of society.

4. Case 2: The Goddess – a figure uniting contrasts

A functional parallel, in the world of the deities, to the soteriological idea of neutralizing oppositions, the overall idea behind the compartmentalisation strategy, can be found in the goddess known as Mahādevī. The Goddess not only unites opposite poles within herself (the tamed and the wild), but she also lives them out or symbolises the coexistence of opposites. She is depicted both as the benign Mother goddess, who creates benevolence in the social world, securing that the śakti-energy will manifest itself through the women bearing children, and as the horrific Virgin goddess who slays humans. She manifests the śakti-energy in a form that takes human lives in the service of a greater cause as, for example, the maintenance of the Universe in order that people may attain mokṣa, etc. Mandakranta Bose (2010, 12–13) differentiates between the Goddess' function as an image of either a philosophical/metaphysical or a social archetype. This fits my differentiation between the Goddess understood in relation to either the soteriological or the social world. Behind her archetypes typology, Bose intends to understand the kinds of ideals that nurture the conceptions of the female gender, their assigned role, and the treatment of women in Hindu society. This of course influences the archetype model which distinguishes between the following four types: 1) mother/nurturer; 2) wielder of power/protector; 3) wife/helper/daughter; 4) destroyer. O'Flaherty (1986), who is inspired by Ramakrishna, differentiates between the mother of tooth and the mother of breast. The goddess of breast embodies the material virtues as motherhood, generosity and subservience to her husband. The goddess of tooth is, in many ways, the contrast to the goddess of breast. She is, first of all, independent and furious, and she can be both erotic and dangerous (O'Flaherty 1980, 90–91).

The divisions of Bose and O'Flaherty are fruitful with respect to understanding how different names given to goddesses differ from each other and how they idealize different roles of women and different aspects of women's lives. In relation to Mahādevī, it is important to understand that she encompasses all aspects in one and the same figure. Wilderness and tameness are here united; in Mahādevī opposites coexist. The paradoxical nature and the multiple identities of the goddess have made a clear definition of Śaktism as an entity in itself, and not just as a part of either Śaivism or Vaiṣṇavism, difficult to formulate. From my point of view, the genius of Śaktism is that it is both an autonomous division as well as an important part of the other two divisions, which is in line with the paradoxicality found in other parts of Hinduism. It is also in line with the theme of this book and the arguments in this article about wilderness both understood as a place, as a state of being and as a state of mind. Wilderness is, from one perspective, oppositional to order, and from another, it is a necessary condition for the attainment of order.

I suggest that Śāktism can be seen as an exploratory field, in which we can approach an understanding of the general paradoxicality within Hinduism, while also coming to an understanding of the multilayered idea of wilderness. First of all, wilderness can be understood as the necessary counterweight to tameness and the tamed, both as a place and as a state of being. Secondly, wilderness is a necessary feature in the process through which order (understood as the condition in which dharma rules) replaces disorder (understood as antidharma). In Śāktism, we find the representation of wildness, as well as tameness, in one and the same figure, namely the goddess herself. On the one hand, she is the creator and upholder of life, the nurturer, the devoted wife and mother, representing the ideals for upholding the social order/the tamed world. On the other hand, she is described as the independent, wild, tameless and furious demon killer, and bloodthirsty destroyer, symbolising the need to destroy the demons (personifying the wild as a negative power) in the world as well as in the individual. (Brockington 1981, 123; Samuel 2008, 248). Clearly, the wild can both be seen as a positive, as well as a negative, power. When respect to the goddess, we see clearly how she is a figure encompassing seeming oppositions within herself, how she herself encompasses paradoxicality. The many double aspects of the goddess, here in the form of Kālī, are described as follows in the Mahānirvāṇa Tantra from around the sixth century:

> Though you have form, you are formless,
> Through your māyā you assume many forms,
> You are the beginning of all, but you have no beginning,
> You are the mistress, the destroyer and the preserver
> (Mahānirvāṇa Tantra, 4.34, translated by Mandakranta Bose, 2010, 32).

4.a. The goddess as both tamed and wild

The narratives about Hindu goddesses and their double-bound nature can be traced back to the Vedic literature,[12] but it is especially in the Tantras and in the Puranic literature from around the fifth century and on-

12 Even though she played a peripheral role and only a few hymns are dedicated to her, she played an important role as being both a life-giver and a mother or nur-

wards that we find a thorough Devī (goddess) mythology. The most prominent figure for an understanding of the double-bound nature of the Devī is the goddess Durgā, who is both depicted as a wild, independent warrior goddess slaying demons and as a loving, caring mother and wife idealising domestic life. In the first aspect, she is depicted as an independent force, the reason behind, as well as the upholder of, the world. As Rachel McDermott (2005) emphasises, Durgā "offers her devotees either worldly enjoyment (*bhukti*) or liberation *(mukti)*". This supports the argument of this article, because it demonstrates that she supports not only the man "living in the world" (gṛhastha and the gṛhini), but also the ascetic, who is oriented "out of this world" (McDermott 2005, 3607).

The description of Durgā as a demon slayer and as an omnipotent entity can be found in a part of Mārkaṇḍeya Purāna called Devīmāhātmya, in the Devībhāgavata Purāna and in various tantric texts (Flood 1996, 174–176). When depicted as the idealised mother and wife, she is most often described in relation to Śiva and can be related closely to Pārvatī as one of her many names. This aspect is played out in the Skanda Purāna (Bose 2010, 30). When depicted as an independent self-reliant power, she can be described as Durgā herself or as Kālī,[13] who – despite her being depicted as a devourer of blood offerings – is also an embodiment of pure consciousness and the omnipotent power of the universe. This aspect is given a profound expression in a Tantric text from the Kālīkula school (Bose 2010, 35).

One of the most interesting ambiguous and ambivalent goddesses is the goddess for smallpox and other skin maladies mostly called either Śītalā or Māriyamman. She represents lay or village Śaktism, and is not only understood as the reason for the disease (disorder), but also as the cure of the disease (order). Thus, it is obvious how she both represents order and disorder, and it is only by pleasing her that order will be attained or secured. Śītalā or Māriyamman also has the ability to do both (Samuel 2008, 246–248).

---

turer as Uṣas, Aditi, Vāc, Pṛthivī or Srī or as a life-destroyer as Nirṛti who represents death and anger (Boose 2010, 15–25).

13 As Boose (2010, 33) rightly stresses, Kālī emanates from Durgā and can be seen as her antithesis – at least when Durgā is depicted in her creating and sustaining aspects. What is of interest is that they both emanate from the same primeval energy known as either śakti, prakṛti, or māyā and sometimes understood as Brahman. That means that the same energy can manifest itself in at least two presumably opposite figures.

While Axel Michaels argues that the understanding of ascetism is the key to understanding Hinduism, Brockington argues that the key is found in the study of Hindu goddesses. "In short, a study of Hindu goddesses is not so much a study of one aspect of the Hindu tradition as it is a study of the Hindu tradition itself" (Brockington 1996, 123). I have pointed to the necessity of studying both for understanding Hindu religion. They function as mirror images of each other with inverse transcendental/profane indicators. In relation to the theme of wilderness, the study of Hindu goddesses offers another necessary layer for understanding the paradoxicality within Hinduism and how this paradoxicality is closely related to the understanding of wilderness as both benign and as destructive.

5. Conclusion

When a person decides to become either a sanṃyasin or a sādhu, a ritual of renunciation will be conducted, in which it is communicated to everybody that this person is departing from all relations to the social world. Sometimes the renouncer symbolically performs his own funeral rite in front of the sacrificial fire, which means that he consumes his old socially related self. At the same time, he is freed from the relativistic part of Vedic religion which has the Brahman-priest as its most important figure. His liberation from this category of Vedic religion is also shown by his removal of his holy thread and abandonment of the sacred fire. He is no longer part of the tradition that has been his overall guiding principle through his life in the world. He is therefore out of place and space, and he must live in the wilderness as a wild man, and in all matters uphold the opposition to ordinary social life. Yet, by doing so, he demonstrates an obligation to the social or profane life, where he functions as its counterweight in a system that aims for equilibrium between oppositions. At the same time, by living through wilderness as a place, and by taking in the wilderness both as space and as a condition for himself, he can accomplish a state of equilibrium within. The religious geography of wilderness is therefore twofold. It is both placed on the outskirts of the profane world, in the mountains and in the jungle, and it is placed within the person who seeks wilderness as the necessary opposition to tameness within himself. Only then can a person strive for transcendence and mokṣa. This is a condition of opposition to the created world, but also a state of equilibrium and undifferentiation. It is the epitome of paradoxicality, but also where paradoxicality stops. As shown in the quotation from the Kena Upaniṣad, Brahman is only a paradox if the seeker of mokṣa tries to encompass it from a profane or worldly perspective. Seen from a soteriological perspective, Brahman is undifferentiated and therefore an entity or a condition in which paradoxicality is dissolved. The conclusion follows that the interwoven paradoxicality within Hinduism is only paradoxical as long the Hindu seeker is part of the world; while in the world the seeker requires the paradoxicality as a necessary stepping stone.

The ascetic supports the individual soteriological path, but he simultaneously also supports the social relativistic world: Firstly, by having lived through the life stage of the householder (gṛhastha). Secondly, by living as a beggar he offers the man "living in the-world" the possibility of fulfilling his dharma by offering food to the ascetic. Thirdly, by functioning as the necessary counterweight in the world so that equilibrium may be attained in the micro-cosmos as well as in the macro-cosmos. Therefore, wilderness is, in Hindu religion, not only a dangerous and chaotic place, but also a sanctioned space that leads, ultimately, to cosmos. It is important to note that wilderness in Hinduism is thus not only understood as a place, but also as way of living, where the person looks and lives as a wild, and as a stage of mind, where the wild and the tamed have equal importance. As a geographical place in India, wilderness is most often related to the mountains, the desert, and the jungle, as well as the cremation ground. As a way of living, it can be seen in the way the person looks, lives and eats; as a stage of mind it shows in the way the person relates himself/herself to the surroundings, assigning all things equal importance.

The same form of ambiguity in relation to wilderness can be noted in the understanding and the worship of the Devī. She idealizes the tamed and ordered world, and, at the same time, the wild and unordered world. As the archetype of womanhood in the grihini-stage, she is seen as the upholder of ordinary life in the world, securing the basis of the social world by manifesting the śakti-energy in forms that uphold life. At the same time, she is depicted as the opposition or counterweight to the social world. She is untamed, wild, furious and, most importantly, independent. This figure, most often depicted as Kālī, lives in the unordered wilderness, eating flesh and drinking blood while slaying demons. She symbolises the disorder necessary to uphold or maintain order. While the Goddess epitomises the unity of opposites that is striven for in Hindu religion, the ascetic functions as the opposition to the man "in the world". Looking at the ascetic in isolation, though, his primary goal is to encompass the oppositions within himself. This can be done by following the aśrama system, as a one-at-a-time system. Or it can be done by means of rituals, mostly known from Tantrism, where oppositions are united before they are dissolved.

Fundamentally, the ascetic cannot live without the householder and vice versa; the furious aspect of the śakti-energy cannot exist without the benign aspect, and vice versa. The same goes for wilderness in relation to tameness, the unordered world in relation to the ordered world: together they are each other's antidotes in a corresponding system striving for equilibrium. Seen from that perspective, the socially based order or culture cannot be legitimised, understood or even lived through, without the necessary counterweight of wilderness. From the perspective of the relativistic world, this is the necessary function of the idea of wilderness in Hindu religion. This does not only provide insight to the paradoxicality which can be found within Hindu religion; it also contests the general understanding of wilderness as seen in opposition to culture or the domesticated world in a conventional dualism. Wilderness can both be seen as a place in and out of space, as a geophysical place, as well as a state of being and a state of mind. It is not in opposition to the ordered world, but a necessary counterweight which has to be lived through in order that one may encompass it.

6. References

Bloch, Esther, Hegde, Rajaram, Keppens, Marianne (ed.). *The Colonial Construction of Hinduism.* London: Routledge, 2010.

Bose, Mandakranta. *Women in the Hindu Tradition. Rules, roles and exceptions.* Oxon and New York: Routledge, Hindu Studies Series, 2010.

Bowen, Paul (ed.): *Themes and Issues in Hinduism.* London and Washington: Cassell, 1998.

Brockington, J.L. *The Sacred Thread. Hinduism in its Continuity and Diversity.* Edinburgh: University Press Edinburgh, 1981.

Brockington, J.L. *The Sacred Thread. Hinduism in its Continuity and Diversity.* Edinburgh: University Press Edinburgh, 1996.

Coburn, Thomas B. *Devī-māhātmya: The Crystallization of the Goddess Tradition,* Delhi, 1984.

Coburn, Thomas B. *Encountering the Goddess: A Translation of the Devī-māhātmya and a Study of its Interpretation.* Albany, New York: State University of New York Press, 1991.

Creel, Austin B. & Narayan, Vasudeva. *Monastic life in the Christian and Hindu traditions: A comparative Study.* New York: Edwin Mellen Press, 1990.

Doniger, Wendy. *The Hindus. An Alternative History.* New York: The Penguin Press, 2009.

Dumont, Louis. *Homo Hierarchicus: The Caste System and its Implications.* Chicago and London: University of Chicago Press, 1980 [1966].

Heestermann, J.C.. *The Broken World of Sacrifice. An Essay in Ancient Indian Ritual.* Chicago & London: Chicago University Press, 1993.

Fibiger, Marianne Qvortrup. "Den hinduistiske institutionalisering af den kvindelige sakti-energi". In Schjødt, J.P. , Geertz, A.W. ; Jensen, H.J.L. (eds): *Det brede og det skarpe: Religionsvidenskabelige Studier.* København: Anis, 2004, 95–108.

Flood, Gavin. *An Introduction to Hinduism.* Cambridge: Cambridge University Press, 1996.

Inden, Ronald. "Orientalist constructions of India". In: *Modern Asian Studies,* 20.3 (1986), p. 401–46.

Inden, Ronald. *Imagining India.* Oxford: Basil Blackwell, 1990.

Jacobsen, Knut A.: *Hyllest til Gudinnen. Visjon og tilbedelse av hinduismens store Gudinne.* Oslo: Emilia forlag, 2007.

King, Richard. *Orientalism and Religion.* London and New York: Routledge, 1999

Kingsley, David. *Hindu Goddesses. Visions of the Devine Feminine in the Hindu Religious Tradition.* Delhi: Motilal Barnarsidass, 1987.

Kingsley, David: *The Sword and the Flute. Kālī and Kṛṣṇa. Dark Visions of the Terrible and the Sublime in Hindu Mythology.* Berkeley, Los Angeles, London: University of California Press, 1975.

Klostermaier, Klaus K. *Hinduism: A Short History.* Oxford: Oneworld Publications, 2000.

Leslie, Julia, ed. *Roles and Rituals for Hindu Women,* London: Pinter, 1991.

Lévi-Strauss, Claude. *The Structural Study of Myth.* Harmondsworth: Penguin 1963.

McDermott, Rachel. "Goddess Worship: The Hindu Goddess". In: *The Encyclopedia of Religion* p. 3607–3611. From *Gale Virtual Reference Library.* Gale: Macmillan Reference USA, 2005.

Menon, Usha. "Making śakti: Controlling (Natural) Impurity for Female (Cultural) Power". In *Ethos 30, 1/2* (2002), *American Anthropological Association,* 140–157.

Michaels, Axel. *Der Hinduismus. Geschichte und Gegenwart.* München: Verlag C. H. Beck, 1998.

Nandy, Ashis. *The Intimate Enemy: Loss and Recovery of Self under Colonialism.* Delhi: Oxford University Press, 1983.

O'Flaherty, Wendy Doniger. Women, *Androgynes and Other Mythical Beasts.* Chicago and London: University of Chicago Press, 1980.

O'Flaherty, Wendy Doniger. Women, *The Rig Veda – an Anthology.* Harmondsworth: Penguin Books, 1981.

O'Flaherty, Wendy Doniger. *Textual Sources for the Study of Hinduism.* Manchester: Manchester University Press, 1988.

Samuel, Geoffrey. *The Origins of Yoga and Tantra. Indic Religions to the Thirteen Century.* Cambridge: Cambridge University Press, 2008.

Singer, Milton. "Beyond Tradition and Modernity in Madras". in: *Comparative Studies in Society and in History vol. 13*, Cambridge: Cambridge University Press, 1971, 160–195.

Singer, Milton. *When a Great Tradition Modernizes.* New York: Praeger, 1972.

7. Notes on Qur'ānic Wilderness – and its absence

Thomas Hoffmann

1. Introduction

'Wilderness' has never been a topic in the curriculum of *'ulūm al-Qur'ān*, the traditional Islamic sciences of the Qur'ān. Neither has the topic gained access to standard Islamological encyclopedias or dictionaries as an entry in own right[1], as compared to Biblical studies and the study of religion (e.g. Ryken, 1998, 948–951; Taylor, 2005, 858–862). Wilderness, nonetheless, does merit scholarly attention for Qur'ānic studies; not only as a heuristic tool and general phenomenological category to be investigated historically, analytically, and theoretically,[2] but also for reasons intrinsic to the Qur'ānic semantic universe. Among these intrinsic reasons, most importance pertains to the Qur'ānic conceptualization and articulation of nature and natural phenomena. Despite the highly theo- and anthropocentric emphases of the Qur'ānic universe, this text is shot through with images of nature: from the grand cosmologies of starry firmament to the ocean and earth's expanses with their towering waves and mountains. From the serene and pleasant images of fecund cultivated nature with its gardens, fruits and crops to the dreadful and – for humans – unpredictable nature with its earthquakes, pulsating storms, chaotic flights or crowded gatherings of animals and humans.

---

1 Though related topics like "Nature", "Geography" and "The Animal Kingdom" certainly figure in works of Islamology. Ambros and Procházka's *The Nouns of Koranic Arabic Arranged by Topics* have recently provided us with a very helpful and encyclopaedic overview of the universe of Qur'ānic lexemes, divided into analytical categories like, for instance, "11. Geography → 11.1. General terms; flat land → 11.2 Elevated land → 11.3 Depressed land; cavities." Wilderness, however, does not appear as a category. It should also be added that in order to go beyond such analytical categories, we believe one must also pursue a more intuitive and tentative strategy of synthesis and interpretation. Thus, we rely on such ultra-systematic works but maintain the legitimacy of a, as it were, textual unconscious.

2 As outlined by Feldt in this volume.

The experience and conceptualization of animate and inanimate nature serve as a template for visions and forms of the transnatural world, such as transnatural topographies like Paradise and Hell, and an intervening eschaton with fantastic animals, earthquakes, terrified behaviour and chaotic behaviour *tout court.*[3] In this sweeping transnatural and natural cosmos, we must subsume wilderness/wild nature and – equally important, its absence – as a category with significant functions. If we venture beyond Qur'ānic semantics proper and approach neighbouring semantics such as pre-Islamic poetry or various post-Qur'ānic hagiographical accounts,[4] then wilderness as a topic acquires additional significance.

At least two restrictions deliberately curtail this article. A first restriction regards the scope of wilderness: wilderness will primarily be analysed in relation to concrete natural phenomena (whether in a natural or transnatural setting). Our intention is not to provide an exhaustive overview, but rather to go into detail with some examples that illustrate how the Qur'ān verbalises and exploits ideas of wilderness and wild nature. This means that we will curb our explorations regarding figurative wilderness. Themes like death, rebellion, and bodily loss of control can often be associated with wilderness and 'deep' nature, but such associative and figurative extensions will be outside the scope of this article.

In our methodology, we have opted for synchronic analyses of the relevant Qur'ānic passages, as they appear in the 1924 King Fu'ād-edition of the Qur'ān, one of the most accepted and disseminated Qur'ānic editions in modern time. The synchronic analyses to be pursued here consider etymology, root-semantics[5] and Semitic cognates, structural semantic components such as contrasts and hierarchies, the use of imagery (in its concrete and literal sense as well as in its connotative and metaphorical sense) and its rhetorical arrangement. These synchronic considerations will be framed and complemented by select diachronic observations, such as to be obtained from what we know from historical, anthropological and ar-

3 For a general introduction, see Toelle, 2007, 589–591.

4 E.g. the foundational story of Muhammad's reclusive retreats in Mecca's mountainous environs where he, in the cave of mount Ḥirā', eventually received his first revelation. Another wild cave-narrative concerns Muhammad's flight from Mecca during which he and Abu Bakr hid from their hostile pursuers in a cave. The enemy subsequently chose to ignore the obvious hideout because a newly spun cobweb covered the opening of the cave.

5 Arabic words are usually built around a morphological 'skeleton' of three consonants, i.e. the so-called roots, which together represent an ostensible basic meaning.

chaeological research into the Qur'ānic environment. What we will discover, then, from these approaches is not so much a wilderness of clearly demarcated contrasts or dichotomies vis-à-vis a universe of order and design. Rather, wilderness in the Qur'ānic lifeworld is – to quote from Feldt's introduction – "an epitomical example of the muddiness of the interface of world and humanity." Wilderness seems to take on a more complex functionality than merely one of staging cosmos and spatiality in neat structuralist binaries. To be sure, structuralist contrasts do crop up throughout our analyses, but wilderness does also appear as faint echoes or vestiges of a dying lifeworld, i.e. that of the pre-Islamic Arabs.

In order to pursue our topic, we will first cast a glance at some of the pre-Qur'ānic etymological, poetic and mythological contexts for Qur'ānic wilderness imagery.

2. The Qur'ān and the Wild Arabs

In much early modern scholarship, an air of 'wilderness' was attributed to the Arabs as the purported originators and first recipients of the Qur'ānic message. Thus, the word 'Arabs' has often been explained as meaning 'desert-dwellers', deriving from the words *'arabat-/'arabā*, which are said to mean 'desert.'[6] Arabs have been depicted as people stemming from 'Arabia', being associated in particular with the steppes and deserts bordering ancient Egypt and the Fertile Crescent (Hoyland, 2001, 1). The Arabs in antiquity have been classified as nomadic pastoralists and Bedouins (plural *a'rāb*; no singular), often seen as being of a somewhat aggressive bent, cultivating an ethos of tribal heroism, egalitarian autonomy and a strong sense of honour and shame. This ethos was partly expounded by the Arabs themselves in their poetry and prose, especially in the so-called 'battle-day' motifs ("We'll tell you of the days long and glorious, we rebelled against the king and would not serve him"), as it was partly mentioned in non-Arab sources, e.g., pre-Islamic texts like the Hebrew Bible (see further below) or post-Qur'ānic reports on the 'Saracens'.[7] Though

6 For a much more complex philological decipherment of the root, see Retsö, 2003, 24–62; 591–599.

7 E.g. the Greek source, Maximus the Confessor (d. 662): "…To see a barbarous people of the desert overrunning another's land… to see civilization itself being ravaged by wild and untamed beasts whose form alone is human." Quoted from Hoyland, 1997, 77–78.

the early Islamic communities also displayed a propensity of dynamic aggressiveness and forged intricate (though sometimes hostile) relations with Bedouin culture and worldview (emphasis on genealogy and virtues like honour, generosity, courage, loyalty, endurance etc)[8], it would be a major mistake to view early Islam as the epitome of Bedouin or nomadic religion. Emergent Islam, rather, arose in urban or semi-urban communities engaged in commercial and agrarian activity (cf. the importance of Mecca and Medina) and the early Muslims were, quite often, at odds with the capricious and aggressive manoeuvres of the Bedouins and their supposed stubbornness regarding conversion to Islam, e.g., Q 9:90 & 97: "And the Bedouins came with their excuses… The Bedouins are more stubborn in unbelief and hypocrisy, and apter not to know the bounds of what God has sent down on His Messenger." Of course this fact should not tempt us to conceive of these milieus as completely insulated from each other or the surrounding natural environments – a major factor like climate does of course not differentiate between city and wasteland.

What seems to unite a substantial amount of the different scholarly interpretations of 'Arabness' (pre-Islamic and Islamic) is the quality of aggressiveness and war-like ethos. As the Swedish Arabist Jan Retsö sums up in his recent monumental study on the Arabs in antiquity, the Arabs (and early Muslims) constituted "a community of people with war-like properties, standing under the command of a divine hero, being intimately connected with the use of the domesticated camel" (Retsö 2003, 623). Retsö, however, rejects both the nomadic and desert-anchored etymology, but this etymology does not preclude a later 'wild' semantic development. In that respect Arabness is only 'wild' in a somewhat figurative sense (i.e. war-like) albeit betraying some deep bonds to an arid and barren ecosystem associated with the camel.

If we turn our attention to the Qur'ān's nearest scriptural kin, namely (part of) the Tanakh (and some Mishnaic texts), we notice these scriptures' overlapping topic concerning Arabs as the descendants of the Ishmael, the son of Abra(ha)m and Hagar, Sara's slave girl. The Biblical story about Sara's expulsion of Hagar and her subsequent flight into the desert, the divine promise of Ishmael's birth and his future conflict-ridden life (Gen 16:11–12: "as a wild donkey of man whose hand will be against

8 For sustained arguments for overlaps and continuities between the Qur'ān and pre-Islamic culture, see Izutsu, 2007, 74–104; Bravmann,1998; Bauer, 2010, 699–732.

everybody and live to the east of his brothers"), Hagar's second flight into the desert and Ishmael's miraculous deliverance from dehydration (Gen 21:8–21), seems to confirm the abovementioned theme of Arab aggressiveness, but it, more importantly, puts emphasis on various forms of wilderness, namely desert wilderness as 1) a zone of social exclusion, 2) a zone of numinous encounter, 3) a zone of death in the form of dehydration, and 4) zone of miraculous survival and zone of fertility, *in casu* pregnancy. Furthermore, wilderness is added a strong metaphorical, metonymic and idiomatic profile regarding the prophecy on Ishmael as 1) a metaphorical wild donkey, 2) metonymically representing conflict and strife, and 3) bound to live to the east of his brother, i. e., an idiomatic expression that probably alludes to the desert and is usually translated as a topographic metaphor symbolizing conflict with others. These Ishmaelite wilderness features, however, do not figure in the Qur'ān proper but only reemerge in the post-Qur'ānic Islamic literature that lends from and elaborates on Biblical motifs, i. e., the so-called *isrā'iliyāt* literature (Firestone, 1990).

Moving from the murky etymological and mythological origins, we notice the sharp profile of the Bedouin rider taking off into the desert horizon as depicted in the stock desert journey section, known as the *rahīl*, of the pre-Islamic ode, *al-qasīda* (see for instance Jacobi, 1971). Though the emphasis in pre-Islamic *qasīda* is perhaps less on the harsh desert surroundings than on the beauty and strength of the riding camel, we nevertheless always come across detailed and perceptive vistas of the deserts and steppes and their organic and inorganic elements, sometimes furnished with observations on the exhaustion, hunger and thirst of the poetic persona and his mount thereby calling attention to the Bedouin virtue of *sabr*, endurance in hardship.[9] It is important to note that the wilderness of the *qasīda* often concerns a very nomadic concept of wilderness, that is, wilderness taking over the Bedouins' abandoned encampments or – seen from another angle – culture sliding back into wild nature.

The topic of wilderness is also quite explicit in the so-called *su'lūk* poetry which excels in motifs of night, danger, hunger, thirst, loneliness and wild beasts. The *ṣa'ālik* (sing. of *ṣu'lūk*) was an outcast character, ex-tribals (often due to excessive, dishonourable violence) roaming alone in the des-

9 A semantic shift away from the proud and detailed camel descriptions to the dangers and exhausting nature of the desert journey, *al-rahīl*, is elaborated in the post-Qur'ānic Umayyad qasīdas.

ert, bereft of camels and tribal protection, sustaining his life by stealing, while composing praises to anti-social and belligerent behaviour. The *su'lūk* prided themselves as *dhu'bān al-'arab*, 'the wolves of the Arabs', and their poetry often contains dialogues with the lone wolf.[10]

Given our relatively certain knowledge about the harsh climates and landscapes of the Arabian peninsula, and given our (less certain) knowledge of the pre-Islamic etymological, mythological and poetic context, we are justified in presupposing that these wildernesses function as a meaningful backdrop for the appropriation and development of Qur'ānic wilderness, whether as something to contrast and swerve away from (especially the garden of Eden, *'adnun*, or Paradise, *al-Firdaws*) or as something to develop and elaborate on (e.g., apocalyptical sceneries).[11] In order to understand these elaborations, we will have to outline pre-Qur'ānic notions of wilderness.

3. Before Qur'ānic Wilderness: Outline of Pre-Qur'ānic Conceptualisations of Wild Nature

As already mentioned, the notion of wilderness in the Qur'ān must be subsumed by a wider notion of nature and it should therefore prove worthwhile to outline the pre-Islamic and Qur'ānic presentation of nature. In an article on the subject of *Naturgefühl* (by which the author means the wild, untamed nature) in Arabic poetry, Gustave E. von Grunebaum concludes that "[o]n the whole, when portraying his environment, the classical poet emphasizes its harsh, rough, and fearful aspects. His descriptive gifts are quicker to respond to the appeal of the overwhelming, be it desert or storm, than to the charm of the idyllic" (von Grunebaum, 1945,140). Thus, pre-Qur'ānic poetry seems to have had a certain penchant for the more wild, dangerous and exceptional elements of nature, such as rain- and thunderstorms, the noon-day heat and its deceptive mirages, the terrible chilliness of the winter night, the wild beasts, the barren wastes and the odd appearances of dreadful, fantastic creatures like the *ghūl*, the desert-dwelling, shape-shifting demon that lures the traveller into the wilderness in order to slay and devour him. These dan-

10 For a brief overview, see Borg, 1998, 670–671.

11 This author brackets the early and late revisionist proponents (John Wansbrough, Michael Cook and Patricia Crone) and their radical scepticism concerning an original Meccan/Medinan setting.

gerous elements functioned as appropriate landscapes into which the poet could install himself (or his patron or his tribe) discursively as a masculine and tough figure. We should, however, not overestimate the harshness of landscapes and protagonists. Indeed, the harsh, monotonous and desolate settings honed the Bedouin's ability to notice the smallest signs of fecundity, softness, fragrance and colourfulness. In short: beauty. In pre-Islamic poetry's description of spring and flowers we view the first sprouts of what later develops into a genre of its own in, namely the garden- and flower poetry of urban and dynastic Islamicate civilization, the so-called *rabb'iyyāt* and *zahriyyāt.*[12] In her monograph *The Mute Immortals Speak: Pre-Islamic Poetry and the Poetics of Ritual,* Arabist Suzanne Pinckney Stetkevych interprets the wilderness elements within a theoretical framework that draws heavily on the analytical concepts of the tripartite *rite de passage* and its orchestrations of liminality, both of which are proposed as phases or stages in a human negotiation of the perennial divide of and bridging between nature and culture.The notion of the tripartite rite of passage was first proposed by the French anthropologist Arnold van Gennep (1873–1957) and later to be developed by the British anthropologist Victor Turner (1920–83) in his landmark monograph *The Ritual Process: Structure and Anti-Structure.* Turner explicitly associates liminality with wilderness, (Turner, 1977, 95). As to the negotiated fluctuations between nature and culture, Stekeveych' debt to Claude Lévi-Strauss (1908–2009) and his grand theory of 'wild' thinking, *la pensée sauvage*, is patent. A small number of quotations (with a few comments added) from the pre-Islamic poetic corpus, in casu the famous anthology called *al-Mu'allaqāt,* convey the reader the prominence of wilderness motifs and themes.

From Labīd (dead circa 661)

Verse 1
Effaced are the abodes,
  brief encampment and long-settled ones;
at Minā the wilderness has claimed
  Mount Ghawl and Mount Rijām.

Verse 2
The torrent channels of Mount Rayyān,
  their tracings are laid bare,
Preserved as surely as inscriptions are
  Preserved in rock

12 For an overview of this genre, beginning with the pre-Islamic traces, see Schoeler, 1974.

In these famous opening verses, the poetic persona describes the disintegrating traces of a former tent camp as well as nature's rock-like inscriptions in the form of dried out channels. In Stetkevych's commentary on these two verses, she emphasises how wild nature or wilderness has occupied and almost absorbed the former remains of a temporary cultural setting, the tent camp:

> Essential to this dialectic is the ephemeral and transitory quality of all that is cultural or cultivated – that is, human – as opposed to the permanence and perpetuity of the natural. *Ta'abbada*, to become wild, derives from the root *'-b-d*, which means to become wild, unsocial (*'abda/u*), but also to last, to remain, to dwell permanently, as in *abad*, lasting, everlasting, eternity (*'abada/i*). Thus, *ta'abbada* also conveys at some level the sense of perpetuity or eternity… Thus in verse 2 the tracings, outlines of drainage ditches, have been stripped, laid bare (*'uriyya*) – the opposite of clothing, the preeminent symbol of culture – by the torrents (Stekevych, 1993,18–19).

We should also notice how wild, destructive nature is presented as a linguistic phenomenon to be deciphered by the poet – Stetkevych translates *al-waḥyu* as inscriptions, but the word can also mean 'message'. Furthermore, we should pay attention to the notions of something laid bare (*'uriyya*) and its contrast to clothing and shrouding; we will deal with this motif later on. As of the remaining part of the poem, it consists in 88 couplets, many of which describe the changing milieus of the desert with its animal life, topography, and weather conditions. Michael Sells, in another translation and interpretation of Labīd's poetry, also emphasises the poem's subtle 'structuralist' interplay between "Edenlike domesticity" and images of barrenness, ruins, flooding, death, and dread (Sells,1989, 32–33).

From Imru' al-Qays (circa sixth century):

> Verse 73:
> Between Dārij and al-'Udhayb I sat with my companions
> *to watch the storm…*

Then follows a number of extended similes describing different fatal and 'wild' consequences of this (rain)storm, concluding with a morbid tableau of the drowned animals:

> Verse 82:
> As if the wild beasts drowned at evening
> *in its remotest stretches*
> Were wild onions'
> *plucked-out bulbs.*

Many Western readers would perhaps be surprised to know that the Arabian peninsula is subject to heavy rain storms and subsequent floods.[13] Though these storms are extraordinary, they nevertheless strike on a fairly regularly basis. Again we must cite Stetkevych:

> [t]he closing line expresses perdition, the destruction of wild and savage nature... Thus the corpses of the drowned wild beasts [...], lie bloated on the edge of the floodwaters at nightfall and likened to the plucked-out roots of wild onion (*anābīshu 'unṣulī*). The first term derives from the verb *nabash*, which means to pull forth or draw out (as a leguminous plant), but also to disinter a corpse. The second, which appear to be related to the trilateral *asl* (root, origin), is said to mean wild onion (*basal barrī*), and thus, like the wild beasts, a symbol of what is savage and uncivilized as opposed to the tame and cultivate (Stetkevych, 1993, 283).

What these few samples first and foremost should point to is this poetry's deep existential engagement with wild nature, being sometimes pleasurable, abundant and idyllic, but most often dreadful, harsh and fatal. An almost structuralist interplay between the cultured and the untamed, between idyll and its opposite, is constantly at work in these texts, as is also suggested by recent scholarship. Thus, both Suzanne and Jaroslav Stetkevych have argued that these patterned interplays seem to perform a tripartite *rite de passage*-like semantic trajectory. As to the specific semantics displayed in this pre-Islamic poetry (or sometimes poetry in transition to the Islamic period, i.e., the so-called *mukhaḍram* poetry), we find a fine-grained vocabulary and rhetoric attentive to natural details as well as descriptions of the desert and its landscape (sand dunes, short-lived meadows sprouting after rain, rocky plains, wadis, mountains etc.), atmospheric phenomena like the thunderstorm, clouds, and even snow, animal life in the form of beautiful animals like the shy onager and the swift oryx, dangerous predators like the wolf and the lion, and hideous scavengers like the vulture. Wilderness in a broader sense, however, does not only present itself through descriptions of discrete objects or landscapes, but also through descriptions of actions and sensations carried out or felt by the poetic persona in the wilderness, for instance the journey, the hunt, the hunger etc.

Thus, as a preliminary conclusion, we maintain that this poetry, from which we have only presented a most minimal amount of samples and

13 A favorite example of this author is an early photography depicting pilgrims swimming around the Ka'ba in Mecca after a heavy rainstorm. This does not happen anymore due to technical solutions provided by the modern Saudi state.

which was held in highest esteem among the Arab tribes both prior and subsequent to the time of Muhammad, indeed justifies a wilderness-approach to the Qur'ān. This poetic wilderness universe must therefore be regarded as an important backdrop for the Qur'ānic development hereof, but we should also bear in mind that the Qur'ānic text betrays signs of a significant change of priorities vis-à-vis wild nature, probably attributable to Islam's urban (Meccan)/semi-agrarian (Medinan) *Sitz im Leben* as well as some scriptural precursors, especially various Biblical texts, that display several important semantic universes besides that of wilderness. Wilderness, in the Qur'ān, seems to be relegated to a semantic periphery. Though roaming around in the periphery, motifs of wilderness never disappear completely but erupt sporadically throughout the text. We might even venture to suggest that wilderness in the form of being astray and off track (root Ḍ-L-L), occupy one of the main metaphorical dystopian motifs of the Qur'ān. This motif is borne out in the most prominent ritual dimension of Islam, namely the ritual prayer (the *ṣalāt*), which is performed five times a day and involves an obligatory recital of 'The Opening', i.e. the first sūra of the Qur'ān, which again includes the pious wish to belong to 'those who do not go astray', *wa-lā aḍ-ḍāllīna* (Q 1:7; my translation) and to be guided to/in 'the straight path', (Q 1:6), the opposite to wilderness.

Having now put forward these references displaying intricate relations to wilderness, let us now proceed to the Qur'ānic text proper.

4. Qur'ānic Nature: on Scales and Sublimity

By merely browsing the titles of the sūras in the Qur'ān, it should be evident that nature plays an important role. Of its 114 sūras, 21 hold titles referring to animals, insects, a plant, landscapes and geological phenomena, astronomical phenomena, phenomena related to the passing of day and night (although the themes and motifs of the sūras cannot be reduced to the semantics of the titles): "The Cow" (number 2), "Cattle" (6), "Thunder" (13), "The Bee" (16), "Light" (24), "The Ant" (27), "The Spider" (29), "The Sand-Dunes" (46), "The Mount" (52), "The Star" (53), "The Moon" (54), "The (star)Constellations" (85),"The Night-Star" (86), "The Dawn" (89), "The Land" (90), "The Sun" (91), "The Fig" (95), "The Blood-Clot" (96), "The Earthquake (99), "The Elephant" (105), "Daybreak" (113).[14] When reading through the sūras we encounter a textual universe shot through with a variety of images of nature, some of a macro-cosmological scale (the firmament, the earth, the mountain, the ocean), some of a micro-cosmological scale (specks of dust, the sperm-drop, the foetal blob, the ant and the louse, the date-thread et cetera) and in between these two scales a world of more human- and animal-sized scale. These scales all have a semiotic function as 'signs', *āyāt* (sing. *āya*) pointing to and 'proving' God's existence and sundry qualities (creation, majesty, punishment, guidance, power etc).

As a piece of prophetic literature it is not that surprising that images of warning, pain, and catastrophe take on a crucial role in the 'natural' discourse of the Qur'ān. To use a term going back to antiquity we could speak of the sublime or simply sublimity. This notion should not be taken in the apologetic tradition of *iʿjāz*, which refers to the dogma of the inimitability of the Qur'ān. The definition to be used here derives from the Irish philosopher and politician Edward Burke (1729–1797) who, in his *A Philosophical Enquiry of into the Sublime and Beautiful*, argued that

> Whatever is fitted in any sort to excite the ideas of pain and danger, that is to say, whatever is in any sort terrible, or is conversant about terrible objects, or operates in a manner analogous to terror, is a source of the *sublime*, that is, it is productive of the strongest emotion which the mind is capable of feeling (Burke, 1757,86)

Burke's definition is first and foremost psychological: the sublime is less a category in and of itself than an emotional response, an affect of the mind. However, the phrase "conversant about" is revealing because we want to emphasize that our proper subject of study, the Qur'ān, is a communicational artifact using language to communicate various matters. Sublimity, therefore, should be regarded as partly an effect of the Qur'ānic language.

Thus, examples of pain and suffering abound: from inevitable biological pain as in Maria's labour pains when she gives birth to Jesus (Q

14 If we were to choose a more systematic approach to nature-related terms, we could consult Ambros and Procházka's *The Nouns of Koranic Arabic Arranged by Topics* with its lists of semantic "fields" (see footnote 1), but this would, however, in many cases fail to notice 'wild' passages, partly because the dictionary does not operate with wilderness as a category of its own and partly because the concept of wilderness can yield a more maximalist definition or a more minimalist.

19:23) to the blazing agonies of *an-Nār*, the Fire, i. e. Hell (for example Q 4:56; 9:35; 11:106; 18:29; 23:104). Pain and scenarios of terror are also central in the so-called *Straflegenden* or punishment narratives (episodes involving mistreated prophets and conventionally ending with divine eradication or punishment of the stubborn and transgressive unbelievers) as well as in the many eschatological and apocalyptic passages that paint multiple images of pain and terror. A rhetorically breathtaking example from Q 81:1–14:

> When the sun shall be darkened,
> when the stars shall be thrown down,
> when the mountains shall be set moving,
> when the pregnant camels shall be neglected,
> when the savage beasts shall be mustered,
> when the seas shall be set boiling,
> when the souls shall be coupled,
> when the buried infant shall be asked for what sin she was slain,
> when the scrolls shall be unrolled,
> when heaven shall be stripped off,
> when Hell shall be set blazing,
> when Paradise shall be brought nigh,
> then shall a soul know what it has produced.[15]

These syntactic repetitions, evoking a series of cataclysmic and almost inconceivable events that turn the natural world topsy-turvy and reverse its natural order, build up a sweeping crescendo which nevertheless reaches its climax in the ultimate realisation of each and every human soul. The most obvious reference to wilderness is found in the fifth verse, which evokes an image of wild animals crowding together. Instead of taking away in flight, as would such animals normally do when threatened, they do exactly the opposite and turn into a great multitude.

However, terror and pain are not the only phenomena that prompt the sublime: notions of privation, obscurity, vastness, light and darkness, sound and loudness, which Burke also deals with, are vital to the Qur'ānic sublime as well. To go through all passages that could be construed as sublime is outside the scope of this article and a couple of examples must suffice, e. g. the following passages from the Light-sūra. We are here presented with two similes on the futile and misguided projects of the unbeliever, Q 24:39–40:

---

15 The poetic enjambement-like layout of Arberry's translation has been retained.

And as for the unbelievers,
their works are *as a mirage in a spacious plain* [*ka-sarāb bi-qīʿatin*][16]
which the man athirst supposes to be water,
till, when he comes to it, he finds it is nothing;
there indeed he finds God…

or they are as shadows upon a sea obscure
covered by a billow
above which is a billow
above which are clouds,
shadows piled one upon another,
when he puts forth his hand, wellnigh he cannot see it…

The contrast between the two similes is striking: the first simile evokes a hot desert expanse upon which a thirsting man is deceived by a mirage. This graphic image also prompts a strong vision of almost blinding sunlight. An image of utter privation and water deceptively turning out not to be water comes to our minds. Like the abovementioned apocalyptic passage, this passage is also rounded off with the evocation of the unbeliever's ultimate and fatal realisation: "there indeed he finds God." In contrast to this passage the second simile abounds in water – though undrinkable seawater – and darkness. We are presented with an almost claustrophobic image of a dark ocean to which is added a double billow which again is part of yet another accumulated variety of amorphous dark materials like shadows and clouds.[17] Claustrophobia and hyperbolic accumulation combined. And whereas the unbeliever in the former verse is deceived by a mirage in the horizon, the latter verse's unbeliever is stripped of anything like a horizon or an external point of reference. What is left is only the uncanny and vague view of a groping hand in the darkness, a strong poetic image of form and yet not form.

To sum up these few examples, we may conclude that the primary function of images of nature is semiotic, to point to God's existence and power. A very prominent dimension of this semiotic system is concerned with nature in its most painful and catastrophic manifestations, often of a fantastic character. Nature has become hyperbolically wild in order to accomplish an eschatological scare campaign aimed at evoking

16 The notion of a spacious plain (*qiʿa* is grammatically plural but should probably be translated as a singular noun) is semantically associated with Arberry's favorite translation of *ʿarāʾ* as wilderness since both nouns refer to something flat and barren.

17 The oceanic and maritime – in contrast to the aridity of the steppes and deserts – constitutes a particular interesting wilderness topic calling for further studies.

God's fatal and spectacular interventions if not heeded by people. One of the most striking aspects of this kind of Qur'ānic wilderness is the images of haphazard cosmological movement, deceptive orientation, straying, darkness and even partial privation of corporeal coherence. These images function as images of true wilderness over against those positive, divinely sanctioned notions and images of orderliness, guidance and straightness that are connected with Islām.

5. Naked Qur'ānic Wilderness

The primary Qur'ānic equivalent to wilderness rests on the trilateral root '-Y-R, which literally connotes 'nakedness', that is, a basic human experience and feeling. Accordingly, the verb *ya'rā* means 'to go naked.' Semantically, the noun *'arā*, 'wilderness' in Arberry's translation, must thus be viewed as a spatial and environmental extension of the original 'naked' root. Wilderness in this sense should of course not be associated with the kind of abundant, burgeoning wilderness that a tropical forest displays, but rather be taken in the sense of a barren, naked, flora-deprived land. In other words, the kind of arid, infertile wilderness that was and is the dominant type of landscape on the Arabian Peninsula, whether as mountains or dunes or steppes. Hence, 'nakedness' should be taken as the primal sense of the root for two reasons: the first reason derives from the fact that the cognate sense is attested in the majority of related Semitic languages, as tabulated from Martin Zammit's comparative lexical study of the Qur'ān (Zammit 2002, 286):

Table 1

Arabic root: '-R-Y	**Lexeme**	**Glosses**
Ge'ez	*'araya*	? 'aequum, planum esse/ fieri'[18]
Syriac	*'rāyūtā*	'nakedness'
Aramaic	*'iryt'tā*	'nakedness;' BA *'arwāh* 'dishonour'
Hebrew	*'ārā*	'to be naked, bare'

18 Zammit does not translate from Italian and German dictionaries he consults.

Table 1 *(Continued)*

Arabic root: ʿ-R-Y	**Lexeme**	**Glosses**
Phoenician	*ʿry*	pi. 'to lay bare'
Ugaritic	*ʿry*	'nackt'
Akkadic	*erū, arū*	'nackt sein'

The second reason is that the notion of nakedness seems to be rooted in a bodily sensation so basic that it has the potential to be transferred to language and subsequently developed metaphorically, *in casu* as a description of landscape. In that sense, the argument is beholden to the embodied cognitive semantic studies in the vein of pioneers like George Lakoff, Mark Johnson, and Eve Sweetser. These scholars developed the idea that thought (and subsequently language) works by means of metaphorical operations. Metaphors, they state, are not merely stylistic devices but provide insight into mental schemas, ways of categorization and constraints of (intelligible) creativity, which again draw heavily on bodily/physical experience and sensations. To mention one example; in a phrase like 'Can you see what I mean', the notion of abstract understanding is immediately metaphorized by means of a basic visual domain, namely 'see', which ultimately spring from our eyes and their connection to our cognitive faculties of image-processing. In other words: in our bodies and our interaction with the world around us. Etymology and cross-linguistic studies are often used to back up this point, as Sweetser, for instance, has demonstrated (Sweetser 1980). These new insights also allow us to reclaim etymology as a viable method not to be wholly dismissed as 'etymological fallacy.'[19] Furthermore, researchers from the school of cognitive semantics reject the notion of dead metaphors and suggest instead that we think of them as being capable of 'resurrection' or – in a less theologically loaded expression – 'revitalization.' Ignaz Nasalski, in his monograph on Arabic political metaphors, forwards the following quotation in the defence of revitalization:

> The forgotten meaning can remain a latent power under the manifest surface described in dictionaries. These survival meanings, though forgotten and repressed, remain alive in the no man's land that is called preverbal, prelogical, or subconscious. (…) The lexical meanings have once been called the "cem-

19 See the classic critique of excessive etymological inferences in Barr, 1961, and recently applied in Qur'ānic studies by Saleh, 2010, 649–98.

> etery of dead metaphors". But verbal signs are indeed often dead symbols that can become alive and reappear with great vigour on the surface of language (Thass-Thienemann 1967, 9 as cited in Nasalski 2004, 99)

If we accept these basic premises of cognitive semantics we realize that religious texts are permeated with metaphors and metaphorical structures – because our minds are. Metaphors cease to be local rhetorical violations of an alleged standard plane of language but reveal themselves as an intrinsic part of the deeper textual dynamics. To a certain degree this deep metaphorical ubiquity transcends genre and historical trajectories because it is bodily and environmentally anchored, but this recognition also allow us to get a grasp of those pre-linguistic motivations and dynamics, which sometimes fade away in our emphasis on lexical language. For these reasons we must regard 'nakedness' as a fundamental aspect of the Qur'ānic imagination of wilderness.

Despite the overall relevance of applying wilderness-thinking to the Qur'ān, the actual root only appears three times in the text: one time as a verb (Q 20:118), and two times as a noun (Q 27:145; 68:49), the latter two being verbatim repetitions referring to the same context.

> Q 20:118
> It is assuredly given to thee
> neither to hunger therein, nor to go naked [*ya'rā*],
> neither to thirst therein,
> nor to suffer the sun.

The verse is taken from a context referring to Adam and Eve's primordial dwelling in the Garden (of Eden). The authorial voice belongs to God who has just warned them about their archenemy Iblīs (later called Shaytān, i.e., Satan) who might be interested in leading them to expulsion from the Garden. If they reject the schemes of Iblīs, they will be guaranteed further residence in the Garden where they will enjoy these basic privileges: no thirst, hunger or nakedness.

The mythological status and aetiological implications of this verse are certainly of importance. The verse seems to suggest that the pre-Fall couple enjoyed a basic life in which their needs were met (there is no depiction of paradisiacal overabundance). Indeed, the words on hunger and thirst could be interpreted aetiologically as referring to those basic deprivations that human beings are likely to suffer in an Arabian wilderness milieu. The promise of paradisiacal no-nakedness is slightly puzzling since the following verses narrate how Shaytān tempts the couple to eat of the Tree of Eternity in course of which they realize their shameful nakedness. This implies of course that they were already naked when God promised them not to go naked, but since they had not yet transgressed they do not seem to feel shameful about their nakedness.

> ***Q 37:139–146***
> Jonah too was one of the Envoys;
> when he ran away to the laden ship
> and cast lots, and was of the rebutted,
> then the whale swallowed him down,
> and he was blameworthy.
> Had he not been of those that glorify God,
> he would have tarried in its belly
> until the day they shall be raised;
> but We cast him upon *the wilderness*[20] [*bi-l-arā'i*],
> and he was sick,
> and We caused to grow over him
> a tree of the gourds.
> Then We sent him unto a hundred
> thousand, or more,
> and they believed..."

These verses narrate in typical pericopic, brisk style the story of Jonas, one of the few Qur'ānic prophets who is actually designated as a prophet in the Tanakh. Jonah is one of the more ambiguous – *malgré lui*, as it were – prophets who, by trying to escape his prophetic charge, first finds himself on a ship, then later in the belly of a big (whale)fish, until he is finally thrown up on a foreign shore where God lets a gourd tree grow over him and – unlike the usual Qur'ānic stories about prophets – let him be successful in his mission to convert the foreigners, traditionally interpreted as the citizens of Nineveh.

Structurally the Qur'ānic story exploits a number of contrastive effects, first and foremost the basic contrasts between the sea and the dry land, between vain prophetic flight and successful prophetic mission, and between damnation and salvation. But also the contrast between a pitch black, solitary containment, that is, in the belly of the fish in the ocean,[21] and then the abrupt exposure to the barren surface/wilderness, where he first appears lonely and sick, but then seems to be restored to health due to a somewhat miraculously growing plant. The Qur'ān

20 See alternative translations like Paret: "kahlen Stelle (an Land)"; Bobzin: "weite Land"; Bell: "a desert"; Bausani: "spiaggia deserta". Ellen Wulff's Danish translation is perhaps the most apt: "den nøgne landjord", meaning "the naked dry land".

21 See also Q 21:87: "then he called out in the darkness."

does not reveal anything about Jonah eating of the plant, but it seems to be an edible plant and in that respect it resembles the miraculous provision of manna to the Israelites. Perhaps the plant's shading quality is even more important, the Qur'ān describes how it grows *over* him, providing him with an intermediate station between the previous darkness and the intense sunlight (that has scorched all other vegetation). The final contrast is that between the solitary confinement in the fish and Jonah's encounter with an abundant sociality. To sum up: what we witness in this passage is a variety of 'wildernesses' efficiently contrasted with each other as if to push forward the narrative which is held together by a majestic divine We.

Q 68:49
So be thou patient under the judgment of thy Lord,
and be not as the Man of the Fish, when he called,
choking inwarding [*makẓūmun*]
Had there not been overtaking him a blessing from his Lord,
he would have been cast upon the wilderness [*bi-l-arā'i*]
being condemned.
But his Lord had chosen him, and He placed him
among the righteous.

The second and last instance of wilderness-terminology reiterates elements of the Jonah story which we have encountered in Q 37:139–146 and Q 21:87: the fish, Jonah's desperate claustrophobic scream, a conditional phrase, Jonah's rough landing ("cast upon") on the wild and barren mainland and finally the 'happy end', which is quite unusual for these kind of prophet pericopes.

Q 68:49 is somewhat of a contradiction to the previous Jonah pericopes since this verse seems to present the idea that Jonah in fact never was condemned to the wilderness: "Had there not been overtaking him a blessing from his Lord,/ *he would have been* cast upon the wilderness/ being condemned." Wilderness is here presented as a place of condemnation and punishment, but it subsumed by a didactics of admonishment directed to the Qur'ānic listeners: do not try to escape God like Jonah. In the previous verses Jonah's wilderness was staged as a 'naked' stage upon which God's miraculous interventions could take place. Perhaps Q 37:139–146 touches on the motif of the holy man who appears from the desert wilderness and addresses a gawking, urban population.

To sum up, this form of wilderness ultimately derives from a notion of nakedness as human vulnerability (being exposed to the scorching weather conditions) and nakedness as desiccated land. It is staged in two primary ways. On the one hand it expresses sheer punishment, physical and psychological duress exercised by God, on the other hand it functions as the ideal stage upon which God can deliver his subjects. Both types of wilderness are subsumed by a strong didactic functionality.

There are a few other Qur'ānic words beyond *arā'* which could be considered references to wilderness as a barren landscape. The following nouns refer either to land in general terms, but specifically to flat land, this being important in connection to the liaisons between nakedness and barren land: The noun *jurūz* means "lacking vegetation, without herbage (due to lack of water or other reasons)" (Ambros 2004, 58) or 'barren dust,' and holds only two Qur'ānic instances, the apocalyptic Q 18:8 ("We shall surely make all that is on it [i.e. the earth] barren dust.") and Q 32:27 which expounds on God's bountiful intervention in a barren desert landscape ("Have they not seen how We drive the water to the dry land and bring forth crops therewith..."). The noun *badw* means "desert" or perhaps "land of the desert-dwellers" and appears only once (Q 12:100). The last noun that specifically refers to flat and naked ground is *qā'* (pl. *qī'a*) and appears two times. We have already come across the word in the section on sublimity, that is, in Q 24:39 ("And as for the unbelievers, their works are as a mirage *in a spacious plain* [*bi-qī'atin*]). The other instance is the apocalyptic Q 20:105–108:

> They will question thee concerning
> the mountains. Say: '*My Lord will*
> *scatter them as ashes* [*yansifuhā rabbī nasfan*];
> then He will leave them *a level hollow* [*qā'an ṣafṣafan*[22]]
> wherein thou wilt see no crookedness
> neither any curving.'

The three short verses conjure up a natural image of elevation and solidness, namely mountains, but despite these "natural" qualities they can be reduced to flat ground and sterile particles (some translate 'dust' instead of 'ashes') with the apocalyptic intervention of God. The root N-S-F does not only refer to the idea of scattering something, but also suggest the act of uprooting something, as if the mountains were some kind of plant. It remains, however, ambiguous whether the end result, i.e., the level hollow, should be conceived as a threatening and barren instance of wilderness or – as imaginatively proposed by Yusuf Ali in his commentary to his translation – a strange topological 'no-place': "We can imagine the scene

22 The adjective *ṣafṣaf* can also be treated as a noun, meaning "plain, level, even (tract of land)"

of judgement as a level plain, in which there are no ups and downs and no places of concealment. All is straight and level, without corners, mysteries and lurking doubts" (Ali, 1938,786).

The reference to mountains and ashes in this verse also leads us to the root Ṣ-ʿ-D, which connotes ascension, something rigorous and difficult, and – more importantly in this case – elevated ground with no vegetation, and dust (cf. ashes), as in Q 18:40, which refer to God's destruction of a fertile garden and his transformation of it into a "slope of dust." Though this does not refer to wild nature as such, but rather to cultivated nature, we could interpret this verse as referring to a kind of God-inflicted wilderness, a relapse into what was wild before it became cultivated. This again leads us to another root, Ṣ-R-M, which connotes cutting or plucking off, something stripped bare, as in Q 68:20 (the same sūra that refers to Jonah being cast upon the wilderness), which also presents a divinely destroyed garden scene: "…and in the morning it was as if it were a garden plucked [*ka-l-ṣarīm*]." In these verses God is not a god dwelling in the wilderness, but rather a god who can wreak his wilderness on human attempts at agricultural cultivation – sometimes as a specific punishment for transgression, sometimes as part of a global eschatological scenario.

6. Qur'ānic Wilderness as a wider Ecological Category: Flora and Fauna

If we widen the category of wilderness as an arid, barren form of landscape and take into consideration other ecological categories like flora and fauna, the scope of wilderness naturally widens but it also becomes more fuzzy and dependent on interpretational focus. Thus, only a very restricted number of examples will be presented here.

6.a. Flora

The generic Qur'ānic word for vegetation is *nabāt.* Mostly, the word refers to cultivated vegetation, like the garden, or plants that can be harvested and which command cultivation, such as corn, vegetables, fruits, date palms, olive trees, herbs et cetera. In general, there are only a few references to wild flora and they do not play a significant role in the semantic universe. However, one form of flora-like wilderness, which should be mentioned, is the devastated field (in contrast to the reaped field) leaving only stubble, *hashīm*, as in the elaborate parable of Q 18:45:

> And strike for them the similitude of
> the present life:
> it is as water that We send down
> out of heaven, and the plants of
> earth mingle with it; and in
> the morning it is *straw* [*hashīm*] the winds
> scatter...

What cognitive semanticists would label as the target domain 'life' (human and animal life and vegetation) is here transferred from the so-called source domains of climate, seasonal change and its effects on flora (see also Q 54:31) A similar instance, also put forth in a simile, is found in connection with the noun *a'jāz*, meaning stumps of tree, trunks. It only appears twice in the Qur'ān, as in Q 54:20, "...We loosened against them a wind clamorous in a day of ill fortune continuous, plucking men as if they were stumps of uprooted palm-trees" (Q 69:7) In both of these instances (the other is Q 69:7) wilderness exposes the ephemeral state of natural abundance, *in casu* life and vegetation.

6.b. Fauna

The Qur'ān does not employ the standard generic word for animal, *hayawān*, but once and in this case only in the sense of 'present life' (Q 29:64). The closest equivalent for a generic word is *dābba* (plural *dawābb*), which could be translated as 'land animal.' The various species of animals mentioned in the Qur'ān are mammals like the herd animals of camels, cattle, sheep, goats, horses, and donkeys, but also carnivorous animals like lion, wolf, and dog. Other animals mentioned are apes, birds, fish and insects. As for wild animals, *wuḥūsh*, which appear only in the plural and probably only in reference to mammals, we could operate with those that inhabit wilderness but do not present a substantial danger to human beings (for instance the wild ass) and those who are intuitively perceived as dangerous and somewhat awe-inspiring.[23] Of these we only come across the wolf and only in connection with the story of

23 The Qur'ān does operate with the option that wild animals can be somewhat domesticated and used for hunting purposes, see Q 5:4, which probably refers to hunting hounds.

Joseph (Q 12:13, 14, 17). Other dangerous and awe-inspiring animals are mentioned, such as the lion, but in the relevant verses it is not depicted as dangerous to humans but to other animals. In this instance, the wolf, *al-dhi'b*, is mentioned by Joseph's brothers in order to trick their father into believing that his son has been eaten. Birds appear several times in the Qur'ān, but as dangerous animals they are only mentioned once and in connection with a somewhat legendary event, namely in the five verses short and probably very old sura 105, titled *The Elephant*. Here birds are depicted as dropping deadly rocks or stones of baked clay (exegetes and philologists have not come to a final conclusion) on "the Men of the Elephants", usually explained in Islamic exegesis as a reference to a Jewish Yemenite attack on Mecca in 570, the same year as Muhammad was born. The story about the fish/whale that swallows Jonah might have been able to produce some goose bumps among the most imaginative of the initial Qur'ānic recipients, but its role as a wild and ferocious animal seems not to be of particular importance in the text. Rather, the fish seems to have a more prop-like status. Thus, as also observed by Ambros, animals are rarely presented as wild and dangerous adversaries, but are mostly depicted as animals exploitable for human purposes or for making plain that they are a part of God's powerful, wise and bountiful scheme (Ambros, 1990, 309).

7. Conclusion

In "Simple Negative Remarks on the Vocabulary of the Qur'ān", an article written by the French orientalist Robert Brunschvig (1901–1990), the author sketches out a somewhat precarious project, namely to approach the Qur'ān "the *negative* way, i. e. stating not what is but rather that which *is not* in the Qur'ān" (Brunschvig, 1956, 285). Brunschvig's methodological problem is of course that we could pile up a huge number of so-called negative remarks but that these in many instances would fall prey to the fallacy of *e silentio* arguments. Nevertheless, Brunschvig's best intentions are both reasonable and viable if pursued with prudence, taking into consideration our present knowledge about the life world of the Arabs in late antiquity. Indeed, absent categories and words can indeed tell us a great deal about cultural and religious priorities as well as economic and ecological conditions. Many Qur'ānic scholars have thus wondered about the striking paucity of words relating to the camel or the desert in the Qur'ān, both of which command numerous words in pre-Islamic poetry. Compared with the immediate literary precursor of the Qur'ān, i.e., pre-Islamic poetry and its great emphasis on natural phenomena, including the more wilderness-like aspects hereof, the Qur'ān's relative neglect of wilderness landscapes and wildlife is indeed striking. As for the other eminent precursors and great reservoirs of intertextual inspirations and loans, namely the oral and written literatures of the Jewish and Christian traditions, we do not notice a particularly strong wilderness influence either. Why these absences, we might ask. Why are the rough mountains, the barren deserts and its inhabitants so imminent and potentially threatening for the Muslims in Mecca and Medina nevertheless so distant and semantically scanty in the Qur'ān? Our answer to Brunschvig and his 'negative remarks' is provided by Ambros who states that the Qur'ān

> reflects an audience of text recipients, who is at "civilizational quantum leap's" remove from the aristocratic warrior and hunter in the desert steppes of jāhilīya-poetry [pre-Islamic poetry]: a settled population who is dependent on cultivated plants for their daily bread and for whom things "outside" can only be a bugaboo for the constant threats of failure to their endeavours (Ambros, 1990, 321).

Ambros' remark on a settled population mainly dependent on agricultural produce is, however, most relevant for the Medinan society, which first became relevant after Muhammad and his followers' emigration from Mecca to Medina (Yathīb) in 622, the so-called *hijra.* Mecca, in contrast to Medina, was much more dependent on caravan trading and pilgrimage economy (as Mecca represented a pilgrimage centre) than agricultural produce. Despite these differences in urban milieus, both cities represented significantly different life worlds as compared to nomadic life. This probably explains the relative disregard for wildlife and wilderness. Wilderness remained a backdrop and part of the cultural memory for the Qur'ānic Arabs, but as a zone for lived experience and a repertoire for further literary endeavours, it seems to have been superseded by the more urban-like milieus and priorities of late antique Arabia. What is left are vestiges of wilderness. Wilderness, then, is not simply anathema to civilization, a marked contrast to be played off against a world of orderliness – to be captured in neat but somewhat worn-out dichotomies of nature and culture, cosmos and chaos. Wilderness also appears as an absence or trace, as a vague collective memory of a lifeworld that is about to be superseded by a new one, that of Islamic monotheism. Despite these curtailments and vestigial remains, wilderness and wild-like phenomena still play a significant role sporadically (being all the more conspicuous in the sūras then) – as a cultural bugaboo and a theological warning of God's wrath and frightful omnipotence and of his mercy and generosity when he holds wilderness in abeyance.

8. References

'Alī, 'Abdullah Yusuf. *The Holy Qur'ān. Text, Translation and Commentary.* New Revised Edition. Brentwood: Amana Corporation, 1989 (original 1938).

Ambros, Arne A. and Stephan Procházka. *A Concise Dictionary of Koranic Arabic.* Wiesbaden: Reichert, 2004.

Ambros, Arne A. and Stephan Procházka. *The Nouns of Koranic Arabic Arranged by Topic.* Wiesbaden: Reichert, 2006.

Ambros, Arne A. "Gestaltung und Funktionen der Biosphäre im Koran", in: *Zeitschrift der Deutschen Morgenländischen Gesellschaft,* Band 140, Heft 1, (1990), 290–325.

Barr, James. *The Semantics of Biblical Language.* Oxford, Oxford University Press, 1961.

Bauer, Thomas. "Relevance of Early Arabic Poetry for Qur'anic Studies Including Observations on Kull and on Q 22:27, 26:225, and 52:21", in Angelika Neuwirth, Nicolai Sinai & Michael Marx (eds.), *The Qur'ān in Context. Historical and Literary Investigations into the Qur'ānic Milieu.* Brill: Leiden, 2010, 699–732.

Bravmann, Meir M. *The Spiritual Background of Early Islam: Studies in Ancient Arabic Concepts,* Brill 1998 (org. 1972).

Borg, Gert. "*ṣa'ālik*" in: Julie Scott Meisami and Starkey, Paul (eds.). *Encyclopedia of Arabic Literature.* Vol. 2. London: Routledge, 1998, 670–671.

Bravmann, Meir M. *The Spiritual Background of Early Islam: Studies in Ancient Arabic Concepts.* Leiden: Brill, 1998 (orginal 1972).

Brunchvig, Robert. "Simple Negative Remarks on the Vocabulary of the Qur'ān", in: Andrew Rippin, ed. *The Qur'ān: Style and Contents.* The Formation of the Classical Islamic World. Farnham: Ashgate, 2001, org. 1956, 285–295.

Burke, Edmund. *A Philosophical Enquiry into the Sublime and Beautiful.* London: Penguin, 2004 (orginal 1757).

Firestone, Reuven. *Journeys in Holy Land: the Evolution of the Abraham-Ishmael Legends in Islamic Exegesis.* New York: State University of New York Press, 1990.

Grunebaum, Gustave E. von. "The Response to Nature in Arabic Poetry", in: *Journal of Near Eastern Studies,* Volume IV, no. 3, July 1945, 137–151.

Hoffmann, Thomas. "Koranisk/kaotisk: Om Koranens semantik, retorik, ritualisering, reception og konstruktion i lyset af begrebet 'kaos'", in: *Chaos: Skandinavisk Tidsskrift for Religionshistoriske Studier* nr. 49, (2008), 75–94.

Hoyland, Robert G. *Seeing Islam as Others Saw it. A Survey and Evaluation of Christian, Jewish and Zoroastrian Writings on Early Islam.* Studies in Late Antiquity and Early Islam 13. Princeton: The Darwin Press, 1997.

Hoyland, Robert G. *Arabia and the Arabs: from the Bronze Age to the coming of Islam.* London, Routledge, 2001.Izutsu, Toshihiko: *Ethico Religious Concepts in the Qur'ān.* Toronto: McGill-Queens University Press 2007, orginal 1959).

Izutsu, Toshihiko. *Ethico-Religious Concept in the Qur'ān,* McGill-Queen's University Press, 2007 (org. 1959).

Jacobi, Renate. *Studie zur Poetik der altarabischen Qaside.* Wiesbaden: Franz Steiner Verlag, 1971.

Lakoff, George and Mark Johnson. *Metaphors We Live By.* Chicago: University of Chicago Press, 1980.

Nasalski, Ignacy. *Die politische Metapher im Arabischen. Untersuchungen zu Semiotik und Symbolik der politischen Sprache am Beistpiel Ägyptens.* Wiesbaden: Harrassowitz Verlag, 2004.

Retsö, Jan. *The Arabs in Antiquity: Their history from the Assyrians to the Umayyads.* London: RoutledgeCurzon, 2003.

Ryken, Leland et al. (eds). "Wilderness", *Dictionary of Biblical Imagery,* Downers Grove, InterVarsity press, 1998, p. 948–951.

Saleh, Walid. "The Etymological Fallacy and Qur'anic Studies: Muhammad, Paradise and Late Antiquity", in: Angelika Neuwirth, Nicolai Sinai, Michael Marx (eds.), *The Qur'ān in Context: Historical and Literary Investigations into the Qur'ānic Milieu.* Leiden: Brill, 2010.

Sells, Michael. *Desert Tracings. Six Classic Arabian Odes by 'Alqama, Shánfara, Labíd, 'Antara, Al-A'sha, and Dhu al-Rúmma.* Middletown: Wesleyan University Press, 1989.

Scholer, Gregor. *Arabische Naturdichtung. Die Zahrîyât, Rabî'îyât und Rauḍîyât von ihren Anfängen bis aṣ-Ṣanaubari. Eine Gattungs-, Motiv- und Stilgeschichtliche Untersuchung.* Beirut, Franz Steiner Verlag, 1974.

Stetkevych, Suzanne Pinckney. *The Mute Immortals Speak:Pre-Islamic Poetry and the Poetics of Ritual.* Myth and Poetics. Ithaca: Cornell University Press, 1993.

Sweetser, Eve E. *From Etymology to Pragmatics. Metaphorical and cultural aspects of semantic structure.* Cambridge Studies in Linguistics 54. Cambridge: Cambridge University Press, 1980.

Taylor, Bron. *The Encyclopedia of Religion and Nature.* London: Continuum, 2008.

Thass-Thienemann, Theodore. *The Subconscious Language.* New York: Washington Square Press, 1967.

Toelle, Heidi. "Nature", in: Mohammad Ali Amir-Moezzi (ed.), *Dictionaire du Coran.* Paris: Robert Laffont 2007, p. 589–591.

Turner, Victor V. *The Ritual Process: Structure and Anti-Structure,* Ithaca, N.Y.: Cornell University Press, 1977.

Zammit, Martin. *A Comparative Lexical Study of Qur'ānic Arabic.* Leiden: Brill, 2002.

8. Wilderness, Liminality, and the Other in Old Norse Myth and Cosmology

Jens Peter Schjødt

1. Wilderness: destruction and potential

In this article, I use the term 'wilderness' to refer to the uncontrollable, which I assume is experienced in all cultures, although it is much more pronounced in archaic and "primitive" societies than in those of the modern world. In this way it reminds us of other terms, such as 'chaos' (e.g. Eliade 1957), 'nature' (e.g. Lévi-Strauss 1962), 'the liminal' (Turner 1969), 'the wholly other' (Otto 1917), and no doubt many more, used in the study of religion to characterise that which is opposed to the ordered world in which we live our everyday lives (cosmos, culture, society, 'this' world).[1] All of the former terms, including wilderness, may thus be seen as classificatory constructions of a reality which is different from the one we control,[2] and the various terms point to various aspects of this 'uncontrollable' reality. Thus 'wilderness' is not, of course, a term or a category which is exclusively connected to religion – it is definitely much more profound than that – but since 'wilderness' is 'out there somewhere' in the world constructions of most cultures, it must be addressed in religions, too. It is common knowledge within the general study of religion that religions are multifunctional: they have sociological, psychological, political etc. functions, and so does ritual. These functions may be divided into emic and etic functions, i.e functions which are recognised by the adherents of a certain religion, and those that are not, but have to be made visible by

1 It is obvious that some of these terms are related to the 'sacred' as a fundamental category, although the 'sacred' may be much more than the semantics of these terms. But just as the terms here mentioned connote something which is different from everyday life, the same applies to 'the sacred'.

2 'Constructions' in this context does not signify something which has no basis in reality. 'Wilderness' certainly does exist, but our way of perceiving it is a 'construction' in the sense that we may focus on various aspects of this reality. Thus there is no full-fledged constructivist agenda behind this terminology.

the analysts (anthropologists, historians of religions etc.). Concerning the etic functions a lot of theories have been proposed (see Bell 1997), some of them more likely than others. From an emic viewpoint, however, an important function in all religions is the attempt of controlling the uncontrollable during rituals of various kinds.[3] For that reason it is of utmost importance to have a rather fixed view of the realities which differ from those of everyday life.[4] This is what myths are all about: they construct worlds which are in many ways reflections of the experience of reality, although certainly not 'mirrors' of reality, so that they may be manipulated in rituals. In most religions, several such worlds are imagined, usually because the uncontrollable has several aspects: it may cause victory or defeat in battles, it may cause abundance or famine, it may cause richness or poverty etc. Therefore, we most often meet at least two 'otherworlds' in religion, namely that of the benevolent gods, the positive forces, and that of the demons, the negative forces.[5] But this rough division may be refined in an infinite number of ways. Some demons may cause illnesses, whereas others may cause defeat or bad luck of all kinds; the same applies to the positive forces: some are gods of fertility whereas others are gods of victory etc. In some religions these gods or demons belong to different worlds or categories, which may be only loosely connected, whereas, in other religions, they constitute whole pantheons and belong to the same world and thus more or less to the same category.

In this way, we may see wilderness as one such category, and before turning to what could be called 'wilderness' in pre-Christian Scandinavian religion, I will attempt to characterise the term more precisely, even if one of the characteristics of wilderness is exactly that there is no precise defini-

---

3 Ritual has thus been defined as 'representative acts designed to change or maintain their object' (Podemann Sørensen 1993, 19–20), which seems quite appropriate, although we probably could add '....through the communication with an other world' because this is what *religious* rituals do: they attempt to manipulate and control things in this world by invoking the gods and spirits of the other world. See also Schjødt 2008, 62–72.

4 This is certainly not to say that religious world views are 'coherent' which they are not. But within a certain ritual it is necessary to know which god to turn to, and how to perform the individual rituals, in order to achieve the desired goal. The ideas may well vary from time to time and from person to person, but whether this or that sacrificial victim, for instance, is sacrificed to a certain god in a certain way, is not coincidental but will be dependant of the 'tradition' , i.e. the cluster of ideas concerning this or that god.

5 Often it turns out to be impossible to make such a clear-cut division since many 'otherworld' powers can be benevolent as well as malevolent.

tion of it from an indigenous perspective.[6] Wilderness is what is 'out there'. To be sure, one of its central characteristics, from an emic perspective, is that it possesses unknown qualities; it lacks a recognisable structure although it will always be structured in some way. This is an important aspect, to which I will return in a moment. As mentioned above, most cultures have several other worlds. The specific way in which they are articulated in various cultures comes down to the variety of human constructions, although they are always related, in some way, to the physical reality experienced.[7] Wilderness is one such other world, and it can therefore be seen simply as a subcategory of the 'Other'. This said, it is important to note that it is not simply identical with 'otherness'. In order to make sense, the term should be reserved for a certain category of otherness, namely that spatial aspect of otherness which is particularly ambivalent in relation to human beings. I will thus propose that we reserve the term wilderness to denote that particular otherworld, known from nearly all cultures, which is characterised by being at the same time – and to a higher degree than other 'otherworlds' – a threat and a potential.[8] The 'real' wilderness – that which is not a social or an ideological construction, but a reality placed outside and sometimes far from the village of the 'myth makers' –, is by definition dangerous because no rules, or at least other rules than the ones we know, will apply when we are out there. It is full of beings which may be hostile to humans, and we cannot rely entirely on the skills that will normally keep us safe. On the other hand, the wilderness also carries great potential, since it can be used for improving the qualities of your own world: you may hunt animals or procure timber, or you

6 Wilderness is of course an *etic* category – words with the significance we attribute to the wilderness cannot necessarily be found in all indigenous languages.

7 Thus the gods may be reflections of the physical conditions that are experienced by humans in general, as well as of features specific to the individual cultures. So death, for instance, is always part of religious world views, just as are illness and cure, warfare, food, sexuality etc. They may also, however be reflections of certain sociological or economic conditions of a particular society.

8 We could say that most otherworlds are characterized by ambivalence: the sorceress may activate her magical universe in order to be benevolent, but she is also able to destroy you; a certain god may favour you but can also abandon you, etc. Thus the differences between otherworlds are most often a matter of degree rather than of absolute distinctions: what can be achieved from one otherworld may also be achieved from another. We are here talking about tendencies. The difference between 'chaos' and 'wilderness' would thus also be a matter of degree: both terms would connote threat as well as potential, but chaos, I believe, tends to connote something much more abstract than 'wilderness', although the two terms are definitely related.

may pass through it to get to places that are attractive in some way or another. So far the 'other world' wilderness is characterised in complete agreement with the wilderness of reality – the unknown is unpredictable, and it may be attractive as well as lethal. But in religious universes, this 'other world wilderness' is supplemented with so-called counter-intuitive beings, to use a term from Pascal Boyer (Boyer 2001, 51–91). Whereas those beings which govern the results of our daily activities are known to us – and we can, to a certain extent, predict their reactions although we cannot be certain of anything, the counter-intuitive beings of the wilderness, in contrast, are just as dangerous as are the animals or the plants or whatever is out there. So even if the encounters with gods or spirits will always be somewhat unpredictable (we are, after all, dealing with beings from another world), spirits of the wilderness are unpredictable to a much higher degree. They may be hostile, or they may be friendly, but unpredictability is their defining trait. Since it seems to be a universal cognitive property in humans that we seek to control, and to a certain degree make understandable, what happens in our lives (cf. Geertz 1966, 14–17), it is important to be able to operationalise the factors which create the unpredictability. In religious societies, the basic tools for this process are myth and ritual. Humans attempt to control the unpredictable through ritual, and through myth we try to understand it. The control, of course, is the most important feature, but it is hard to control anything that we do not understand. So, from an etic perspective, ritual is certainly the basis for myth whereas, from an emic perspective, myth, or rather the mythic world view which is constituted by the individual mythic narratives, is the prerequisite for ritual: that which we aim to control must be understood as well as possible.

This is probably the reason why most mythologies across the world are full of beings which represent the positive as well as the negative sides of the wilderness: in a religious or a mythological world view, it is important, in order to cope with the experiences of the wilderness (and the dangers of this world in general), to be able to manipulate such forces by gaining knowledge about them.

2. Cosmological axes in pre-Christian Scandinavia

Turning to the pre-Christian Scandinavian mythology, it may be useful to give a brief summary of some of the important features within the cosmology in order to depict the 'worlds' with relevance to the theme in focus, 'wilderness'.[9] As in most other mythologies, it seems as if the world view was not coherent and as if many more or less incompatible ideas are brought together (Schjødt 2009). Nevertheless, it appears that we can discern some basic structures which, probably with some variations, constitute an important part of the pagan world view. It can of course be discussed to which extent the various worlds apply to the term wilderness, but it seems as if some of the worlds mentioned in the following and the interchange between them at least reveal the basic characteristics of wilderness according to the definition of the term given above.

For the present purpose, a detailed discussion of the source critical problems connected to the various pieces of information is unnecessary. In accordance with most other mythologies, the Scandinavians structured

9 There is no reason here to go into all the problems having to do with sources. Everyone familiar with the field of pre-Christian Scandinavian mythology (or, as it is often called, Old Norse mythology) will know that the main reason for disagreement among scholars is the source situation. To put it very briefly, by far the largest part of information concerning the mythology is transmitted to the present through manuscripts written down during the Middle Ages. The two main sources are the so-called Eddic poems (or the Elder Edda) and Snorri's Edda (or the Younger Edda). The Eddic poems are known from various manuscripts, the oldest one being from the second half of the 13th century. The poems contain, among other things, information about the cosmology and a long list of gods. The basic problem is that, although most scholars nowadays agree that many of the poems were transmitted orally before being written down, we do not know the time of composition: were they composed in pagan times (before c. 1000 AD) or in Christian times, the importance of which is obvious. Besides that, the poems are most often very brief and appear to assume that the audience is familiar with the myths related. In other words, to understand the poems, the audience is required to have a cultural competence which modern scholars generally do not possess. The other main source, Snorri's Edda, is a prose work written by the Icelandic chieftain, politician and author Snorri Sturluson c. 1220 AD. Snorri was no doubt a good Christian, and he wrote his work from an antiquarian interest. He clearly expresses the view that the pagan mythology that he wrote about was wrong (something that Christian people should not believe in) in comparison with the Christianity of his time, and part of his agenda was clearly to show this. Nevertheless, it seems as if he relates many myths in an (arguably) trustworthy manner (although some would disagree about that), especially because he uses some of the Eddic poems as his source. It seems clear that, from a source critical point of view, this situation is far from ideal, and, as mentioned above, a lot of controversies have taken place concerning almost every piece of information – whether seen as an expression of the pagan world view or whether seen as a Christian idea about paganism that cannot illuminate the pagan mythology 'as it really was'. This, of course, is a rough simplification of the state of affairs but reflects, nevertheless, the main argument. For a far more detailed discussion with many references, see Schjødt 2008,85–107.

their cosmos along two axes – vertical and horizontal (Schjødt 1990). Although these two axes were certainly part of the same cosmological framework, the myths connected to each of them seem to deal with different aspects, to which I will return. From the existence of these axes, we naturally get two sets of oppositions, namely centre vs. periphery and up vs. down. Concerning the horizontal axis, we are thus told by Snorri (Faulkes 2005, 12–13) that the world was structured with the gods – who are also called the Aesir (the group in which we find some of the most well-known gods of the Scandinavian pantheon, such as Odin and Thor) – living at the centre in a place called Asgard (literally meaning fence of the Aesir but also connoting the farmstead). Outside this abode of the Aesir, there is the world of the humans called Midgard (the fence in the middle), and outside that again is the world of the giants called Utgard (the outer fence). These worlds are apparently seen as concentric circles – although this is not stated explicitly and although no sort of dogmatic discourse on to how to perceive the world existed. Outside these worlds is the deep sea, at the bottom of which lies the Midgard serpent, encircling the world.

The vertical axis is constituted by the world tree Yggdrasill, which reaches up to the sky and the roots of which reach out to the whole world. Down in the underworld, there are many beings of different kinds, such as the dead (or rather some of the dead) and the dwarfs, who are portrayed as smiths in a wide sense of the word: they create all kinds of goods. In the underworld, we also have the alfar, who were transformed into the elves of the modern world (Gunnell 2007), and other categories of beings who mainly have to do with death and fertility. This agrees well with what we meet in many mythologies, namely that phenomena which are connected to death and fertility are spatially seen as chthonic.

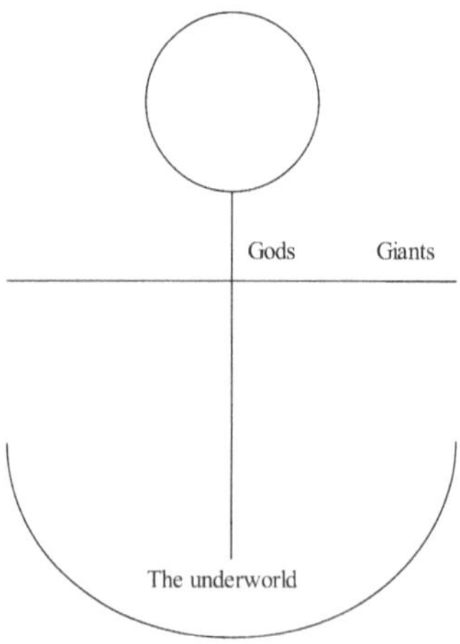

All information concerning the cosmos related in the sources has been the subject of numerous investigations, and some scholars have even attempted to draw a detailed picture, taking into consideration all available information which is supposed to depict the view of the ancient Scandinavians (see for instance Branston 1980, 72–75 for examples). Such attempts, however, are doomed to fail since, as was mentioned earlier, the world view of the pagan Scandinavians was certainly not coherent. This is the case in almost any cosmology, and particularly in those of religions which have no canonical writings (cf. Schjødt 2010). In my opinion, what we can do is to draw the main structures, in which details have undoubtedly varied from time to time, from area to area, and even from individual to individual. We should thus focus on the basic structure constituted by the two axes – horizontal and vertical – and the beings positioned along these axes.

Nearly all the main gods belong to the Aesir group. According to some sources with an obvious Christian bias, such as Snorri (Faulkes 2005, 17), the Aesir are said to live in the sky or in the top of the world tree Yggdrasill. In most sources, however, their location seems to be at the roots of Yggdrasill, in the middle of the world. In this way they are located close to the humans, and it is also conspicuous that, like in many mythologies all over the world, the differences between gods and humans appear to be of a quantitative rather than qualitative kind: the gods are stronger, cleverer and, in general, much more powerful than men, but they are subject to death, and they mingle with humans on many occasions, just as we are told that whole peoples or at least the noble families descended from the gods (Byock 1990, 35 and Jónsson 1911, 10). So, from a certain perspective, humans and gods are of 'the same kind', although, of course, not identical. This is confirmed when we look at the 'societies' of gods and humans respectively. Here it seems clear that the norms among the Aesir and the humans are similar. This is probably reflected in the myth of the peace agreement between the Aesir and another group of gods called the Vanir, who should, in all likelihood, be seen as chthonic. Here we are told that incest, marriages between brother and sister, was the norm among the Vanir; the hostages that the Vanir had to send to the Aesir in connection with the peace agreement thus had to give up this practice because it was not allowed among the Aesir to marry that close (Jónsson 1911, 5–6)). This indicates (and further examples could be mentioned) that the Aesir in some way can be seen as the representatives of the humans of the culture which created the mythology that we usually call the Old Norse or pre-Christian

Scandinavian. The relation between the Aesir and the other groups of beings within the mythology should thus be seen as a reflection of the relation between the 'in-group' and various 'out-groups'. It has thus been argued (Mundal 1996 and 1997; Lindow 1995) that, in some way, the Sami to the north of the Germanic speaking Scandinavians were mythologically reflected in the giants or, as they are often called in the mythological texts, the jötnar.[10] It will certainly not be a surprise for historians of religion that the world of myth and that of the human culture are to some extent imagined as parallel: it is simply one of the basic functions of myth to treat, and possibly explain, problems of all sorts within societies (c.f. Doty 1986, 41–71 with many references), and among these not least the relation with what is foreign, whether it be foreign cultures or the unknown wilderness.

We thus have a cosmos with an 'in-group' and various 'out-groups'. In relation to 'wilderness', it seems that the two aspects mentioned at the beginning of the article, the threat and the potential, are to a certain extent divided into two other worlds, one on the periphery and the other one downwards. Wilderness thus seems to be a much more ambivalent notion than is usually believed. It is certainly not just an equivalent to 'Chaos', since 'civilisation', although of another kind than the one of the in-group, is definitely at stake. We can still see a kind of dualism connected to the opposition between this world and the other world, but the terms that are invested are related in a much more sophisticated way than is usually believed, as we shall see in the following sections. This being said, it must be emphasised once again that mythologies are not coherent, which indicates that there is a lot of overlap in the distribution of the various functions among the various worlds. We are therefore just talking about some tendencies, tendencies that I will attempt to analyse below.

10 There are many native designations for what is usually translated into English 'giants'. Jötnar is one of them, and others are thursar, troll, risar. Whether these names signify the same group of beings is very uncertain. It seems as if the giants have many, sometimes opposing characteristics, which may be distributed to various groups, originally being classified differently. In the extant sources, however, it is not possible to draw any strict borderlines between the groups, or to connect certain characteristics with certain designations (see Schultz 2004, Jakobsson 2006).

3. Wilderness and the giants

The world that immediately comes to mind when talking about wilderness is the world of the giants, which we find mainly on the horizontal axis. This is often described as a world characterised by coldness. It is rocky, and it is usually placed to the North and to the East.[11] Taking the geographical position of Scandinavia into consideration, it seems obvious that the 'natural', geophysical counterparts to the imagined wilderness of the mythology would be the areas of Northern Scandinavia and the vast areas to the east in Northern Russia. It thus appears from the mythological descriptions that the region of the giants must be viewed as 'wild', as a waste land.

However, in order to characterise the worlds of Old Norse mythology, as in any mythology, we must look at the inhabitants of these worlds and at the relation between them and the representatives of the in-group. The relation between giant land and the gods is a theme in many Old Norse myths, and in most cases, the relation is of a hostile nature. Even if several gods are involved in the interchange between these two worlds (to which I will return below), the primary representative of the in-group is Thor. Now, Thor is presented as the great warrior in the mythology; from a Dumézilian perspective, he is characterised as the representative of the second function, that of war (e.g Dumézil 1959, 106–117). But what is interesting is that Thor was apparently not the god who was venerated in connection with wars among humans; he is hardly ever seen as the protector of armies, and he never appears on the battlefield. His role as a warrior seems to exist exclusively in the mythology, where he is the strong champion who throws his hammer Mjöllnir against his enemies and crushes everything with it. His enemies are almost always giants, and he fights them in single combat.[12] Sometimes the giants

---

11 The examples here are legion. For instance at Ragnarök the giant armies arrive to the battlefield from the east, as it is told in the eddic poem *Völuspá*, st. 50 and 51 (Neckel and Kuhn 1962, 11–12).

12 Thor's main enemy is the Midgard serpent representing the periphery of the world. The two antagonists kill each other during the eschatological battle at Ragnarök, at which the beings of the periphery attack the world of gods and humans. And in a certain myth it is thematised how Thor during a fishing expedition attempted to kill the serpent by rowing so far out that he could meet it (the main sources here are Snorri [Faulkes 2005, 44–45] and Hymiskviđa [Neckel and Kuhn 1962, 88–95]). It thus seems as if the relation between Thor and the serpent is analogous to the one between centre and periphery.

will appear in Asgard, but most often Thor travels to the land of the giants to fight them. Thus we often hear that Thor was not at home on this or that occasion because he was in the east fighting giants (for instance in the eddic poems Hárbarðslióð [Neckel and Kuhn 1962, 78–87] and Lokasenna [Neckel and Kuhn 1962, 96–110]). It is important to notice that Thor is unambiguously the defender of cosmos (Schjødt 2004). As mentioned, he was apparently not a war god in the traditional sense of the term. Quite on the contrary, he is often portrayed as having strong affiliations with fertility (e.g. Adam of Bremen IV, 26–27), and in that sense he reminds us of the Roman Mars. This is probably due to the fact that every sort of failure of the crops in this mythological universe is seen as an attack from the powers of chaos (Schjødt 2004, 130–131). And these powers may very well be seen as the giants, who are represented in several myths as stealing important objects from the gods, such as the apples of the goddess Idun, which the gods eat to prevent them from getting old, or Thor's own hammer, which is necessary for the general survival of gods and humans. The giants can thus be seen as chaotic forces which constantly try to destroy the world of gods and humans, a world that is basically good, but which is constantly threatened from the outside. The fights of Thor should probably be seen in this light: he is defending cosmos against the forces of chaos, the demons of the wilderness trying to break into the civilised world. Thus wilderness is certainly a spatial notion, but it is also a notion which characterises a basic condition, namely that the civilised world is always threatened by exterior forces.

This whole exposition about what is happening on the horizontal axis (and therefore in the relation between centre and periphery) shows that the basic code used here is that of antagonism, namely between Thor and the giants. Further, it is characterised by a strong physical bias, which is quite natural in connection with fighting, just as Thor, in his entire appearance, is extremely physical. He is the strongest of the Aesir (Faulkes 2005, 22), and in several myths he is seen as having an enormous thirst and appetite (e.g. Faulkes 2005, 41 and Neckel and Kuhn 1962, 111–115). However, there is another kind of interplay between the Aesir and the giants, namely an exchange of knowledge – numinous knowledge. An entire poem tells us how Odin visited a giant called Vafthrudnir (Neckel and Kuhn 1962, 45–55), with whom he exchanged knowledge, primarily of a mythical kind. Odin also acquires knowledge from various giant women (e.g. Neckel and Kuhn 1962, 1–16 and 277–279. See also Schjødt 2008, 217–224).

This leads to the other aspect of the wilderness – its potential. This potential, when related to giants (which is certainly not always the case), is connected almost exclusively with female giants, and it can be divided into two realms, one physical and one mental. The first one can be seen in the marriages of the Vanir gods brought to the Aesir as hostages. The two male Vanir, Njord and Frey, both marry giant women who are particularly beautiful (which is certainly not the case with all giant women),[13] since they were not allowed to practice the incestuous marriages that characterised the Vanir society (Jónsson 1911, 6). And even if these marriages do not seem to be particularly successful, they show that sexuality was part of the potential of the female giants.[14] I will not elaborate further on this issue although the significance of these marriages is very important for the structure of Old Norse mythology (treated in Schjødt 2008, 382–396). Instead, I focus on the mental aspects of the potentiality which, however, are not one of the primary traits of the giant world. As just mentioned, we nevertheless find exchanges of numinous knowledge in connection with giant women, and in these instances the protagonist on the side of the Aesir is Odin, who in general is portrayed as the intellectual among the Aesir. It is worth noticing that when Odin contacts the giantesses, they are either dead, indicating that they have changed from the horizontal to the vertical axis (because death is strongly linked to the underworld [Schjødt 2008, 410–421]), or, in one instance, they possess an object of knowledge, i. e. the mead of poetry and wisdom, which is captured by Odin, again from the underworld (Faulkes 1998, 3–5), so that he can pass it on to his favourites. Thus, the notion of the feminine seems to be an important characteristic of the underworld. Creating parts of semantic universes is what myth and cosmology do, and this is exactly what we see here: even if we find a lot of elements in the pre-Christian Scandinavian texts that are part of what we usually connect with cosmologies, we also note categories not normal-

---

13 These myths are related and hinted at several places in the sources (e. g. Faulkes 1998, 2 and Neckel and Kuhn 1962, 69–77).

14 Many of the male giants are also represented as sexually obsessed in their lust for some of the goddesses (particularly Freyja and Idun). However, whereas the sexuality of the giant women is not in general viewed in a negative light, it is different with the male giants. Their lust is illegal, which is probably due to exactly their relation to the wilderness, seen from the patriarchal (in-group) society: the sexuality of our women belongs to us, and not to foreign barbarians, whereas female sexuality from these same barbarians may be quite attractive.

ly seen in this connection, such as numinous knowledge, femininity and masculinity etc. as building blocks for the semantic universe.

4. The acquisition of numinous power: myth and ritual

Whereas the exchange along the horizontal axis is thus a matter of hostility, the acquisition of numinous knowledge is most often actualised along the vertical axis. The pertinent spatial opposition here is up vs. down, and the protagonist is not Thor, but rather Odin. The actors of this other world are of many kinds and do not represent a closed group as do the male giants in the hostility code. As mentioned above, on this axis we meet giant women as well as encounters with the dead and female entities of varying descent. Sometimes it even seems as if no actors at all are involved, as for instance in a famous myth from the poem *Hávamál,* stanzas 138–141 (Neckel and Kuhn 1962, 140), about Odin hanging on the world tree Yggdrasill (often having the role of *axis mundi* in Old Norse religion) for nine nights, in a process of dying. The stanzas go like this:

138 I know that I hung on a windy tree
nine long nights
wounded with a spear, dedicated to Odin
myself to myself,
on the tree of which no man knows
from where its roots run.

139 No bread did they give me nor a drink from a horn,
downwards I peered;
I took up the runes, screaming I took them,
then I fell back from there.

140 Nine mighty spells I learnt from the famous son
of Bolthorn, Bestla's father,
and I got a drink of the precious mead,
poured from Odrerir.

141 Then I began to quicken and be wise,
And to grow and to prosper;
one word found another for me,
one deed found another deed for me. (Larrington 1996, 34).

We notice that when Odin has been hanging on the tree, without food or drink and exposed to the weather, for nine nights, which is a sacred number in Old Norse religion, and which connotes a kind of fulfilment (Schjødt 1988); he looks down, and from beneath he picks up the runes (the word 'rune' basically means 'secret'). From then on, he starts growing in words and deeds and is able to acquire new magical tools, such as the mead of wisdom[15] and various magical spells.[16] The myth clearly aims to show how Odin became the god of runes and secrets in general, and it is just as clearly played out along the vertical axis, emphasising the opposition up vs. down: Odin is hanging high up in the tree, and he picks up the runes from below.

The fact that several beings can be seen as providers of numinous knowledge indicates that we are facing a mental operation, classifying several basic categories into a vast semantic system in which the relations between these categories are made homologous. Thus the female, the dead, and the underworld are classified together and in opposition to the male, the living, and the upperworld; other oppositional pairs could be mentioned as well (see Schjødt 2008). It is thus possible to juggle with these terms in order to create and sustain a particular world view. In all cultures such operations take place primarily through myths, an important function of which is to establish cultural relations.

Let us return for a moment to the myths which represent the wilderness as a category of potential. For what we see in these myths of acquisition of numinous knowledge, or numinous powers, is a certain structure, using certain symbolic elements. The structure of the sequence is quite simple, namely that the subject representing 'our' world – often, but far from always, Odin – is brought into contact with the other world. Again, in these myths the other world is most often, but not always, represented as the underworld. The result of this contact is, in all cases, that the protagonist acquires some numinous powers, whether they be magical skills, supernatural tools or something else, which he (the subject is always male) can bring back to the world in which he lives his normal life (Schjødt 2008,398–410).

In this 'underworld' locality on the vertical axis, it is sometimes difficult to see the character of 'wilderness', probably because the ideas of the underworld, as stated earlier, were very imprecise. In the myths in which the protagonists are gods, it causes no problems to create an underworld scenario as an opposition to the upperworld along the vertical axis. However, when it comes to the human sphere, and thus the ritual sphere, it is

15 'Odrerir' at this place is the name of the mead itself and means 'that which moves the spirit'.

16 The son of Bölthorn is unknown, but we know that Bölthorn is a giant and besides that also the father of Odin's mother Bestla (Faulkes 2005, 11). In this way, we may deduce that he is Odin's maternal uncle.

much more difficult to establish an 'underworld'. This is probably why, in such ritual scenarios, we most often see only vague symbolic expressions of what is among the gods a real underworld. Nevertheless, in the human sphere we do meet 'the other world', characterised both as 'potential' and 'danger'. In the following, I will give a couple of examples of narratives (whether they are 'myths' in the traditional sense is not crucial here). They are both chosen from the so-called 'fornaldar sögur'; sagas that were written down during the 13th and 14th centuries, but which at that time had been transmitted orally for centuries.[17] The first example is taken from the saga of Hrólfr Kraki, the name of a Danish king who may have existed as a historical person, but whose existence is only known from legendary sources, definitely without much basis in historical facts. The episode mentioned here, which illustrates the double nature of the wilderness, is about one of the great heroes of the saga, Bödvarr Bjarki, and how he manages to transform a young boy, Höttr, into a great warrior with a new name, Hjalti. A brief summary of the episode goes like this:

When Bödvarr, a brilliant warrior, for the first time arrives at the king's hall, he notices that all the great warriors of the king's retinue enjoy themselves during the banquettes by throwing bones after a small boy, Höttr, who constantly fears that he might be killed and is scared to death. Bödvarr does not care for such behaviour, and he pulls Höttr out of the pile of bones in which he is sitting. Höttr is screaming, but Bödvarr tells him to be quiet, takes him outside, washes him and then places him next to himself on the bench. In the evening, the retainers come into the hall and start throwing bones towards the two, but when Bödvarr kills one of the retainers, the king accepts him (and reluctantly Höttr) as his man. Some time passes and as yuletime[18] approaches, Bödvarr notices that gloom settles over the men. Bödvarr asks Höttr what causes their dejection, and he is told that a huge monster (called a troll –

17 This is not to say that the sagas, as we have them in the written form, had been transmitted in exactly this form. The 'authors' of the 13th and 14th centuries may very well have played an important role in the written products – the final result, so to speak. But I suppose that all scholars of Old Norse studies would agree that an oral transmission of smaller or larger parts of the sagas had taken place for centuries (see Mitchell 1991), reaching back into pagan times.

18 The word 'yule' is of unknown etymology, but it exists in modern Scandinavian languages (Danish: 'Jul') and designates Christmas. So the period mentioned in the saga is definitely the time around winter solstice.

one of the names for giants)[19] has come there for the past two winters, causing much damage to men and cattle. At Yule eve, the king tells everybody to stay indoors because he does not want to lose any of his men. But Bödvarr steals away in the night. He brings with him the terribly scared Höttr, who screams loudly, but Bödvarr tells him to be quiet and throws him down on the moor. Bödvarr succeeds in killing the beast, returns to the place where he left Höttr and carries him to the dead monster. Höttr is now forced to drink some of the beast's blood and eat some of its heart. The two of them start fighting, and after a while Bödvarr tells Höttr that he has now become strong and no longer needs to fear the king's retainers. Höttr answers proudly that from now on he will fear neither them nor Bödvarr. The next passage tells how the two raise the dead beast to make it look as if it is alive. The next morning, they go out together with the king and his men, and Höttr attacks the dead monster and kills it, for the second time, so to speak. The king is a bit suspicious but accepts that Höttr has changed completely, and gives him a sword and his new name, Hjalti. From then on, Bödvarr and Hjalti remain the two greatest warriors in the king's retinue until they die a heroic death while defending the king.

There is no doubt that the wilderness is here verbalized as the space outside the castle, from where the monster will appear, and it is thus constituted along the horizontal axis. It is designated as a dangerous place with threatening beasts which kill cattle and men. So far it reminds us of the home of the giants, but it turns out to be more than that. For this is where the protagonists of the saga go in order to acquire the powers of another world. Höttr is a boy who is absolutely without any kind of power until he drinks the blood and eats of the heart of the beast, and it seems that the whole sequence is a kind of initiation (Schjødt 2008, 312–326), the purpose of which would be to enable him to act, in the rest of the saga, as one of the greatest warriors in one of the most famous warrior bands in Scandinavia. What is important in connection with the subject treated here, however, is that, even if the wilderness scenario seems to be played out exclusively on the horizontal axis, one small detail indicates that the idea of the underworld is not completely lacking. This detail is that Höttr is thrown down on the moor, or perhaps a more correct translation would be 'the bog'. Such watery locations were, for centuries prior to the Viking Age, sacred places for the pagans. They are best

19 In another version of the incident (written by the Danish clerk Saxo Grammaticus), the beast is said to be a bear (Friis-Jensen 2005, 168).

known from the so-called bog corpses, which can be found all over the world of the Germanic speaking peoples (Turner and Scaife 1995). Although it is always difficult to interpret the notions lying behind archaeological findings, most archaeologists would certainly agree that these bogs were perceived as representing some kind of otherworld. In connection with the dead in particular, as we just saw, it appears very likely, in an imprecise sense, that the bog was seen as a sort of entrance to the underworld. All the more so since the idea that water came in wells or streams, which symbolised 'communication lines' between the upper world and the underworld, can be found in other myths among the Scandinavians.[20]

Thus, the area outside the castle contains the two aspects of the wilderness introduced at the beginning: the threat and the potentiality. It is dangerous to go there, but it may prove to be worth the risks. I believe that this 'wilderness' scenario is what we meet all over the world in initiation rituals. Here the novices often have to spend some time in the wilderness, bringing numinous qualities back home to the society, which are of great value to them for the rest of their lives. This double aspect of the wilderness is therefore, particularly in initiations, of fundamental importance.

Another example, which emphasises the same double aspect and which also deals with the phenomenon of initiation, can be found in another fornaldar saga, *Völsunga saga* or the Saga of the Völsungs. This example, however, introduces new elements. In this saga the spatial opposition is not a castle vs. a bog, but a castle vs. a wild forest. Again we have two persons in focus, a father and a son. Sigmund, a son of King Völsung, has been betrayed by his sister Signy's husband, a king named Siggeir. At his wedding party with Signy, Siggeir attacks his father in law and brothers in law treacherously, and only Sigmund remains alive. He escapes into the woods outside the castle, where he is supported by his sister, who is furious about what has happened to her family and strongly desires revenge. Therefore she sends her sons by Siggeir, one by one, when they reach the age of ten, to Sigmund in the woods to assist him in his quest to avenge Siggeir. However, they prove not to be strong enough when their courage is put to the test. Sigmund demands them

20 The most well-known examples are the springs of the mythological beings Mimir and Urd, both characterised as very wise, and both being situated in a well or a spring (the word *brunnr* in Old Norse can signify both) (e.g. Faulkes 2005, 17). The symbolism here seems to be straightforward, namely that wisdom flows from the underworld to the upperworld, where it can be manifested or 'used'.

to bake a loaf of bread, giving them a sack of flour, but the boys do not dare to bake the bread because something is moving in the sack. Because of their lack of courage in this situation, Signy urges Sigmund to kill them, which he does. After some time, Signy takes the shape of a sorceress and goes to see her brother in his 'earth house', where they sleep together for three nights. She returns to the castle, having become pregnant, and she later gives birth to a boy named Sinfjötli. When he reaches the age of ten, he is also sent out in the forest to Sigmund, who does not know that it is his own son. Sinfjötli has no problems passing the test with the sack of flour (in which there was a very poisonous snake), and he proves in every respect to be a true descendent of Völsung. The following incidents should be quoted here in full:

> It is now to be told that Sigmund thought Sinfjotli too young to seek vengeance with him, and that he first wanted to accustom the boy to hardship. During the summers they traveled widely through the forests, killing men for booty. It seemed to Sigmund that Sinfjotli took much after the Volsung race. Nevertheless, he believed the boy to be the son of King Siggeir and to have the evil disposition of his father along with the fierce zeal of the Volsungs. Sigmund felt that Sinfjotli did not put much store in kinship, for the boy often reminded Sigmund of his grievances, strongly urging the man to kill King Siggeir.
>
> One time, they went again to the forest to get themselves some riches, and they found a house. Inside it were two sleeping men, with thick gold rings. A spell had been cast upon them: wolfskins hung over them in the house and only every tenth day could they shed the skins. They were the sons of kings. Sigmund and Sinfjotli put the skins on and could not get them off. And the weird power was there as before: they howled like wolves, but understanding the sounds. Now they set out into the forest, each going his own way. They agreed then that they would risk a fight with as many as seven men, but not with more, and that the one being attacked by more would howl with his wolf's voice. "Do not break this agreement," said Sigmund, "because you are young and daring, and men will want to hunt you."
>
> Now each went his own way. And when they had parted, Sigmund found seven men and howled in his wolf's voice. Sinfjotli heard him, came at once, and killed them all. They parted again. Before Sinfjotli had traveled very far in the forest, he met with eleven men and fought them. In the end he killed them all. Badly wounded, Sinfjotli went under an oak tree to rest. Then Sigmund came and said: "Why didn't you call?" Singjotli replied: "I did not want to call you for help. You accepted help to kill seven men, I am a child in age next to you, but I did not ask help in killing eleven men." Sigmund leapt at him so fiercely that Sinfjotli staggered and fell. Sigmund bit him in the windpipe. That day they were not able to come out of the wolfskins. Sigmund laid Sinfjotli over his shoulder, carried

> him home to the hut, and sat over him. He cursed the wolfskins, bidding the trolls to take them.
>
> One day Sigmund saw two weasels. One bit the other in the windpipe and then ran into the woods, returning with a leaf and laying it on the wound. The other weasel sprang up healed. Sigmund went out and saw a raven flying with a leaf. The raven brought the leaf to Sigmund, who drew it over Sinfjotli's wound. At once Sinfjotli sprang up healed, as if he had never been injured.
>
> Then they went to the underground dwelling and stayed there until they were to take off the wolfskins. They took the skins and burned them in the fire, hoping that these objects would cause no further harm. Under that magic spell they had performed many feats in King Siggeir's kingdom.
>
> When Sinfjotli was fully grown Sigmund thought he had tested him fully. (Translations by Byock 1990,43–45).

After this they try to avenge themselves, but are captured by Siggeir, who locks them up in a mound, from where they, nevertheless, manage to escape with the help of Signy. Then they go to Siggeir's hall and set it alight. Signy now tells Sigmund how Sinfjötli's conception came about. But she wants to die and so enters the burning hall. Sigmund and Sinfjötli return to the country of their kin, and Sigmund becomes a famous king, and Sinfjötli a great warrior.[21]

The sequence has for almost a century been recognised as an initiation ritual transformed into narrative (e.g. Weiser 1927 and Höfler 1934). In relation to the phenomenon of wilderness, it is clear that the forest is strongly opposed to both the castle of Siggeir and that of their own kin. This is truly a wilderness, emphasised by the fact that both father and son are transformed into wild wolves, Sinfjötli being the 'wilder' one. The danger of the wilderness is, of course, emphasised by the risks taken by both protagonists, and particularly for Sinfjötli, this risk leads to his death: he is killed by his own father, but is resurrected by the leaf which is brought by a raven, no doubt indicating the relation he has to Odin, who is strongly connected to ravens in several ways. So, wilderness, in this story,[22] is strongly related to the 'uncivilised' and to 'nature' as opposed to civilisation and culture, but what makes it particularly 'wild' is

21 In relation to the initiation theme, it is important to notice that, when Sinfjötli dies, he is taken back by Odin himself, who is said to be his great-great-great grandfather.

22 I wish to stress that wilderness, like many other phenomena which, in certain situations, are attributed a symbolic value, can function in various ways: what is decisive is its function within a certain structural framework, and it thus varies from one narrative or ritual to the next.

the fact that it cannot be controlled: when our two protagonists have put on the wolf skins, they cannot get them off. They have to fight men and, of course, run the risk of being killed, and they even get into a fight between themselves, resulting in the death of the 'initiate'. On the other hand, it is also obvious that the wilderness, like in the story of Höttr, is a resource and a potential, for what especially Sinfjötli learns is that he has to be brave, but not foolhardy, and both are qualities without which he cannot fulfil his function as a warrior. It is also worth noticing that they gain an insight into the nature of the wild beasts, for they learn the language of wolves. Finally, we notice that again we have a certain combination of a scenario, played out along both the vertical and the horizontal axes. The forest is outside the castle, but the protagonists also live in an underground dwelling, probably a reflection of the fact that rituals have to use a more 'realistic' background scenario than is the case with myth proper (in this connection understood as narratives that take place exclusively in the 'other world').

The two examples analysed briefly here show that, at least in initiation scenarios, wilderness is used exactly as something threatening as well as a potential. The two qualities which, in the cosmological framework, are divided – although not consistently – into two different realms, the periphery and the underworld, blend here in order to emphasise the necessity of the combination of both when it comes to the preparations for becoming a warrior: on the one hand, you must be able to withstand threatening dangers and, on the other hand, you have to acquire skills in order to survive.

5. Conclusion

A couple of examples of how the wilderness is verbalized both as a threat as well as a potential obviously cannot be taken as a universal feature, but it is certainly characteristic of the way 'wilderness' was perceived in Scandinavia. As mentioned in the introduction, wilderness is also related to reality, to the world 'as it is', and should thus be seen as an emic construction. Although it is 'real' (in the sense that in opposition to the known, the so-called civilised world, it is uncontrollable, foreign and unknown), it is also a construction in the sense that its content will vary from one culture to the next: what is 'wild' in one culture may well be part of civilisation in the other. It is difficult to imagine a culture, however, until a couple of centuries ago perhaps,[23] which would have no notion of wilderness, understood in this double sense. In pre-Christian Scandinavian mythology, it is characteristic that the wilderness is part of a cosmological structure which is orientated along both a horizontal and a vertical axis and that the 'threat aspect' is mostly organised along the horizontal axis, whereas the 'potential aspect' is organised along the vertical axis. In the 'human world', on the other hand, the potential aspect may also be acted out along the horizontal axis. This certainly indicates that wilderness is part of a cosmological structure. It is no doubt based on experiences, but to cope with such experiences it is necessary to place them within a structure than can be manipulated by myth as well as ritual. This cosmological structure, however, involves much more than just those elements that we usually connect with the term 'cosmos', understood as the spatial framework within which our everyday life is played out. As we have seen, the dualism that is often thought of in the relation between 'us' and the wilderness is both more complex and ambivalent than is usually thought, and it is at the same time part of a larger semantic universe which has to do with much more than just the spatial notions of wilderness and cosmos.

6. References

Bell, Catherine. *Ritual. Perspectives and Dimensions.* New York, Oxford: Oxford University Press 1997.

Boyer, Pascal. *Religion Explained. The Evolutionary Origins of Religious Thought.* New York: Basic Books 2001.

Branston, Brian. *Gods of the North.* London: Thames and Hudson 1990 [1955].

Byock, Jesse (trans.). *The Saga of the Volsungs. The Norse Epic of Sigurd the Dragon Slayer.* Berkeley, Los Angeles, Oxford: University of California Press 1990.

Doty, William G. *Mythography. The Study of Myths and Rituals.* Tuscaloosa and London: The University of Alabama Press 1986.

Dumézil, Georges. *Les dieux des Germains.* Paris: PUF 1959.

23 And even in our modern world, it is obvious that, although our 'wilderness' is not out there, outside the town, behind the next mountain or across the ocean, there is a basic idea, which is quite widespread, that, somewhere in space, behind the stars, or perhaps even at the bottom of the sea or somewhere else, there are dangerous creatures and great treasures. What has changed is probably just that here on earth, even if there are many dangers, they are not exclusively related to a certain space outside the civilized world. The greatest dangers are probably nowadays primarily attributed to the civilized world itself. It is trite to maintain that scientific and other kinds of knowledge have demystified most of the world.

Eliade, Mircea. *Das Heilige und das Profane. Vom Wesen des Religiösen.* Hamburg: Rowohlt 1957.

Faulkes, Anthony (ed.). *Snorri Sturluson: Edda. Skálskaparmál I.* London: Viking Society for Northern Research 1998.

Faulkes, Anthony (ed.). *Snorri Sturluson: Edda. Prologue and Gylfaginning.*London: Viking Society for Northern Research 2005.

Friis-Jensen, Karsten (ed. and Peter Zeeberg [trans.]). *Saxo Grammaticus: Gesta Danorum; Danmarkshistorien, I.* Copenhagen: Det danske sprog- og litteraturselskab & Gads Forlag.

Geertz, Clifford. "Religion as a Cultural System", *Anthropological Approaches to the Study of Religion* (M. Banton, ed.). London: Tavistock Publications 1966, 1–46.

Gunnell, Terry. "How Elvish Were the Álfar?", *Constructing Nations, Reconstructing Myth. Essays in Honour of T. A. Shippey* (A. Wawn *et al.*, eds.). Turnhout: Brepols 2007, 111–130.

Höfler, Otto. *Kultische Geheimbünde der Germanen.* Frankfurt a. M.: M. Diesterweg 1934.

Jónsson, Finnur (ed.). *Snorri Sturluson: Heimskringla. Nóregs konunga sögur.* Copenhagen: G.E.C. Gads Forlag 1911.

Jakobsson, A. "Identifying the Ogre: The Legendary", *Fornaldarsagaerne. Myter og virkelighed,* (A. Ney *et al*, eds.). Copenhagen: Museum Tusculanums Forlag 2009, 181–200.

Larrington, Carolyne (trans.). *The Poetic Edda.* Oxford: Oxford University Press 1996.

Lévi-Strauss, Carolyne. *La pensée sauvage.* Paris: Librairie Plon1962.

Lindow, John. "Supernatural Others and Ethnic Others. A Milenium of Worlds View", *Scandinavian Studies* 67 (1995), 8–31.

Mitchell, Stephen. *Heroic Sagas and Balads.* Ithaca, London: Cornell University Press 1991.

Mundal, Else. "The Perception of the Saamis and their Religion in Old Norse Sources", *Shamanism and Northern Ecology* (J. Pentikäinen, ed.). Berlin: de Gruyter 1996, 97–116.

Mundal, Else. "Kong Harald Hårfagre og samejenta Snøfrid. Samefolket sin plass i den norske rikssamlingsmyten", *Nordica Bergensiana* 14 (1997), 39–53.

Neckel, Gustav and H. Kuhn (eds.). *Edda. Die Lieder des Codex Regius nebst verwandten Denkmälern.* Heidelberg: Carl Winter Universitätsverlag 1962.

Otto, Rudolf. *Das Heilige. Über das Irrationale in der Idee des Göttlichen und sein Verhältnis zum Rationalen.* München: Verlag C.– H. Beck 1917.

Podemann Sørensen, Jørgen. "Ritualistics: a New Discipline in the History of Religions", *The Problem of Ritual* (T. Ahlbäck ed.). Åbo: The Donner Institute for Research in Religious and Cultural History 1993, 9–25.

Schjødt, Jens Peter. "The "Fire-Ordeal" in the *Grímnismál* – Initiation or Annihilation?", *Medieval Scandinavia* 12 (1988), 29–43.

Schjødt, Jens Peter. "Horizontale und vertikale Achsen in der vorchristlichen skandinavischen Kosmologie", *Old Norse and Finnish Religions and Cultic Place-Names* (T. Ahlbäck ed.). Åbo: The Donner Institute for Research in Religious and Cultural History 1990, 35–57.

Schjødt, Jens Peter. "Kosmologimodeller og mytekredse", *Ordning mot kaos. Studier av nordisk förkristen kosmologi* (A. Andrén e, a. eds.). Lund: Nordic Academic Press 2004. 123–134.

Schjødt, Jens Peter. *Initiation between two Worlds. Structure and Symbolism in pre-Christian Scandinavian Religion.* Odense: The University Press of Southern Denmark 2008.

Schjødt, Jens Peter. "Diversity and its Consequences for the Study of Old Norse Religion. What Is it We Are Trying to Reconstruct", *Between Paganism and Christianity in the North* (L. P. Slupecki and J. Morawiec, eds.). Rzeszow: Wydawnictwo Uniwersytetu Rzeszowskiego 2009, 9–22.

Schjødt, Jens Peter. "Aspekter ved kristningen af Norden", *Middelalderens verden. Verdensbilledet, tænkningen, rummet og religionen* (O. Høiris and P. Ingesman eds.). Aarhus: Aarhus Universitetsforlag 2010, 376–385.

Schulz, Katja. *Riesen. Von Wissenhütern und Wildnisbewohnern in Edda und Saga.* Heidelberg: Universitätsverlag Winter 2004.

Turner, Rick C. and R. G. Scaife 1995. *Bog Bodies. New Discoveries and New Perspectives.* London: British Museum Press.

Turner, Victor W. *The Ritual Process. Structure and Anti-Structure.* Ithaca, New York 1969.

Weiser, Lily. *Altgermanische Jünglingsweihen und Männerbünde. Ein Beitrag zur deutchen und nordischen Altertums- und Volkskunde.* Baden: Konkordia 1927.

9. Making a Garden out of the Wilderness: landscape, dwelling and personhood in the encounter between European settlers and the Mi'kmaq in "New France"

Anne-Christine Hornborg

1. Introduction

The encounter with North America was, by the European explorers in the fifteenth and sixteenth century, depicted as a discovery of a New World. They brought a basic storyline of how this meeting should proceed with them from Europe, as we can see from the names given to the provinces in Eastern Canada: New France for the French or Nova Scotia (New Scotland) for the English colonists. It was not only a matter of discovering a new continent; it also implied a transformation of the new land in the meeting. The classification of "the other"[1] found in the older texts about the inhabitants of the new continent written by explorers, travellers and missionaries, reflects what they experienced as their mission from a Western and Christian standpoint. Statements of the "savages'" nature and the image of the "Indian" is clearly drawn from the perspective of the colonizer and must be evaluated from the perspective of which interests were hidden in these statements: With a supposed right and a duty to dominate other peoples, the colonizers nourished an image of the Indian as a savage, or as a child who needed to be controlled or taken care of (Gill 1982, 7). Thus the boundary drawn, by the newcomers, between culture and nature, and their use of the idea of wilderness, did not only function as a means of description, it also comprised a power relation which had devastating consequences for the inhabitants of "New France". Wilderness was understood as the condition offered by nature

1 "The Other" is, in this context, understood along Saidian lines (Said 1978). Said investigated how the "Orient" was constructed from the Occidental perspective in order to legitimate political intervention and domination. The imaginary picture which is drawn attributes to the other that which the Westerner sees as undesirable, which thus makes his own, congratulatory self-image appear much clearer.

itself, if a human hand had not cultivated the land. The "Indian" was classified as a part of nature, as a savage – a child of the wilderness. Christianity and civilization were seen as the means of their liberation. The present analysis investigates the use of the idea of wilderness by the European newcomers to justify domination, while also bringing out the voices of the Mi'kmaq people appearing in the material. The aim of the paper is to trace, through the texts of the explorers and missionaries, the contours of the encounter between two differing perspectives on landscape, dwelling and personhood.

2. The Wilderness Idea and the Agricultural Ideal

In the older texts and the writers' early explanation of the new continent and its inhabitants we find one, key metaphor (Ortner 1973), and that is wilderness. The newcomers used their concept of wilderness as a way of explaining the differences they saw in the landscape, culture and in the personhood of the indigenous peoples. Thereby, they justified the creation of a new order. According to the Christian myth of Genesis 2–3 in its contemporary 16th century interpretation, God had planned Eden for humans, but the Fall changed the conditions. Adam and Eve were driven out from the well-organized Garden with its fruits of plenty and had to fight "thorns and thistles" in order to eat the plants of the fields. The key to humankind's survival was, thereafter, to till the land. "The father of modern science", Francis Bacon (1561–1626), who was a prolific writer and philosopher in Europe during the early colonisation of America, advocated that man's mission was to recover the master kingdom over nature, which was lost in the Fall (Farrington 1949, Merchant [1980] 1994, 191–192, Merchant 1989,127). To the minds of the newcomers, the New World they encountered was in need of cultivation, since there was no single trace in the landscape of a human, caring hand. For the newcomers, the local Mi'kmaq nomadic hunters resembled animals roaming in the forests more than humans settled in organized communities. Christianity and civilization were connected to the agricultural landscape, and so the wilderness became the landscape of heathens. The agricultural landscape was seen as similar to the order in Eden, created by God, who had assigned man to be given dominion over the earth – or in a milder form – being the steward or loyal viceroy of the earth (Callicott 1994,14). Wilderness was, in the early texts of the explorers and missionaries, thus also given a moral connotation: dwelling in the wilderness, the nomadic way of life of the Mi'kmaq Indians, was seen as morally defunct.

In the early explorers' and missionaries' texts we can read that the newcomers were inspired by the biblical story of the Israelites, who after the exodus from Egypt had to suffer in the desert wilderness before they reached the goal: to till the land of Canaan. The transition to agriculture was, in the European texts, seen as an effort to return everything to the divinely sanctioned order in the cultivated landscape. Yet, the texts also show how the Mi'kmaq resisted the European "cultivation". Certainly, their narratives depict not wilderness but a structured and ordered world. Since they lived in an oral culture, there are no Mi'kmaq texts from the time of the early encounter, so we have to look at the meetings through the lens of the explorers. In order to discuss this meeting between cultures, I will mainly use texts by missionaries, explorers and travellers that are compiled in *The Jesuit Relations and Allied Documents*, a rich collection of documents in 71 volumes with invaluable notes from 1620 to 1791. The main texts I use are from the Jesuits Joseph Jouvency and Pierre Biard, the Recollect missionary Chrestien Le Clercq, the lawyer Marc Lescarbot and the tradesman Nicolas Denys, whom all of them stayed in New France, the Eastern Province of Canada, in the seventeenth century. To look into the Mi'kmaq hunters' world, I also have to use later compilations, well aware that time passed between the first meetings and when Mi'kmaq oral culture was documented. For this purpose I will use Silas Rand's compilation of Mi'kmaq oral culture, *Legends of the Micmacs* (1971 [1894]), since he was the first who, in a more systematic way, wrote down the Mi'kmaq stories. Rand was a Protestant missionary whose mission was to convert the Mi'kmaq from Catholicism to Protestantism and in order to succeed he collected their stories to better understand the Mi'kmaq life world. Although the Mi'kmaq were forced into the reserves during the years Rand started his compilation (beginning in 1850) and they had lived with the missionaries for about two hundred years, it is through his work we still can read and interpret the earlier life of the Mi'kmaq as hunters, and learn of their cosmology. These stories provide the necessary background information to better understand the encounter between ways of dwelling depicted in the early texts.

I will show, by using examples from these texts, how the concept of wilderness was employed as a key metaphor for establishing the moral right and duty of Europeans to colonize the land, settle the nomadic hunters in Christian communities, and to "civilize" or tame "the savages", and I will try to trace the contours of this encounter between very different perspectives on landscape, dwelling, and personhood. I start with the newcomers' interpretation of the landscape as an uncultivated and inhospitable wilderness. Then, I offer an image of how the Mi'kmaq structured the landscape in a different way in their narratives of the mythical being Kluskap. Afterwards, I discuss the differences between the settled and the nomadic ways of dwelling in the environment, how the newcomers had difficulties in accepting and understanding the nomadic way of life as a legitimate way of dwelling and constructing culture, as well as the glimpses we get, from the newcomers' reports, of how the Mi'kmaq responded to the newcomers' view. Finally, I will discuss how the newcomers and the Mi'kmaq had different ways of understanding what it is to be a person. In the Mi'kmaq narratives, humans can marry animals – and animals are addressed as "persons", an alien view for the newcomers who laboured to maintain a clear border between the cultivated man, an image of God with a soul, and the wild animals, lacking souls and culture.

3. Landscape

According to Tim Ingold (2008,168), people create meaning in the world through engagement and practical activity: "meaning is immanent in the relational contexts of people's practical engagement with their lived-in environment"(Ingold 2000,168). When the Europeans encountered the New World, their prerequisites were different from the people who already inhabited and dwelled the land. The meaning of the landscape was not generated from a day-to-day activity as hunters. They entered the land with models and narratives in order to navigate in an uncertain world, and these models were drawn from their interpretation of the Christian Bible. As Moses had led the Israelites out of Egypt into the wilderness with an aim to return and settle in Canaan, so had they, as Europeans, left their land to be tested in the wilderness, but they were eventually to build a New World. In the Bible interpretations of the day, the Lord showed mercy for the chosen people in the wilderness, feeding them manna from heaven until they reached the Land of Canaan, which they had an obligation to cultivate into a land filled with milk and honey (Chamberlin and Feldman 1950, Merchant 1989: 100). The parallel was clear for the newcomers; the Wilderness was an area of transit on their way towards later rewards. But the wilderness was also a place of temptation, since this place appealed to the chaotic inner nature of man. If the settlers endured and defeated the chaotic landscape, they could, like the Israelites, build a world similar to that of Canaan (Merchant 1989,101, Jaenen 1976,191). With this perspective, wilderness was often displayed as a place which had to be endured, as the opposite of the Promised Land, which was depicted as an ordered place to dwell in and fulfil thc mission of man on Earth.

The journeys of exploration to the New World in the fifteenth and sixteenth centuries did create a genre of wilderness stories in the European tradition (Merchant [1980] 1994,153). The "savages", dwelling in the wilderness, were seen as warnings, as bad examples of the chaotic and threatening nature or of wilderness within man, lest they could be settled and tamed by Christianity and reason. The texts by the missionaries or travellers from the seventeenth century reflect this literary genre, and the new province, New France, is often depicted as God's punishment, meant to be reorganized and turned into fields and gardens by the settlers (Merchant 1989,100ff, Jaenen 1976, 191 ff.). The Jesuit Biard describes how a terrifying wilderness meets the newcomers in New France and starts his Relation 1616 by stating that "Satan's malevolence" must reign in this horrible wilderness:

> *A Garden of delight lies before him, behind him is a solitary wilderness.*[2] For verily all this region, though capable of the same prosperity as ours, nevertheless through Satan's malevolence, which reigns there, is a horrible wilderness, scarcely less miserable on account of the scarcity of bodily comforts than for that which renders man absolutely miserable...Where can the glory of a Christian more successfully ennoble him, than there where it brings both bodily and spiritual happiness to his brethren; and where, as one of God's great instruments, he would make a Garden out of the wilderness: where he would subjugate satanic Monsters, and would introduce the order and discipline of heaven upon earth (Biard [1616] 1959, 34–35).

The Christian mission was thus seen not only as spreading the Christian gospel, but also as forming a Garden out of the wilderness in the new country. The cultivated landscape was a symbol of God's order ordained in creation and by tilling new land and sowing it with European plants the colonizers could restore God's order. To banish the wilderness, Biard found, would also be to extinguish "Satan's malevolence" and "subjugate satanic Monsters" (Biard [1616] 1959, 35, Vecsey 1980,33). As the text tells us, humans were seen as "one of God's great instruments", su-

2 Biard says he is here quoting of one of the earlier Prophets, it is probably Joel 2:3.

perior to nature. Progress would come by forming "a Garden out of the wilderness".

For the Mi'kmaq, on the other hand, what the Europeans experienced as wild nature was not dichotomized as a counterpart to culture, but as a dwelling place they established a humble relationship to (Jaenen 1976,191, cf. Ingold 2000). In Biard's and other missionaries' notes we find indications that for the inhabitants of this "wilderness", the place was far from imagined as chaotic or wild. Firstly, it seems that the Mi'kmaq had organized the land into well-defined units. Biard writes that it was the responsibility of chiefs to divide the province into different territories, each with a local leader. Often the divisions coincided with natural boundaries such as a bay or a river. For example, there was one chief for Pentegoet River and another for St. Croix (Biard [1616] 1959, 89). The division of land was one of the important tasks for the summer councils. Both the missionary Le Clercq and the tradesman Denys confirms, in the seventeenth century, a structuring of the landscape by the Mi'kmaq. Le Clercq is surprised at their skill in drawing maps. With meticulous precision and an intimate knowledge of places they drew on birch bark the exact outlines of different areas (Le Clercq [1691] 1968,136, Denys [1672] 1908, 4). Denys described how familiar the Mi'kmaq were with the landscape and how this familiarity was shown when the prey was left in the forests for the women later to fetch to the camp:

> The man will tell only the distance of the road, the woods that must be passed, the mountains, rivers, brooks, and meadows, if there are any on the route, and will specify the spot where the animal will be...This is enough to enable them to find it, to such a degree that they never fail, and they bring it back (Denys [1672] 1908, 404–405).

For the explorers, the nomadic life was a life of animals, roaming the woods, but for the Mi'kmaq it did of course not mean that they simply moved around haphazardly in the forests or along the shorelines. They had names for the places they returned to, and many of these names were lost during colonization and thereby ceased to be "places". Some Mi'kmaq names that were included in English and French maps were also given to "wrong" places that had no connection with the place that they were originally attached to (Hewitt 1908, 3).

Secondly, in Rand's collection of Mi'kmaq stories from the nineteenth century, we find a rich collection of names attributed to places which the Europeans saw as wilderness. A prominent figure in the Mi'kmaq narratives is the mythical being Kluskap and it is through his arrangement and movements that the Mi'kmaq landscape is created and explained. Thus the European interpretation of "Wilderness" was, in a Mi'kmaq perspective, a structured place and reflected a highly symbolic universe. The water area Minas Basin is likened to a huge beaver pond with its opening at Cape Split. The Mi'kmaq named Cape Split *Pleegum*, which is the opening made in the beaver dam to let water pass (Rand [1894] 1971, xliv; cf. his different spelling 1919,69 *pleegun*). The Mi'kmaq narratives thus explained why the valley of Annapolis and Cornwallis, previously inundated according to their stories, had been drained by the opening at Cape Split and Annapolis Gut. Aylesford swamp is described as a big lake with a huge beaver lodge, which explains the Mi'kmaq name *Cobeetek*, the home of the beaver. Kluskap chases a little beaver from this beaver lodge all the way to Bras d'Or Lake in Cape Breton. There it tries to hide in another lodge, but is killed by Kluskap, who turns the lodge upside down and transforms it into an island. When he opens another beaver lodge at Cape Chignecto, a piece of land floats away. From this piece he creates a moose, which he lets his dogs chase. As a treat they get food from the dead moose (probably the entrails). Large pieces of this dog food are transformed into stones, "and remain there to this day; the place is called Oopunk" (Rand [1894] 1971: 293). Oopunk is Mi'kmaq for "the lights (lungs)" (Rand 1919, 63). The hunt ends when Kluskap transforms the moose into an island, Isle of Holt (Rand [1894] 1971, 236). Parts of this story are repeated further on, but then the island is named Isle of Hant (p. 293, cf. Parsons 1925, 38). It is not a coincidence that the beaver hunt is chosen as a metaphor in the narratives for transforming the landscape. Since the Mi'kmaq often hunted beavers it was a natural metaphor to employ in their narratives. The beaver was also the animal that most radically affected the environment in Nova Scotia. Breaking beaver dams was literally a way of transforming the landscape. When the giant beaver, in retaliation, builds a dam across from Blomidon to the Cumberland side, Kluskap's medicine garden is destroyed. The Minas Basin area has the highest tide difference in the world. The great difference between high and low tide in this place could be metaphorically spoken about and likened to the effects of a huge beaver dam being broken and rebuilt every day (Hornborg 2008, 86) These and similar narratives assisted in providing orientation in the landscape for the Mi'kmaq.

It is not only when hunting animals that Kluskap travels and transforms the landscape. In one of Rand's legends, *A wizard carries off Glooscap's housekeeper*, Grandmother is stolen away and Kluskap travels all around the province searching for her (Rand [1894] 1971, 284–293.With the help of some whales, he goes to Cape Breton (*Oonumage,* The spelling varies; *Onumage* in Rand 1919, p. 61) by the Strait of Canso and arrives at Cape North (*Uktutun*, "highest mountain", Rand 1919, 81), only to find that the one he is looking for has been brought to Newfoundland (*Uktukamkw*, "a large body of land", Rand 1919, 80). After a while he returns to Nova Scotia and is stranded with his whales in Pictou (*Piktook*). *Piktook* means "an air explosion" (Rand 1919, 68), possibly referring to the whale, blowing water in the air, when stranded. His travels then continue to Partridge Island, Cape Blomidon, and Spenser's (Spencer's) Island. On Spenser's Island Kluskap has good luck in hunting. He cuts the meat from the prey in pieces to be dried, a common Mi'kmaq practice. He then puts the left-over bones in a large kettle and boils them, in order to get the coveted marrow. When the food is ready and there is no more use for the kettle, he turns it upside down so that it becomes an island. The Mi'kmaq call this island *Ooteomul*, which means "his kettle", that is Kluskap's kettle. According to Rand another Mi'kmaq name for Spencer's Island is *Wochuk* (Rand 1875, 100). Kluskap then stays near Cape d'Or during the winter. To make it easier for people to travel between Partridge Island and Cumberland Bay he makes a passage to walk on, *Awokun* (Rand [1894] 1971, 292). Rand (1919, 17) says that the name for Parrsboro in Nova Scotia is *Awokun*, which means "a portage, a short cut, over the boar's back"; cf. Rand (p. 96) on *Owokun* – a crossing-over place.

The above examples show how places and natural forms were incorporated into Mi'kmaq mythical stories. What the newcomers saw as pure wilderness were familiar places to the Mi'kmaq, embedded in a storyworld, which undoubtedly functioned as a means of orientation and navigation (Ingold 2000, 219–242). Two classificatory systems collided. The Mi'kmaq view of the landscape certainly did not correspond to the colonizers' perspective. It was a way of talking about the landscape, where natural features and events were transformed into signs, reflecting the Mi'kmaq life world, and organized into a larger symbolic representation of the environment (Hornborg 2008, 86).The wilderness was thus not wild or chaotic for its inhabitants, but a structured world. In their narratives, places were woven together in a meaningful web that enabled them to dwell in the landscape and which provided orientation in it.

4. Dwelling

Tim Ingold defines the engagement in a landscape as "dwelling" and a "dwelling perspective" is a "perspective which situates the practitioner, right from the start, in the context of an active engagement with the constituents of his or her surroundings" (2000, 5). It is obvious, when we read the texts by the Jesuit Pierre Biard, that according to him culture and dwelling is defined in close relation to tilling the soil. He saw the great forests as pure wilderness, lacking civilization and not a place worthy for human dwellings:

> It now remains to tell you that the conversion of this country, to the Gospel, and of these people to civilization, is not a small undertaking nor free from great difficulties; for in the first place, if we consider the country, it is only a forest, without other conveniences of life than those which will be brought from France, and what in time may be obtained from the soil after it has been cultivated" (Biard [1611] 1959, 171, 173).

Biard clearly express his concept of culture when he says that the country "is only a forest". Culture was for him a structured area like a city or a village with houses and "other conveniences". It seems fair to say that for the new settlers, transcending the alienation and the discontinuity of this new place was of vital importance. One way of banishing alienation and maintaining the experience of continuity was to recreate familiar, European landscapes in the New World. By imposing the English landscape ideas on "the wilderness", it could be tamed and transformed into a cultivated and familiar dwelling (Reid 1995, 30). The Anglican Church in Fredericton was thus built as a copy of Portland Chapel in London (Reid 1995, 30). The colonial cartographers tried, as systematically as possible, to create a sense of being at home. They conjured up an imagined British community in the new colony by deliberately avoiding representations of the "strange" regions beyond the British settlement. The first map to show Halifax, drawn in 1749, was widely circulated in England and showed an English ship in the harbour, a fortress, a palisade, a church, army barracks, the governor's residence and a court house. Behind the city's palisade were depicted two fruit trees, a beetle, and a porcupine. The forests and the animals were excluded and the presence of an aboriginal group with hunting and fishing grounds in the area was not even hinted at (Reid 1995, 32). The British colonialists thus elided Mi'kmaq culture from the map. The Mi'kmaq were seen as part of the wilderness, part of alterity, and thus of no interest for the cartographer. If the Mi'kmaq themselves were mentioned, it was as children of the forest characterised by the features of wild nature: inhospitable and fierce (Reid 1995, 31–32). In building a New France or a New Scotland (Nova Scotia), the newcomers sealed themselves off from the Mi'kmaq and from North American nature. It became their strategy of feeling at home in their new country.

There are many examples in the Mi'kmaq landscape narratives of a "Mi'kmaq dwelling perspective" and engagement-in-the-world as described by Ingold. What the European cartographer saw as 'wilderness' was the home of the Mi'kmaq. We get a glimpse of the encounter between the two ways of life in the writings of the missionary Le Clercq, who relates an early example of a Mi'kmaq hunter's appreciation of their life as nomadic hunters. He compared the Mi'kmaqs' nomadic life with the life of tortoises, which also carried their houses on their backs (Le Clercq [1691] 1968, 100). He wrote that the French may think that life in the forest only offered simple and miserable conditions, but in fact the Mi'kmaq appreciated their nomadic dwellings more than French houses. To leave the forest for them was the same as abandoning their home. Many times the missionaries also complain over the difficulties they had in baptizing and settling "the heathens", since they refused to abandon their nomadic lives. This Mi'kmaq man had, at that time, grown tired of hearing colonizers telling him to abandon his own traditions. He expressed his criticism of the attempts to assimilate his people into the European world. He delivered the harshest critique against the settlement, expressing his preference for the nomadic lifestyle:

> 'I am greatly astonished that the French have so little cleverness, as they seem to exhibit in the matter of which thou hast just told me on their behalf, in the effort to persuade us to convert our poles, our bark, and our wigwams into houses of stone and of wood which are tall and lofty, according to their account, as these trees. Very well! But why now,' continued he, 'do men of five to six feet in height need houses which are sixty to eighty? For, in fact, as thou knowest very well thyself, Patriarch - do we not find in our own all the convenience and the advantages that you have with yours, such as reposing, drinking, sleeping, eating, and amusing ourselves with our friends when we wish? This is not all,' said he, addressing himself to one of our captains, 'my brother, hast thou as much ingenuity and cleverness as the Indians, who carry their houses and their wigwam with them so that they may lodge wherever they please, independently of any seignior whatsoever? Thou art not as bold nor as stout as we, because when thou goest on a voyage thou canst not carry upon thy shoulders thy buildings and thy edifices. Therefore it is necessary that thou preparest as many lodgings as thou makest of residence, or else thou lodgest in a hired house which does not belong to thee. As for us, we find ourselves secure from all these

> inconveniences, and we can always say, more truly than you, that we are at home everywhere, because we set up our wigwams with ease wheresoever we go, and without asking permission of anybody' (Le Clercq [1691] 1968, 103).

The Mi'kmaq speaker goes on by criticizing those French who despise the Mi'kmaq life. He thinks his people have been wrongly characterized. He strongly rejects the notion that they lack religion, traditions, honour, and an organized society, and that they live more like animals in the forest, without bread, wine, and other products of which the French said that they had abundance. But, he continues, the French cannot fully understand the feelings that a Mi'kmaq has for his country. He affirms that they are fully content with the life they live. If France was like a celestial paradise, why then would the French leave their country and risk their lives to visit a strange and barbaric country in the West? Everything that makes life worth living – wives, children, relatives and friends – they leave behind them. They do not even manage to cope in the new country but must either eat a monotonous diet like cod or ask the Mi'kmaq to hunt for them: "We see also that all your people live, as a rule, only upon cod – cod in the morning, cod at midday, cod at evening, and always cod" (Le Clercq [1691] 1968,105). If a Mi'kmaq appears to be short of anything, the reason is that he has adopted the French lifestyle. The speaker asserts that the one who lives the longest and best life among the Mi'kmaq is the one who holds on to their traditional customs and is fully content with eating beaver and moose meat, seabirds, and fish. He is a thousand times happier with his life in the forest and his wigwam than if he stayed in the most grandiose palace and ate with the mightiest princes in the world (Le Clercq [1691] 1968, 107). But the later history of the encounter teaches us that this Mi'kmaq man spoke to deaf ears. Overall, the newcomer's negative view of wilderness and the Mi'kmaq dwelling remained unchanged. The missionary reports tell of their clear ambition to get the Mi'kmaq to settle. The Mi'kmaq protests are often, as in this case above, mostly put in the texts as what the writer sees as "the Other's" protests of holding back the mission, and doing so, being hindrances for the "savages'" possibilities of progress and becoming "civilized". But they are, at the same time, valuable passages for catching a glimpse of the "little points of views" in a text, dominated by the colonizers point of view and perspective (Hornborg 2008, 57).

Central for the European farmer was the settlement, where generations before him had left traces in the form of other buildings, cleared fields, or household utensils. Settlement was thus defined as culture, the forest wilderness as nature. Animals roam the land, while humans are settled. From the Christian missionary and the settlers' view, living in the forest wilderness could never generate a moral duty to care for the land:

> ...they must be made sedentary by the cultivation of the land, thus obliging them to remain in one place. For anyone who has taken the trouble to cultivate a piece of land does not really abandon it, but struggles valiantly to keep it (Lescarbot [1610] 1959, 87).

Biard too sees the necessity for the Mi'kmaq hunters to be settled in order to be civilized and not "savages" and thereby get attached to the land:

> They are, I say, savage, haunting the woods, ignorant, lawless and rude: they are wanderers, with nothing to attach them to a place, neither home nor relationship, neither possessions nor love of country (Biard [1611] 1959, 173).

The texts show that the conflict between the two different ways of dwelling was clearly discernible also in the alimentary area. Biard experienced New France as a place where people had to struggle hard for their daily bread. Yet, researchers have documented how the Europeans' statements about starvation among hunters, and their struggle for daily food, tell us more about the colonizers than about the actual conditions of the hunters (Martin 1992, 48, cf. Sahlins 1972). A cultural myth concerning hunting cultures and starvation seems to have been in existence; one grounded in the radical difference between the hunters' consumption of food (what they ate and how much) and the European way of treating food. The Mi'kmaq ate intestines, bone marrow and whole, boiled fish, something the colonizers regarded as inedible or barbaric – or "wild food", but which, measured in nutritive value, is not unfit to eat. The cultural classification of food is very obvious when Lescarbot complains ([1609] 1928, 225) about the lack of bread among the Mi'kmaq, stating "that bread is a food very natural for man".

The Mi'kmaq also practiced a kind of ritual eating, the so called "eat-all feasts", when they consumed enormous amounts of food, and then fasted for some days, a practice alien to Westerners (Martin 1972, 80–81). Both Lescarbot and Le Clercq commented on this practice. Lescarbot writes ([1609] 1928, 226): "at home in their houses (as the Brazilians) they stretch out their bellies as much as they can, and do not leave eating as long as there is any meat; and if any of ours be at their *Tabagie* [feast, including ritual eating, my comment], they will bid him do as they do". He also draws parallel with ancient cultures, to stress the progress Westerners have undergone. Eating properly and modestly is equated with high moral qualities: "I see no gluttony like to that of Hercules, who alone did eat whole oxen" (ibid.). Le Clercq describes ([1691] 1968, 291–292) the "eat-all feasts", as feasts "in which it is necessary that everything be eaten before anybody goes out from the wigwam...It is, however, permissible for those who cannot finish their portions to present them to their companions, each of whom takes whatever he desires thereof". The latter thus ate moderately with the result that they had a hard time to keep pace with the Mi'kmaq when they did not have access to food for several days. The days of fasting, which they experienced as starvation, were not experienced as such by the Mi'kmaq. Le Clercq says about this way of eating: "It is not difficult for them to go three or four days without food, especially when they are hunting" (Le Clercq [1691] 1968, 116). Lescarbot confirms ([1609] 1928, 242) Le Clercq's observation: "And, although they sometimes exceed in their *Tabagies*, or feasts, they diet themselves afterwards well enough, living very often eight days more or less with the smoke of tobacco, not returning to hunting until they be hungry". Thus, to have regular meals everyday was not a necessity for the non-agricultural Mi'kmaq (se also Cronon [1983] 1997, 41–42).

Affluence in food is a cultural notion – and depends on what is classified as food and when and how to eat meals. The Mi'kmaq managed quite well in the "wilderness" by hunting and gathering their food instead of clearing land from "thorns and thistles" in order to cultivate the forest into fields. As hunters, they were living in a truly affluent society in the New World (Sahlins 1997). Many of the reports from New France in the seventeenth century, like Biard's *Relation*, tell about a plenty of food:

> From the month of May up to the middle of September, they are free from all anxiety about their food; for the cod are upon the coast, and all kinds of fish and shellfish (Biard [1616] 1959, 81).

The Jesuit Joseph Jouvency describes the affluence of food in the province: moose, beavers, different kinds of birds and fishes, most of them edible for the Mi'kmaq but not culturally preferred food for the French (Jouvency [1710] 1959, 245–255). Also, the French were not familiar with the landscape and what it had to offer. Lescarbot tells us that during severe scarcity of food among the French: "They heard of some roots which the Savages eat in their time of need, and which are as good as Truffles" (Lescarbot [1612] 1959, 169).

What the Europeans had a hard time to understand was why the Mi'kmaq did not store the food surplus. Why go hungry in wintertime, when there was an amount of good food to collect and save for coming, harder times? For the French settlers the Mi'kmaq seemed to be unconcerned of winter shortage of food, even if this meant days without anything to eat. There are various explanations for this. Firstly, as mentioned above, the practice of eating differed between the newcomers and the native population. Secondly, the nomadic life did not, as the settled life, permit a surplus to be stored. Thirdly, the nomadic life was in one way adjusted to food surplus; not in one place, but in several different places. The seasonal patterns of Mi'kmaq wanderings followed the variation of access of food. The lack of food in one place or in one season, as experienced by the settled Europeans, could mean a surplus in another place. If the products from the "wilderness" only supplemented the food the settlers produced, it held, for the Mi'kmaq, all the products necessary for survival.

The permanent settlements affected the supply of game. As hunters, the Mi'kmaq were dependent on a mobile life. The settlers soon became aware of the mobility of the game and the necessity of following the routes of the game in order to find food. Half of the French population who first stayed in the province during winter 1604 died from scurvy and the rest of them would not have survived the next winter if the Mi'kmaq had not visited them and provided them with food from their hunting. Starvation became so severe that some of them had to stay in Mi'kmaq families during winter to survive. Evidently, the French depended on the Mi'kmaq for their survival from time to time (Hornborg 2008, 6). With this in mind, it is fully understandable that the utterly exhausted French experienced their surroundings as threatening. It is, however, also reasonable to surmise that real, human encounters between the Europeans and the Mi'kmaq must have taken place during those early years of settlement.

5. Personhood

The lack of order and cultivation which the colonists saw in the 'wilderness' was also transferred to Mi'kmaq persons. Jennifer Reid, in examining the British colonists' image of the indigenous group in Nova Scotia (1995, 38,98), describes this image thus: "Like the uncultivated land that was without meaning, the Mi'kmaq were regarded as lacking human significance..." The chaotic, devilish forces were thought to pervade the souls of the nomadic Mi'kmaq, reflecting the wilderness and challenging the moral order that was characteristic for having human qualities. To meet a real "savage" could be a lesson to learn for the Europeans. As early as 1508, Captain Thomas Aubert, of Dieppe, brought back "some of the Natives, whom he exhibited to the wonder and applause of France" (Biard [1616] 1959, 39). During the seventeenth century many exhibitions were organized to allow the French to visit and experience the "savages" – the children of the wilderness.

The reports from New France to France of the "savages'" nature vary depending on the author's background and aims with his travels, which put a personal mark on the documents. The lawyer Marc Lescarbot, who visited the province in 1604, is one of the first travellers who published texts from the first encounters. As an author, he generally describes the Mi'kmaq in a positive way. Part of the reason for this may have been that he had lost a court trial in France and had become disappointed with his own country, and had desired to flee from what he depicted as a corrupt country (Lescarbot [1609]1928, x). He even criticized the use of the word "savage": "So if we commonly call them savages, the word is abusive and unmerited" (Lescarbot, quoted from Smith 1977, 22). Yet, his idealization of the Mi'kmaq must be read as a way of criticizing French society. He did not want to see New France become a replica of the misfortunes of the old country, and in his view, Christianity would be the belief that would unite the French colonists with the Mi'kmaq (Smith 1977, 35). The Jesuit Pierre Biard was far more pessimistic than Lescarbot concerning the cultural potential among the "savages" and did not think that France had anything to learn from the Mi'kmaq (Smith 1977, 45). Biard complained a lot about the persistent customs and traditions among the Mi'kmaq, but since he was dependent on his financiers for his work he stressed, in his letters, how successful the mission was.

During this time, the discussion of the "savage's" nature was also highly debated in Europe. The cultural elite of the Renaissance spent much time discussing and demarcating the boundaries between themselves and the uncivilized children of nature. The question had, of course, one root in the colonial encounter. The colonists' descriptions of the "new" world also introduced a new issue for Europe to discuss: were the "Natives" humans or animals? Had Jesus died for these poor creatures and did they have the capacity of understanding the Gospel? Already in 1537, Pope Paul III had produced the bull *Sublimus Deus*, in which it was stated that the natives were humans, as rational beings with souls, who had the capacity of understanding the Christian message (Jaenen 1976, 18).

Further, in 1550–1551, the Spanish king Charles V organized a debate on the subject of how the Indians should be Christianized, inspired by the European encounter in Middle and South America. The discussion concerned how the mission should proceed - with the aid of violence or peacefully - and the Spanish philosopher and theologian Sepúlveda and the friar Las Casas participated. The latter preferred a gentler mission: the Indian was like fertile soil, ready to till, and he could be integrated into the Christian world according to the divine calling (Green 1995, 13, Gill 1982, 7). In this debate, Westerners were portrayed as civilized, learned, and artistic, in short a people of culture. "The Others" - the Indians - were pagan, uncivilized, incapable of deeper theoretical insights; they were inhuman (Green 1995, 13; Gill 1982, 7).

The Jesuit Biard attributes qualities to the Mi'kmaq, but these qualities needed the helping hand of the missionary in order for the Mi'kmaq to be able to live in a new social order, otherwise: "...these miserable people...will always remain in a perpetual infancy as to language and reason" (Biard [1612a] 1959, 13). This is but one example of how the Mi'kmaq were seen as children of the 'wilderness', as belonging to the sphere of nature, for the purposes of domination. The limit that was drawn between nature and culture did not only function as a description of the world and society, it clearly comprised a power relation. The "savages" were like a raw material, like the soil or like wild plants, demanding cultivation and civilization in order to be liberated and remade, transformed into 'real' humans, Christians (Miller 1989, 8, Merchant 1992, [1980] 1994, Ingold 1996, Evernden 1985, Vecsey 1980, 4). Biard writes:

> For in truth this people, who, through the progress and experience of centuries ... is like a great field of stunted and ill-begotten wild plants, a people which ought to have produced abundant fruits in philosophy, government, customs, and conveniences of life; which ought to be already prepared for the completeness of the Holy Gospel, to be received in the house of God (Biard [1612b] 1959, 111).

After Biard has described the savages of New France as "wild plants" and wanderers in the wilderness, he attributes them with uncivilized vices: "As a people they have bad habits, are extremely lazy, gluttonous, profane, treacherous, cruel in their revenge, and given up to all kinds of lewdness, men and women alike" (Biard [1611] 1959,173). To Biard's annoyance, the Mi'kmaq constantly contested these descriptions and, in Biard's words ([1611] 1959, 175): "they consider themselves more ingenious: 'For,' they say, 'you are always fighting and quarrelling among yourselves...and are neither generous nor kind; as for us, if we have a morsel of bread we share it with our neighbour'". Also, Mi'kmaq religion was seen by the French as a religion that had similarities with the ancient religions in Europe – another way of designating it as inferior. Their religious specialist, the shaman, was experienced as a juggler and did not belong to the modern world (i.e., the seventeenth century) but to a heathen history: "Similar criminal maxims and ridiculous observations are still to-day in force among the Gaspesians, who observations and superstitions which the Romans recognized of old in the ministry of their diviners" (Le Clercq [1691] 1968, 215). Thus the conquest of the "savages" also implied turning the heathen history into "modern beliefs".[3] The wild savages should be liberated from their more ancient and historical lifestyle – also found in the past of European history – and brought into the modern era, to the enlightenment (Reid 1995, 41).

The missionaries' frustration with the difficulties of communication with the Mi'kmaq was based on their divergent conceptions of personhood. This bore on the classification of human-animal relations especially. The French experienced the Mi'kmaq perspective on classification and, especially, personhood as a mental shortcoming, reflecting the wilderness, and an obstacle for the progress of reason among the Mi'kmaq: "likewise they will name to you a wolf, a fox, a squirrel, a moose and so on to every kind of animal they have, all of them which are wild, except the dog; but as to words expressing universal and generic ideas, such as beast, animal, body, substance, and the like, these are all together too learned for them" (Biard [1612a] 1959, 11).

The differences between the Amerindian and the Western cosmology have been examined by the social anthropologist Eduardo Viveiros de Castro, and he sees them as structural inversions (Viveiros de Castro 1998, 1999). In West, following the Cartesian tradition, it is the body that unites us with the animals, and what separates humans from the rest of creation is the gift of being a person, a subject and able to build cultures. Viveiros de Castro dubs the Western outlook on the world a *multiculturalistic* ontology. The Amerindian ontology is, on the contrary, *multinaturalistic*, since it is grounded in the assumption of spiritual (cultural) unity (all beings are subjects and thus united), underlying a bodily diversity (the body is what separates us). Thus, the Mi'kmaq ascription of personhood to animals challenged the idea of being human and humankind's unique position as God's viceroy or steward on earth in the European perspective.

---

3 Gaspesians is the name of the Mi'kmaq group Le Clercq visited.

Viveiros de Castros did field work in South America (Araweté in Brasil). He argues that there is a structural similarity between the cosmology of the Araweté and that of other Amerindian peoples (Viveiros de Castro 1992). His analysis of Amerindian ontologies can clearly be useful when examining the Mi'kmaq cosmology. The Mi'kmaq narratives drew boundaries between humans and other beings with whom they shared their dwelling, both animals and spirits, but in different ways than did the newcomers. Their mythology emanates from the concept that all beings share the same essence. They are all subjects, persons, and all part of a greater subject/power that they named *buoin.* Since everyone is part of this cosmological essence as subjects, they share the same experience of dwelling in the world (Hornborg 2006, 315). This is one explanation why humans as well as animals are depicted as persons in the stories, dwelling in "cultures" and why they can communicate with each other "cross-culturally". Whether the culture of the Mi'kmaq entertained an idea of an 'otherworld' outside of their normal world, which could be seen as more or less equivalent to 'wilderness', is difficult to gauge. To be sure, the environment classified as wilderness by the newcomers was a cultured environment for the Mi'kmaq, and the Mi'kmaq ascription of personhood to animals was an issue of great contention. In the contemporary, European outlook of the day, animals/nature were sharply demarcated from humans/culture. In Mi'kmaq cosmology, the separation between different beings' dwelt-in-worlds and humans consisted in their bodily differences (Hornborg 2008, 35; compare with Viveiros de Castro 1998, 470).

In Mi'kmaq stories, humans may marry animals. This is one way of metaphorically showing that alliances between humans and animals are possible; that is, also animals are persons to be communicated with. However, the bodies that animals (and humans) inhabit differ from each other, so they have to dwell in different ways. That is to say, "the limit of my world is my body" (Hornborg 2006, 317). This is one reason why marriages between animals and humans always fail – humans cannot forever inhabit other beings' niches due to bodily differences. The limit of the body means that every person exists in the world according to his or her special condition. Out of the diversity of objects in the world, every being has to choose from his or her perspective those that suit his or her body's requirements (for example what to eat). This does not change the fact that the subjective experience of being in the world is shared by all.

The Mi'kmaq stories abound with examples of people crossing temporarily into the niches or worlds of other beings. Without the notion of all beings as parts of a greater subject, the beings in different dwellings could never communicate with each other. They would be forever captives of their body's specific outlook on the world (Hornborg 2006, 318). But such visits are by no means without problem, since the body of one person is not adjusted to the worlds of other beings. Hence, no visit may be permanent and the journeys between worlds must be conducted with many precautions, in order for a person to be able to return to his/her true element and regain the perspective which goes with the natural body of the person. Only the Mi'kmaq religious experts fully master the ability to transcend the borders between species, i.e., they manage more than one perspective of the world at the same time.

It was difficult for the newcomers to cope with the complexity involved in the encounter with other boundaries between humans, animals, culture and nature than their own. For them, humans were created in the image of God and only humans could be attributed personhood, a soul and the capacity to build cultures. Animals were not persons and they lacked souls. The Europeans misinterpreted the Mi'kmaq narratives which depicted animals as persons with human qualities, and their ritual communication with animals, when they understood them as exemplifying how the Mi'kmaq failed to distinguish between humans and animals. This, of course, only strengthened their idea that the Mi'kmaq belonged to the wilderness.

6. Conclusion

For the early Western explorers of the New World, the idea of wilderness implied opposition to civilization and to an agricultural ideal understood as the divinely sanctioned order. The wilderness was seen as a place to be defeated and not a place of dwelling for humans. When the European travellers and missionaries met the nomadic Mi'kmaq people of the Eastern coast of Canada, they encountered a different structuring and understanding of landscape, a new mode and understanding of dwelling, and other ways of understanding personhood. The European view of what constituted a human being vis-à-vis other species was challenged. In this context, the idea of wilderness became a useful key metaphor to apply on the new continent's landscape, mode of dwelling and understanding of personhood. The idea of wilderness thus played a crucial role in the colonization of the land, in justifying the perspective of the newcomers, in enabling suppression and the eradication of traditional culture and in "civilizing" the "savages". The "wilderness" was ascribed moral connotations and this justified domination. The newcomers, at first, felt alienated from the uncultivated landscape that was the home of the Mi'kmaq and which they used for hunting and gathering. To encourage other Europeans to inhabit the new land, they framed themselves as "new Israelites", defeating the wilderness before entering Canaan's land. The nomadic culture of the Mi'kmaq was described, in the early sources, as immoral and the "savages" as animal- or plant-like. The texts show clearly that the wilderness idea was used, by the newcomers, as a means of justification in their attempts to force the nomadic Mi'kmaq to be settled and baptized, which, according to the rhetoric of the settlers, rendered them human and cultivated. Yet, the texts also clearly show that the concrete, bodily meetings between Europeans and Mi'kmaq hunters involved a considerable amount of contestation of the European perspective, of negotiation, and, occasionally perhaps, of exchange between the European and the Mi'kmaq perspectives.

It is interesting to note that Biard's idea, from the beginning of the seventeenth century, that the nomadic Mi'kmaq could never develop an attachment to the land was turned upside down at a later time and in the Romanticist perspective. Here, the "Indian" was by definition deeply involved in the land (Francis 1992, Parkhill 1997). Charles Leland, who collected Mi'kmaq and other Algonquin stories in the nineteenth century, is a good example of this. He was greatly inspired by the German Romanticists and ascribed the "Volk" soul of America to the indigenous groups. Leland criticizes other Romantic authors like Wordsworth, Byron, Tennyson and Thoreau for their distanced view of nature. In his view, the mystical bond between the country and man is ascribed to the "Indians":

> A few second-hand scraps of Byron and Tupper, Tennyson and Longfellow, the jingle of a few rhymes and a few similes, and a little second-hand supernaturalism, more 'accepted' than felt, and that derived from far foreign sources, does not give the white man what the Indian *feels* (Leland 1884, 339).

It does not help that Thoreau is assisted by Penobscot Indians in his descriptions of nature. The famous nature philosopher also becomes a target of Leland's criticism:

> This writer passed months in Maine ...Yet he could in the same book write as follows: 'The Anglo-American can indeed cut down and grub up all this waving forest, and make a stump and vote for Buchanan on its ruins; but he cannot converse with the spirit of the tree he fells, he cannot read the poetry

> and mythology which retires as he advances.' ... Such a writer can, indeed, peep and botanize on the grave of Mother Nature, but never evoke *her* spirit (Leland 1984, 66).[4]

However, in the 17th century, the opposite was the case. The texts studied here show that the differences between the European settlers and the Mi'kmaq hunters and gatherers in ways of dwelling in an environment generated, in the best cases, meetings in which the voices of the Mi'kmaq contesting the European perspective were heard. Sometimes it generated relatively harmless misunderstandings, but most often it ended in dominance, suppression and violence. The dichotomous image of the wilderness as opposed to civilization was, as we have seen in this article's case study, a product of a specific era's specific type of bible interpretation used for purposes of domination. It was contested and challenged on the ground, and cannot be seen as a universal feature of religious cosmologies.

The new settlements, the cultivated land and the fencing of private property challenged the usual ways of interacting with the 'wilderness' for the Mi'kmaq people. Therefore two different ecological systems came to compete in the new provinces. When wilderness was turned into cultivated land, the game disappeared. With one of the basic means of subsistence at stake, Mi'kmaq culture was threatened. Although the early Europeans texts are full of Mi'kmaq arguments of being privileged as nomads and hunters, their way of dwelling – for the Europeans classified as dwelling in the wilderness – was disappearing. Settlement was forced upon them and in the middle of nineteenth century it ended with the Mi'kmaq living in small enclosures in their former province – the reserves.

7. References

Biard, Pierre."Lettre...." In *The Jesuit Relations and Allied Documents*, Vol. I, edited by Reuben Thwaites, translated by John Covert, 138–183. New York: Pageant Book, [1611]1959.

4 Thoreau's *Walden* (1854) is a classic that made the author into a nineteenth century "guru" within the transcendentalist movement. Although he received much of the information in the book from local Native guides, he does not give them credit for this (Jaimes 1995, 274).

Biard, Pierre. "Lettre…" In *The Jesuit Relations and Allied Documents*, Vol. II, edited by Reuben Thwaites, translated by John Covert, 3–55. New York: Pageant Book, [1612a] 1959.

Biard, Pierre, "Biard's Epistola." In *The Jesuit Relations and Allied Documents*, Vol. II, edited by Reuben Thwaites, translated by John Covert, 60–105. New York: Pageant Book, [1612b] 1959.

Biard, Pierre. "Relation de la Nouvelle France." In *The Jesuit Relations and Allied Documents*, Vol. III, edited by Reuben Thwaites, translated by John Covert, 26–301. New York: Pageant Book, [1616] 1959.

Callicott, John Baird. *Earth's Insights: A Multicultural Survey of Ecological Ethics from the Mediterranean Basin to the Australian Outback.* Berkeley, Los Angeles and London: University of California Press, 1994.

Chamberlin, Roy and Herman Feldman, eds. *The Dartmouth Bible: An Abridgment of the King James Version, with Aid to Its Understandings as History and Literature and a Source of Religious Experience.* Boston: Houghton Mifflin, 1950.

Cronon, William. *Changes in the Land: Indians, Colonialists, and the Ecology of New England.* New York: Hill and Wang, 1997 [1983].

Dennis, Clara. *Down in Nova Scotia. My Own, My Native Land.* Toronto: The Ryerson Press, 1934.

Denys, Nicolas. *The Description and Natural History of the Coasts of North America*, edited and translated by William Ganong. Toronto: The Champlain Society, [1672] 1908.

Evernden, Neil. *A Natural Alien.* Toronto: University of Toronto Press, 1985.

Farrington, Benjamin. *Francis Bacon: Philosopher of Industrial Science.* New York: Schumann, 1949.

Francis, Daniel. *The Imaginary Indian. The Image of the Indian in Canadian Culture.* Vancouver: Pulp Press, 1992.

Gill, Sam. *Native American Religions: An Introduction.* California: Wadsworth Publishing Company, 1982.

Green, Michael. "Cultural Identities: Challenges for the Twenty-First Century." In *Issues in Native American Cultural Identity*, edited by Michael Green, 1–38. New York: Peter Lang Publishing, Inc, 1995.

Hewitt, Harry. 1908. "Customs of the Micmac Indians." Manuscript of 33 pages, read before the Nova Scotia Historical Society April 21, 1908, and preserved in the files of that society.

Hornborg, Anne-Christine. *Mi'kmaq Landscapes: From Animism to Sacred Ecology.* Bodmin, Cornwall: Ashgate, 2008.

Hornborg, Anne-Christine. "Visiting the Six Worlds –Shamanistic Journeys in the Canadian Mi'kmaq Cosmology," in: *Journal of American Folklore*, 119:473 (2006), 312–336.

Ingold, Tim. *The Perception of the Environment: Essays in livelihood, dwelling and skill.* London and New York: Routledge, 2000.

Ingold, Tim. "Hunting and Gathering as Ways of Perceiving the Environment." In *Redefining Nature: Ecology, Culture and Domestication,* edited by Roy Ellen and Katsuyoshi Fukui, 117–155. Oxford: Berg, 1996.

Jaenen, Cornelius. *Friend and Foe: Aspects of French-Amerindian Cultural Contact in the Sixteenth and Seventeenth Centuries.* New York: Columbia University Press, 1976.

Jaimes, M. Annette. "Native American Identity and Survival: Indigenism and Environmental Ethics." In Michael Green ed., *Issues in Native American Cultural Identity.* New York: Peter Lang Publishing, Inc, 1995, 273–296.

Jouvency, Joseph. "De Regione et Moribus Canadensium seu Barbarorum Novæ Franciæ." In *The Jesuit Relations and Allied Documents*, Vol. I, edited by Reuben Thwaites, translated by William Giese, 241–317. New York: Pageant Book, [1710]1959.

Le Clercq, Chrestien. "*Nouvelle Relation de la Gaspesie....*", edited and translated by William Ganong. New York: Greenwood Press. 1968 [1691].

Leland, Charles G. *The Algonquin Legends of New England or Myths and Folk Lore of the Micmac, Passamaquoddy, and Penobscot Tribes.* London: Sampson Low, Marston, Searle & Rivington, 1884.

Lescarbot, Marc. *Histoire de la Nouvelle France.* English translation *Nova Scotia: A Description of Acadia, 1606.* London: Routledge & Sons, [1609] 1928.

Lescarbot, Marc. "La Conversion des Savvages." In *The Jesuit Relations and Allied Documents*, Vol. I, edited by Reuben Thwaites, translated by John Covert, 49–113. New York: Pageant Book, [1610]1959.

Lescarbot, Marc. "Relation Dernière..." In *The Jesuit Relations and Allied Documents*, Vol. II, edited by Reuben Thwaites, translated by John Covert, 125–191. New York: Pageant Book, [1612] 1959.

Martin, Calvin. *In the Spirit of the Earth: Rethinking History and Time.* London: The John Hopkins University Press, 1992.

Marting, Calvin. *Keepers of the Game: Indian-Animal Relationship and the Fur Trade.* Berkeley, Los Angeles, London: University of California Press, 1972.

Merchant, Carolyn. *Naturens död: Kvinnan, ekologin och den vetenskapliga revolutionen.* translated by Öjevind Lång. Stockholm: Symposion, [1980] 1994.

Merchant, Carolyn. *Ecological Revolutions. Nature, Gender and Science in New England.* London: The University of North Carolina Press, 1989.

Miller. James. *Skyscrapers Hide the Heavens: A History of Indian-White Relations in Canada.* Toronto: University of Toronto Press, 1989.

Ortner, Sherry. "On Key Symbols"." *American Anthropologist,* 75:5 (1973), 1338–46.

Parkhill, Thomas. *Weaving Ourselves into the Land: Charles Godfrey Leland, "Indians," and the Study of Native American Religions.* New York: State University of New York Press, 1997.

Parsons, Elsie C. "Micmac Folklore." In *The Journal of American Folklore,* 38 (1925), 55–133.

Rand, Silas T. *A First Reading Book in the Micmac Language.* Halifax: Nova Scotia Printing Company, 1875.

Rand, Silas T. *Legends of the Micmacs.* New York: Longmans, Green, & Co, [1894] 1971.

Rand, Silas T. *Micmac Placenames in the Maritime Provinces and Gaspe Peninsula,* collected, arranged and indexed by W. M. P. Anderson. Ottawa: Printed at the Surveyor General's Office, 1919.

Reid, Jennifer. *Myth, Symbol, and Colonial Encounter. British and Mi'kmaq in Acadia, 1700–1867.* Canada: University of Ottawa Press, 1995.

Sahlins, Marshall D. *Culture and Practical Reason.* Chicago: University of Chicago Press, 1972.

Said, Edward. *Orientalism.* London: Routledge & Kegan, Paul, 1978.

Smith, Colin. *The Acadian Classics: Seventeenth Century Views of the Micmac Indians.* National Library of Ottawa: Unpublished thesis.

Thoreau, Henry D. *Walden.* Swedish translation 1998: *Skogsliv vid Walden.* Stockholm: Wahlström och Widstrand, 1854.

Vecsey, Christopher. "American Indian Environmental Religions". In *American Indian Environment. Ecological Issues in Native American History,* edited by Christopher Vecsey and Robert Venables, 1–37. New York: Syracuse University Press, 1980.

Viveiros de Castro, Eduardo. "The Transformation of Objects into Subjects in Amerindian Ontologies", paper held at a session, *Re-Animating Religion: A Debate on the New Animism,* at the 98th Annual Meeting of the American Anthropological Association, Chicago Nov. 1999.

Viveiros de Castro, Eduardo. "Cosmological Deixis and Amerindian Perspectivism." In *The Journal of the Royal Anthropological Institute,* 4:3 (1998), 469–88.

Viveiros de Castro, Eduardo. *From the Enemy's Point of View: Humanity and Divinity in a Amazonian Society,* transl. Catherine V. Howard. Chicago: University of Chicago Press, 1992.

10. William Robertson Smith on the Wilderness

Robert A. Segal

1. Introduction

William Robertson Smith (1846–94) was a celebrated Scottish biblical critic, theorist of religion, and theorist of myth. Smith's accomplishments were multiple. First, he brought higher biblical criticism from Germany to the English-speaking world and then developed it far beyond its Continental origins. Where his German mentors reconstructed the history of Israelite religion from the Bible itself, Smith ventured beyond the Bible to Semitic religion and thereby pioneered the comparative study of religion. Where others viewed religion from the standpoint of the individual, Smith approached ancient religion from the standpoint of the group and thereby helped pioneer the sociology of religion.

As an original theorist of religion, Smith asserted that ancient religion was centrally a matter of ritual, not creed. Practice, not belief, counted most. Religion was initially communion between god and humans, not a pre-scientific account of the world. As an equally original theorist of myth, Smith similarly maintained that myth was initially an explanation of ritual. Since Smith's time, the ritualist theory of myth has found adherents not only in biblical studies but also in classics, anthropology, and literature. Contemporary theorizing on myth and ritual by, among others, Walter Burkert and René Girard harks back to Smith.

Smith's originality is to be found in all of his works, but above all in *Lectures on the Religion of the Semites* (1889). There Smith makes two sets of distinctions that bear on the theme of the wilderness: not only that between the wilderness and civilization but also that between the physical and the spiritual. On the one hand ancients and "primitives" pit the wilderness against civilization. The wilderness is the domain of demons, of evil, of wild beasts, and of physical danger. Civilization, which is my term, is the domain of gods, of good, of domesticated animals, and of safety. The conquest of an uninhabited spot means the driving out of the resident demons, who are displaced by gods.

On the other hand the deeper distinction for Smith is that between the physical and the spiritual. Now the divide is not between one physical domain and another but between the conception of the holy as physical and the conception of it as non-physical. The emergence of this higher conception of religion begins with the Prophets of ancient Israel but is not fully developed till Christianity. The wilderness is no longer a physical state but is now a spiritual one. Now God resides not in any one place but everywhere. God is now a spiritual rather than a physical being. The relationship between humans and God no longer involves the physical ingestion of God, as it does in totemism, the earliest stage of religion for Smith.[1] Now communion means spiritual ingestion.

At the outset of the *Lectures* Smith declares his subject to be "the religion of the Semitic peoples" (Smith 2nd ed. [1894], 1). From the Semites emerged "three of the great faiths of the world": Judaism, Christianity, and Islam. These three religions, which Smith labels "positive religions," arose out of ancient "heathenism" but advanced far beyond it. Smith uses the terms "primitive" and "ancient" (or "antique") almost interchangeably.

In seeking the nature of Semitic religion, Smith turns to "heathen Arabia" as the earliest and therefore presumably clearest case: "In many respects the religion of heathen Arabia, though we have little information concerning it that is not of post-Christian date, displays an extremely primitive type, corresponding to the primitive and unchanging character of nomadic life" (Smith 1894,14). Semites were initially at a "primitive" stage of culture.

2. Primitive and Ancient Religion versus Modern Religion

Smith sharply differentiates primitive and ancient religion from modern religion. Primitive and ancient religion is ritualistic, mythic, partly amoral, and collectivist. Modern religion is creedal, nonritualistic, nonmythic, wholly moral, and individualistic. Most of all, primitive and ancient religion is materialist, where modern religion is spiritual.

In ancient and primitive religion god is conceived of as physically related to worshippers:

1 Totemism as the first stage of religion has long been discredited. See Jones 2005.

> The relation between the gods of antiquity and their worshippers was expressed in the language of human relationship, and this language was not taken in a figurative sense but with strict literality. If a god was spoken of as father and his worshippers as his offspring, the meaning was that the worshippers were literally of his stock, that he and they made up one natural family with reciprocal family duties to one another (Smith 1894, 29–30).[2]

God as father has moral as well as physical consequences. The relationship of blood dictates social obligations, with "the parent protecting and nourishing the child, while the child owes obedience and service to his parents" (Smith 1894, 41). Whatever the source of morality in modern religion, the source of morality in primitive and ancient religion is biological.

The moral link between god and worshippers is wholly beneficent. The god of a clan or nation is friendly, solicitous, and above all loving. Smith puts unfriendly, malevolent forces in magic or sorcery rather than in religion. Because the forces in magic are unfriendly, they can only be controlled, not obeyed. No morality can arise from dealings with them. Smith associates magic with the individual rather than, like religion, with the group. Magic, because private, is anti-social and therefore immoral: "private and magical superstitions were habitually regarded as offenses against morals and the state" (Smith 1894, 55).

One consequence of the materialist conception of god in primitive and ancient religion is that god need not be human-like but can be an animal or even a plant. Smith thus roots what for him is the earliest form of religion, totemism, in this materialist or, better, biological conception. Just as an anthropomorphically conceived god is taken to be the father of his worshippers, so an animal is taken to be their brother: "Every animal of this kind is looked upon as a brother, is treated with the same respect as a human clansman, and is believed to aid his human relations by a variety of friendly services" (Smith 1894, 124). Not only the obligations among worshippers but the obligations between them and their gods derive from kinship, which turns worshippers and gods into one big, happy family. In primitive and ancient religion blood is thicker than water.

Strictly, totems for Smith are not gods, for totems, while kin to humans, are not anthropomorphic. Still, totems are the basis for gods. Like

2 For all his emphasis on God as a father, Smith is prepared to grant that the earliest god was female (see Smith 1894, 51–52, 55–58; see also Jones 1894, p. 40).

gods, who are originally local, totems "are invested with gifts such as we should call supernatural, and of the very same kind which heathenism [later] ascribes to the gods–for example with the power of giving omens and oracles, of healing diseases and the like" (Smith 1894, 125).

For Smith, the Arabian *jinn* are the Semitic version of totems: "This conception of the communities of the *jinn* is precisely identical with the savage conception of the animal creation" (Smith 1894, 127). Yet totemism is not the earliest stage of belief. The belief in demons comes earlier and is magical rather than religious. Smith's chronology is therefore first demons, which are pre-religious, then totems, then local gods, then national gods, and finally universal gods.

Another consequence of the materialist nature of primitive and ancient religion is that the sacredness, or "holiness," of a god, person, or place stems from other than character. To moderns, holiness is "an ethical idea" and is based on character: "God, the perfect being, is the type of holiness; men are holy in proportion as their lives and character are godlike; places and things can be called holy only by a figure, on account of their association with spiritual things" (Smith 1894, 140). By contrast, to ancients, "holy persons were such, not in virtue of their character but in virtue of their race, function, or mere material consecration" (Smith 1894, 141). Persons practicing the worst immoralities could still be labelled "holy." The same Smith who credits primitive and ancient religion with the instillment of morality simultaneously disparages that stage of religion as amoral.

In primitive and ancient religion the attribute "holy" is applied less to persons than to gods, seasons, things, and especially places. While persons, things, and times are, as in modern religion, holy because of their association with gods, their holiness is tied to the locales where gods reside: "Holy persons things and times, as they are conceived in antiquity, all presuppose the existence of holy places at which the persons minister, the things are preserved, and the times are celebrated." In fact, the holiness of the gods themselves "is an expression to which it is hardly possible to attach a definite sense apart from the holiness of their physical surroundings" (Smith 1894, 141). The wilderness is not unholy simply because demons reside there. Demons reside there because it is unholy.

A further or likely consequence of the materialist nature of primitive and ancient religion is the conflation of holiness with uncleanness. True, holiness comes from association with gods and uncleanness from association with demons. Holiness is thus tied to religion and uncleanness tied to magic. But both involve contact. One becomes holy by contact with something holy, and one becomes unclean by contact with something unclean: "Holiness … is conceived as infectious, propagating itself by physical contact" (Smith 1894, 161). Contact with either the holy or the unclean is tabooed, and to make contact with either is to become tabooed oneself.

For Smith, the differentiation between the holy and the unclean marks "a real advance above savagery" (1894, 154), and Smith credits the Semites, or some of them, with forging the distinction. But the distinction between the holy and the unclean was not initially that between the spiritual and the material. Not till Christianity was the holy conceived of as other than material itself. Still, the differentiation between the holy and the unclean does distinguish religion from magic, or at least religion of a more advanced stage than magic.

For Smith, uncleanness was originally assumed to arise from physical contact and was remedied by a physical act:

> Among the Semites the impurities which were thought of as cleaving to a man, and making him unfit to mingle freely in the social and religious life of his community, were of very various kinds, and often of a nature that we should regard as merely physical, e.g. uncleanness from contact with the dead, from leprosy, from eating forbidden food, and so forth…. They were dealt with, where the uncleanness was of a mild form, mainly by ablutions; or where the uncleanness was more intense, by more elaborate ceremonies involving the use of sacrificial blood, of sacrificial ashes, or the like. (Smith 1894, 428)

Where, however, uncleanness stemmed from the murder of a fellow tribesman, the washing away of the blood did not suffice. Semitic religions "provide no atonement for the murderer himself," who thus cannot be restored "to his original place in his tribe." Thus Hebrew law "does not admit piacula for mortal sins." Only the community can be cleansed, and only by collectively "disavow[ing] the act of its impious member" (Smith 1894, 429).

The combination of collective responsibility for "the maintenance of holiness" with "the thought that holiness is specially compromised by crime" led to the conception of the community as "a kingdom of righteousness," which is to say, a kingdom based on ethics. Where ethics rules, "there is no atonement for mortal sin" on the part of individuals (Smith 1894, 429). Smith seems to connect the rejection of the possibility of atonement with the rejection of a physical means of atonement. He certainly connects the rejection of a physical means of atonement with the rejection of sin as physical. While murder obviously involves bloodshed, the real sin was now taken to be ethical.

Smith maintains at once that even the post-exilic Israelite notion of sacrifice as atonement rested on the primitive notion of sacrifice as communion and that even primitive and pre-exilic Israelite notions of sacrifice harbored the supposedly exclusively higher concepts of atonement. Within primitive religion itself misfortunes like plague and famine were attributed to the weakening of the bond between god and the community, and sacrifice served to restore the bond. Eventually, the weakening came to be attributed to sin, and sacrifice came to be taken as atonement. Nevertheless, the ultimate aim of atonement was the restoration of the bond between god and community.

Despite Smith's insistence on the presence of both communion and atonement in primitive and higher religion alike, the gulf between these stages of religion remains wide because primitive religion still conceives of both materially, whereas higher religion conceives of both spiritually. In primitive religion spiritual ideas like atonement, while present, "are very vaguely defined; they indicate impressions produced on the mind of the worshipper by features of the ritual, rather than formulated ethico-dogmatical ideas" (Smith 1894, 439). In "primitive ritual" communion "is grasped in a merely physical and mechanical shape, as indeed, in primitive life, all spiritual and ethical ideas are still wrapped up in the husk of a material embodiment" (Smith 1894, 439). The materialist conception of religion clouds the recognition of spiritual concepts like ethics, sin, and atonement.

Consequently, "to free the spiritual truth from the husk was the great task that lay before the ancient religions, if they were to maintain the right to continue to rule the minds of men." Smith grants that "some progress in this direction was made, especially in Israel" (Smith 1894, 439). The Prophets severed communion from material sacrifice–Smith taking the Prophets as entirely anti-ritualistic. Once communion with god came to be conceived of spiritually, separation from god came to be conceived of ethically, as a matter of the justification for the separation and of the amends needed for overcoming it. But the subsequent, post-exilic restoration of animal sacrifice conflated anew the spiritual with the material. Only with Christianity was the spiritual fully disentangled from the material.

3. The Wilderness

By the "wilderness" Smith means both literal location and religious state. Taken literally, Smith means the land between Egypt and the Promised Land. He means, above all, the desert. The wanderings in the desert for forty years are associated with thirst, hunger, disease, earthquakes, and snakes. Yet wilderness does not mean unoccupied land. It also means land inhabited by the "heathen" cultures that Israel encounters. With the wilderness is associated continual fighting. Also associated with it is the recurrent betrayal of God for pagan gods. The key sections of the Hebrew Bible that present the Israelites' wanderings are Exodus 13.17–22 and 15.22–17.15; Numbers 10.11–14.45, 16.1–17.13, 20.1–25.18, 27.12–23, and 31.1–34.29.

The opposite of the wilderness is civilization. Wherever Israel encounters other cultures, influence is unavoidable, but progress comes only by then rejecting that influence.

> The positive Semitic religions had to establish themselves on ground already occupied by these older beliefs and usages; they had to displace what they could not assimilate, and whether they rejected or absorbed the elements of the older religion, they had at every point to reckon with them and take up a definite attitude towards them. No positive religion that has moved men has been able to start with a *tabula rasa,* and express itself as if religion were beginning for the first time; in form, if not in substance, the new system must be in contact all along the line with the older ideas and practices which it finds in possession. (Smith 1894, 2)

Because of the primitive and ancient conception of religion as materialist, the physical world was divided into two domains: the demonic and the divine, the wilderness and civilization. Generalizing from the case of the *jinn* to that of early Semitic religion, Smith distinguishes the domains:

> In fact the earth may be said to be parcelled out between demons and wild beasts on the one hand, and gods and men on the other. To the former belong the untrodden wilderness with all its unknown perils, the wastes and jungles that lie outside the familiar tracks and pasture grounds of the tribe, and which only the boldest men venture upon without terror; to the latter belong the regions that man knows and habitually frequents, and within which he has established relations, not only with his human neighbours, but with the supernatural beings that have their haunts side by side with him. And as man gradually encroaches on the wilderness and drives back the wild beasts before him, so the gods in like manner drive out the demons, and spots that were once feared, as the habitation of mys-

> terious and presumably malignant powers, lose their terrors and either become common ground or are transformed into the seats of friendly deities. From this point of view the recognition of certain spots as haunts of the gods is the religious expression of the gradual subjugation of nature by man. (Smith 1894, 121–22)

The "triumph" of gods over demons is gradual and is "finally sealed and secured" only in the post-nomadic, agricultural stage, "when the god of the community became also the supreme lord of the land and the author of all the good things therein" (Smith 1894, 122).

The demons who had formerly occupied the land either were driven out into uninhabited land or "were reduced to insignificance as merely subordinate beings of which private superstition might take account" – perhaps as magic – "but with which public religion had nothing to do" (Smith 1894,122–23). In the land now "frequented by the community of men the god of the community was supreme." Now "every place that had special supernatural associations was regarded, not as a haunt of unknown demons, but as a holy place of the known god" (Smith 1894, 123).

If the wilderness for Smith means geographical location and is to be contrasted to the Promised Land, the wilderness for Smith also means "spiritual" location. Now the wilderness is mental. It is the state of separation from God. It is the state of sin. As decisive as the demarcation of the physical world between wilderness and home is for ancient Israelites, once the division between material and spiritual starts to emerge in religion, any distinctions within the physical world become pointless.

4. Legacy

Smith's emphasis on the roots of ancient Israelite religion in "heathenism" has proved exceedingly influential and is now the mainstream view. But Smith's equal emphasis on the transcending of those roots has proved ever more tenuous. Smith sees ancient Israel as casting off the would-be lasting influence of the cultures with which it comes in contact. He accepts as historical the prophetic view of a clash between pure Israelite religion and the corruption of it by its heathen neighbors. Influence for him means corruption and is eventually, if slowly, rejected. He sees the development of Israelite religion and in turn of Christianity as almost a leap of faith, from what is benighted to what is true. This view of the progress of religion as proceeding by rejection is almost anti-evolutionary.

What Smith writes at the end of one of his most famous essays, "Animal Worship and Animal Tribes among the Arabs and in the Old Testament" (1880), sums up his views:

> It is a favourite speculation that the Hebrews or the Semites in general have a natural capacity for spiritual religion. They are either represented as constitutionally monotheistic, or at least we are told that their worship had in it from the first, and apart from revelation, a lofty character from which spiritual ideas were easily developed. That was not the opinion of the prophets, who always deal with their nation as one peculiarly inaccessible to spiritual truths and possessing no natural merit which could form the ground of its choice as the people of Jehovah. Our investigations appear to confirm this judgment, and to show that the superstitions with which the spiritual religion had to contend were not one whit less degrading than those of the most savage nations.... It does not appear that Israel was, by its own wisdom, more fit than any other nation to rise above the lowest level of heathenism. (Smith 1912, 482–83; see also Maier 2009,269)

For Smith, wilderness means heathenism and civilization means revealed religion, so that the rejection of heathenism for the religion espoused by the Prophets sets Israel against its Near Eastern neighbors. However unique Israelite religion is still taken to be by contemporary biblicists, it is no longer seen as the antithesis of the very locale where it arose. Far more influential has been Smith's distinction between primitive and modern religion. Primitive religion is ritualistic, collective, partly amoral, and above all materialist. Modern religion is creedal, individualistic, entirely moral, and above all spiritual.[3]

Yet Smith's influence has proved ironic. His rigid opposition between primitive and modern religion has been dismissed as an expression of his hatred of Catholicism and perhaps also of Judaism. He sees Catholicism as an atavistic revival of the ritualistic, collective, and materialist nature of primitive religion. Thus he regards the Catholic belief in literal transubstantiation as the epitome of primitive materialism and contrasts the Catholic view to the Calvinist concept of symbolic transubstantiation.

In the twentieth century the study of religion has emphasized the similarities rather than the differences between primitive and modern religions. All religions have come ever more to be seen as primarily ritualistic rather than creedal and as primarily collective rather than individualistic. Conversely, all religions have come to be seen as moral and, even more,

3 On Smith's distinction between primitive religion as collective and, in contrast to Emile Durkheim, modern religion as individualistic, see Beidelman 1974, 30–31; Segal 2008, 10–12.

materialist, as the recent focus on the body, on materiality, and on space in religions worldwide evinces. Smith's refusal to see all religions as akin and as equal would today be rejected as a theological bias.

But there is a much bigger and more direct reason for the outdatedness of Smith's claims: his appeal to God as the source of the advance from primitive to modern. Where ancient heathen religions grew up "under the action of unconscious forces operating silently from age to age", the positive religions arose from "the teaching of great religious innovators," who "spoke as the organs of a divine revelation" and "deliberately departed from the traditions of the past" (Smith 1894, 1). Smith contends that "on the whole it is manifest that none of the ritual systems of antiquity was able *by mere natural development* to shake itself free from the congenital defect inherent in every attempt to embody spiritual truths in material forms. A ritual system must always remain materialist" (Smith 1894, 439–40 [italics added]). In other words, Israelites and Christians on their own could never have made the leap from a material conception of sacrifice to a spiritual one. Only God's intercession, undertaken indirectly through prophetic inspiration, can account for the jump (see Rogerson 1995, 146–149; Bailey 1970, 291–293; Bediako 1997, 117–118).

Smith is rightly viewed as a pioneering sociologist of religion. He shifts the focus of the study of primitive and ancient religion from beliefs to institutions and from the individual to the group. For Smith, the function of primitive and ancient religion is the preservation of the group, even if he does not, like the more relentlessly sociological Emile Durkheim (1965), make group experience the origin of religion or make the group itself the object of worship.

Whether Smith can reconcile his sociological account of religion with his theological one is a philosophical issue. Certainly God for him operates behind the scenes, by inspiring religious innovators. The direct causes of religious evolution are for him natural. They are a mix of external factors – political, economic, environmental, and cultural (contact with other cultures) – and internal ones. For example, the beginning of the religion of the individual stems internally from individual Israelites' being left to themselves once sacrifice is centralized.

Smith was the last great theological theorist of religion. With conspicuous contemporary exceptions like John Hick, theorists after him have shied away from appealing to God to account for religion and have restricted God to the *object* of religion. Disputes since Smith's day have been over the natural causes of religion. God is never invoked. Smith straddled the move from a theological to a secular account of religion. His sociological approach to primitive and ancient religion opened the door to the sociology of religion. But his theological approach to higher religion sought to keep the door closed to a wholly secular account of religion.

5. References

Bailey, Warner McReynolds. *Theology and Criticism in William Robertson Smith.* PhD dissertation,Yale University, 1970.

Bediako, Gillian M. *Primal Religion and the Bible.* Sheffield: Sheffield Academic Press 1997.

Beidelman, T. O. *W. Robertson Smith and the Sociological Study of Religion.* Chicago: University of Chicago Press, 1974.

Black, John Sutherland, and George Chrystal. *The Life of William Robertson Smith.* London: Black, 1912.

Brown, Jesse H. *The Contribution of William Robertson Smith to Old Testament Scholarship, With Special Emphasis on Higher Criticism.* PhD dissertation. Duke University, 1964.

Davies, G.I. "Wilderness Wanderings". *Anchor Bible Dictionary* 6, 912–914. New York: Doubleday 1992.

Durkheim, Emile. *The Elementary Forms of the Religious Life* (1912), tr. Joseph Ward Swain (1915). New York: Free Press, 1965.

Jones, Robert A. Introduction to reprint of William Robertson Smith, *The Prophets of Israel and their Place in History.* New Brunswick, NJ: Transaction Publishers 2002, vii–lxi.

Jones, Robert A. *The Secret of the Totem.* New York: Columbia University Press, 2005.

Johnstone, William, ed. *William Robertson Smith.* Sheffield: Sheffield Academic Press, 1995.

Maier, Bernhard. *William Robertson Smith.* Tübingen: Mohr Siebeck, 2009.

Nelson, Ronald R. *The Life and Thought of William Robertson Smith, 1846–1894.* PhD dissertation, University of Michigan 1969.

Riesen, Richard A. *Criticism and Faith in Late Victorian Scotland.* Lanham MD: University Press of America, 1985.

Rogerson, J.W. *Old Testament Criticism in the Nineteenth Century.* London: SPCK 1984.

Rogerson, J.W. *The Bible and Criticism in Victorian Britain.* Sheffield: Sheffield Academic Press, 1995.

Segal, Robert A. "In Defense of the Comparative Method." *Numen* 48 (2001), 339–73.

Segal, Robert A. "William Robertson Smith vis-à-vis Emile Durkheim as Sociologist of Religion". *Journal of Scottish Thought* 1/2 (2008),1–12.

Segal, Robert A. "William Robertson Smith: Sociologist or Theologian?" *Religion* 38 (2008), 9–28.

Smith, William Robertson. *Lectures on the Religion of the Semites.* First Series. 1st ed. Edinburgh: Black, 1889.
Smith, William Robertson. *Lectures on the Religion of the Semites.* First Series. 2nd ed. Edinburgh: Black, 1894.
Smith, William Robertson. *Lectures on the Religion of the Semites.* Second and Third Series, ed. John Day. Sheffield: Sheffield Academic Press, 1995.

11. The Taiga Within. Topography and personhood in Northern Mongolia

Morten Axel Pedersen

1. Introduction

The binary between "the wild" and "the tame" has been out of fashion for some time now among anthropologists, historians of religion, and scholars from cognate disciplines. Like its conceptual parent nature/culture, alongside with a whole number of its more or less distant "cousins" such as the raw/cooked and public/private, wild/tame has been criticized for its inability to capture the diversity of human engagements with and perceptions of their surroundings in a cross-cultural perspective. Indeed, it seems hard to conclude otherwise when faced with the bewildering ethnographic variation documented by anthropologists from all corners of the world, ranging from the Hagen people's "no nature, no culture" in Papua New Guinea's (Strathern 1980) over the Amerindian "multi-naturalism" discussed by Viveiros de Castro (1998) to the landscapes of Northern Mongolia's reindeer-herders, which, to cite a previous work of mine, "is not simply comprised by the physical contours of the environment surrounding the nomadic camp. For [the] camp does not define an enclosure of cultural space on the other side of which its inhabitants encounter a natural wilderness. 'Wild' spaces can be found inside each Duxa camp, just as 'tame' places can be found outside it" (Pedersen 2009, 137)

But perhaps there is reason to pause a bit, and ask whether if is possible, germane even, to try reinstating the binary between nature and culture – and, more specifically between the wild and the tame – as an analytical framework in a manner that seeks to overcome its multiple (Eurocentric, Cartesian, modernist, etc.) biases, but retains its explanatory power. The present article attempts to do just that. More specifically, my aim in what follows is to explore the concept of wilderness as an ethnographically derived analytical concept and, in so doing, demonstrate the scope of a certain form of contemporary anthropology, which may be described as "post-plural" (Holbraad & Pedersen 2009) or "post-relational" (Pedersen 2012). My case is Mongolia's Darhad peoples, with whom I have carried out some 20 months of fieldwork since the mid-1990s.[1]

The Darhads, who number around 20.000 individuals, are a Mongolian speaking group of pastoralists, hunters and village dwellers, who inhabit the north-western corner of Mongolia's Hövsgöl Province 1000 km away from the national capital of Ulaanbaatar and 200 km away from the provincial capital, Mörön. The Darhads originate from a complex mix of ethnic groupings only some of which were Mongolian in cultural and linguistic terms, whereas the rest were Tuvan, Turkic and, possibly, Tungusian tribes. Today these original groupings, many of which at some point were known as clans (*ovog)*, are largely defunct in sociological and economical terms, though people still make reference to certain clans; particularly in the context of the Darhads' shamanist religion, in which they play a crucial role (Badamhatan 1986; Dulam 1992; Pedersen 2011a). The Darhads are clustered around a region bearing their name. Located west of the Hövsgöl Nuur, Mongolia's largest freshwater lake, south-east of the Tuvinian Autonomous Republic of Russia, and south-west of Russia's Buryat Autonomous Region, the Shishged Depression is a geographically isolated lowland situated between three large mountain complexes, south-east of Mongolia's border with the Tuvan Autonomous Republic of Russia. In topographical terms the Shishged Depression is quite unique in Mongolia, for it is comprised of a largely uninterrupted flat steppe zone surrounded on all sides by coniferous forests and high alpine lands (collectively known as the *taiga*).

Ever since the heyday of the Mongolian empire in the 13th and 14th centuries, if not before, the Shishged Depression has fluctuated between

1 Fieldwork was carried out from October 1995 to January 1996; June 1998 to September 1999; June to August 2000, and from May to June 2009. I would like thank the Danish Research Academy; King's College; the William Wyse Foundation; The Department of Anthropology, Aarhus University, Mindefondet; King Christian X's Foundation; HH Queen Margrethe and Prince Henrik's Foundation, and Crown Prince Frederiks's Grant for Scientific Expeditions for financial support that has made this research possible. This article builds in some parts on Morten Axel Pedersen's previous article "Tame from Within: Landscapes of the Religious Imagination Among the Darhads of Northern Mongolia" in *The Mongolia-Tibet Interface*, edited by Uradyn E. Bulag and Hildegard G.M. Diemberger, Leiden: Brill, (2007). Gaby van Rietschoten and the Brill publishing house are gratefully acknowledged for the permission to re-use parts of this previous work.

periods of social, cultural, and religious incorporation during which this region and its inhabitants were fully part of large-scale polities, and periods of social, cultural, and religious fragmentation, where it was not governed by any single sovereign, subject instead to constantly changing alliances and feuds between different tribes and their chiefs. Indeed, the Shishged Depression is situated right on the border between two major geographic, political and cultural regions. Not only is it located on the present-day border between Mongolia and Russia, which, before that, was the border between the Mongolian Peoples' Republic and the Soviet Union, and, still further back in time, the Qing empire and the Russian empire. It is also on the geographical border between the forests of Siberia and plains of Inner Asia; on the religious border between the predominantly shamanic indigenous peoples of Siberia and the predominantly Buddhist peoples of Mongolian societies; and on the sociological border between the traditionally egalitarian hunting societies of Northern Asia and the traditionally hierarchical pastoral societies of Inner Eurasia (Ingold 1980; Hamayon 1994; Humphrey 1994; Pedersen 2001, Pedersen 2011a).

Indeed, "the Darhad people" (*Darhad yastan*) is a product of complex political dynamics unfolding at the fringe of the Qing empire from the late seventeenth to the early twentieth century. With the collapse of the Mongolian empire in 1368, Northern Mongolia entered a tumultuous period in which, for nearly four hundred years until the final defeat of the Jungar (Oirat) federation 1757, it was caught in endlessly shifting alliances and antagonisms between different polities, including the rising imperial powers of Czarist Russia and Qing China, who simultaneously exercised power as part of a wider political outmanoeuvring game. In the mid-18th century, the Shishged entered a period characterized by relative stability, which was to last until the chaotic decade of political turbulence culminating in the revolution of 1921, and, with the death of the "Holy Emperor" (Bogd Khan) in 1924, full Communist takeover. During this period, which people nostalgically refer to as the "golden age of Buddhism", the Shishged was home to the Darhad Ih Shav' ("The Great Darhad Ecclesiastical Estate"), a Buddhist estate belonging directly to the Jebtsundamba Khutuktu, the leading reincarnation or, in popular parlance, Living Buddha of Mongolia's Buddhists. It was this Buddhist estate that defined Darhads as a people of ecclesiastical disciples (*shabinar*) corresponding to a sacred landscape and administrative structure, and it was this institution that established Gelupka Buddhism firmly in a region, which, by all accounts, had until then been a shamanist stronghold (Sandschejew 1930; Zhamtsarano 1979).

This chapter explores certain contemporary ramifications of this historical Buddhist missionizing or, to adopt the lamas' own term, "domestication" project that took place in the shamanic hinterlands of Mongolia in the centuries up until the communist takeover. My argument proceeds from the ethnographic premise that Darhad concepts of the person cannot be understood without a careful exploration of how people are thinking through the landscape. Indeed, many aspects of Shishged social life revolve around two overarching dichotomies, namely a spatial contrast between the taiga and steppe, and an existential contrast between what is widely understood to define the basic substances of Darhad persons, namely the so-called "yellow side" (*shar tal*) and the so-called "black side" (*har tal*). The black side comprises everything that is violent, uncontrolled, harmful, morally ambiguous, and – the logic goes – shamanic. The yellow side, conversely, is all peaceful, balanced, benevolent, morally unambiguous, and – therefore – Buddhist. Indeed, people often refer to shamanism as the "black religion" (*shar shashin*) and to (Gelugpa) Buddhism as the "yellow religion" (*shar shashin*).

Some might wish to think of these two "sides" as ethnic identities, for they spring from a cultural imaginary shared by Darhads and non-Darhads as to what distinguishes the former as a *yastan*. However, the concept of ethnic identity may be misleading in the present context, if not generally (Pedersen 2011b). Not only does it easily give rise to the dubious notion that the Darhads subscribe to only one subjectivity in their relations with significant others, but it also conceals that Darhads see themselves as internally differentiated into "black" and "yellow" persons. For, although both of these "sides" are understood to reside within each and every Darhad, they are also employed to distinguish between different kinds of Darhads. Every person is simultaneously black and yellow on the inside, but not on the outside, where some persons (such as potential future shamans) stand out by being particularly "black", while other persons (like respected male elders) are prominent in a "yellow" way. Thus agency becomes a question of how these two sides are extracted from land, things and people, by different kinds of persons recognized to have the capacity to do so (Pedersen 2011a).

The central claim of this chapter is thus that the two "sides" of Darhad persons are folded into the geographically-cum-historically generated spatial contrast between a Shishged *taiga* zone and a Shishged steppe zone, and between shamanism and Buddhism. What we seem to be faced with, then, is classic line-up of homologous binary oppositions, which, to adopt the old structuralist convention (Levi-Strauss 1962), may be depicted as follows: wild: *taiga:* shamanism: "black side" :: tame: steppe: Buddhism: "yellow side". The point, however, is that there is an inherent asymmetry at work in these homologies; and it is the nature and the implications of this homological asymmetry that defines the thrust of my argument. I contend that the Shishged *taiga* zone constitutes a "multiple margin" in asymmetrical opposition to which the Shishged steppe zone represents a "singular centre"; and that this contrast between a wild or natural dimension that is "black", "shamanic" and "multiple" and a tame or cultural dimension that is "yellow", "Buddhist" and "singular" is replicated across a range of different objects and scales, including, above all, mindscapes and landscapes.

In what follows, I outline the key dynamics of the process by which the "outer topography" of the Darhad landscape has become replicated in the "inner topography" of Darhad persons. I then present in some detail the Buddhist attempt to domesticate of the Darhads and their landscape in the 18th and 19th centuries, and end by discussing some local legends suggesting an alternative view of Darhad cultural history. But first, let me begin by describing the prevailing Darhad stereotype entertained by most people in Mongolia.

2. The black side and the yellow side

How do other Mongolians view the Darhads? Tellingly, little attention is paid to the Darhads' notable Buddhist history. Indeed, the Darhads are widely known to be *the* shamanist ethnicity (*yastan*) in Mongolia. Thus non-Darhads,[2] when asked to characterise the Darhads, tend to single out three traits: 1) The Darhads, due to the extreme remoteness of their homeland, are wild, crude, and poor. 2) The Darhads are deeply shamanic, and they all have the ability to curse (*haraal hiih*). 3) The Darhads are inveterate jokers, about whose intentions one can never be sure. Taken together, these traits signify a more general concept of a significant other in opposition to which non-Darhads can identify themselves. If They are shamanists, then We are Buddhists; if They are wild, then We are civilised; and, since They are positioned at the very margin/border

2 By this term I here refer mainly to Mongolia's Halh majority, whose stereotype of the Darhads many of Mongolia's other ethnic minorities seem to share.

(*zah, hil, hyazgaar*) of the Mongolian nation-state, We are positioned closer to its centre (*töv*).

But the Darhads are not only seen to occupy the margin in geographical, economical and political terms. They are also known to be marginal in sociological and psychological terms. Darhads are thus not held to be as *shuluuhan* ("straight, direct") as other people, notably Mongolia's Halh majority, who, at least in Northern Mongolia, are known to be especially *ilen dalangüi* ("open, earnest'", *töv sanaatai* ("balanced", lit. "to have centred feeling-thoughts") and *töviig baridag* ("balanced", lit. "to be holding the centre").[3] More generally, as Bulag (1988: 70–76) have argued, where the Halh originally constituted one *halh* ("flank, shield") of the Mongolian heartland, they are now widely considered to be positioned at its very *gol* ("core, centre"). Thus Halh's peoples' "balanced" and "centred" *inner* position correspond to their *outer* position in Mongolia's political economy. Darhads, conversely, are "unbalanced" (*töviig baridaggüi*, lit. "not holding the centre"), and, far from being "straight and direct" (*shuluuhan)*, Darhads are feared for "always to speak in a roundabout way" (*dandaa toiruu yaridag*). Indeed, Darhads are widely believed to have "layered minds" (*davhar uhaan*). Interestingly, *davhar*, in addition to meaning "double", "layered" or "stratified", is also used to denote processes of impregnation (e.g. *biye davhar boloh* – "(for a body) to become pregnant") (Hangin et. al. 1986). From the non-Darhad point of view, then, the Darhads' political-economic marginality corresponds to an internal state of multiplicity: every Darhad person is understood to constitute an existential manifold within.

In many ways, the negative ethnic stereotype of the Darhads is similar to their own notion of the "black side", as outlined above. Typically, when people talk about the "black side", it revolves around the famous hostility of Darhad shamans towards the Buddhist *lamas* of the Darhad Ih Shav'. A multitude of legends (*domog*), curses (*haraal, zühel*) and shamanic invocations (*duudlaga*) play on this theme. Common for such narratives is the ability of Darhad shamanic spirits (*ongon*, pl. *ongod*) to undergo constant and unpredictable metamorphoses (*huvilgaan*). In one legend, for example, the spirit of a female shaman (*udgan*) transmutes into a rainbow hovering across a valley, where a caravan of Buddhist *lamas* is passing through. The *lamas* are terrified: the rainbow is defiling them, for it is also the underneath of a menstruating shaman. Eventually,

3 C.f. Hangin et. al. (1986). As Sneath (2000, 144) points out, *töv* – whose general meaning is that of "centre" or "middle" – also means "orthodox and righteous".

a high-ranking *lama* (in some versions the Jebtsundamba Khutuktu himself) says a powerful prayer, and the rainbow dissolves. We have here a fine example of how the historical conflict between shamans and *lamas* takes the form of a contest over control of the Darhad landscape (just as it is strongly gendered). However, an additional notion underwrites the idea of the "black side", namely one on which the Darhad Depression plays the role of a topographical index of the Darhads' uniqueness. Consider, for example, the following explanation, offered to me by a prominent Darhad hunter:

> There is a kind of uranium (*uraan*) around here. Nature (*baigal*) contains it, and the flowers, wild animals and so on receive it, and pass it on to humans. A wild goat may have *uraan* and rest where the blueberries are growing. The blueberries will then receive the *uraan*, and their taste and colour will become excessively nice. So, humans will eat all the berries and receive the harmful things (*hortoi yum*). Their eyes will become light blue and they sight will deteriorate. I think, if Darhad people were able to avoid this influence from nature (*baigalyn nölöö*), we would live for 2–300 years. Darhads are different because the nature is different around here. We are receiving many things from nature – too many things – and this makes our minds very powerful (*hüchtei*) and very strange (*hachin*). What are these things? They are the many different things of nature (*yanz büriin baigalyn yum*), different things that influence peoples' minds and render them powerful. This is why we have shamans here, and why we have the ability to curse. Some people will not admit this, but they ought to, for it is part of them [...] People who have left the Darhad Depression generally have success; their heads are working very well. During their time here, they received enough energy for the rest of their life. People who stay usually do not have good lives – they are receiving too much power from nature. So why don't we leave? Because we cannot: Nature is pulling us back.

Not surprisingly, migrant Darhads seem to agree with this characterisation. An elderly Darhad man living in the provincial capital of Mörön made the following observation: "Why, I am a real Darhad, but I also did right in leaving my homeland. The Darhads living in our native land are smart enough, but their problem is that they can't see to the end of things" (*etsest n' hürgehgüi orhidog*).

The native Darhads, it would appear, are not "straight" enough in their heads, at least not in the superior manner the Halh are known to be. Certainly, this is the message of the following observation, made by a middle aged businessman from the ethnically mixed Arbülag district, located some eighty kilometres south of the Shishged Depression. The man classified himself as *erliz* ("of mixed breed"), half Darhad and Halh, and is (as I was later told) very rich. "You see", he said, "in our district, the Darhads are doing extremely well. We are the best in school, we are highly skilful (*chadvartai*), and we are hardworking. This is because we are living in close proximity to the Halh. In the Darhad Depression, people do only what they feel like doing, which sometimes is very much, but usually is very little. But, here in Arbülag, the Halh leaders know to praise the Darhads so that they will work all the time". (For more on the Darhad minority districts in the Hövsgöl province, see Lacaze 2000).

Again, we are presented with the idea that the Darhads are out of balance. Their minds are hazy and unclear. They cannot see properly. They are unable to carry things through. Too many things are distracting them, influencing them. But the quotation from the hunter also shows that, when Darhads talk about the "influence from nature", they tend to do so with reference to the *taiga*. Thus the vehicles of *uraan* are associated with the *taiga* (blueberries and mountain goats), and not with the steppe zone. Darhads, it seems, are not at risk of receiving *uraan* from the life forms of the steppe, such as the domestic livestock or the grass on which these animals feed. In fact, the steppe is not associated with shamanist activities. Rather, the steppe is known as a peaceful place. It is where the pastoralists nomadise; it is where the wild predators (ideally) do not come, and it is where everyone, man and beast alike, enjoys a carefree life during the lush and plentiful summer.

We can now return to the discussion of marginality and multiplicity. For if we look at the first quotation again, we realise that nature (*baigal*) here is evoked to express the idea of a manifold. There are "many different things" in the Shishged Depression, "too many different things", in fact. Diversity, not unity, is what the hunter's explanation is about, apart from "power" (*hüch*). Certainly, *uraan* is used as an all-encompassing metaphor for the "influence from nature"; yet it is surely no coincidence that this term is adopted for this purpose. Uranium, after all, is known for its inherently unstable nature – and for the harmful power that springs from this instability. *Uraan*, it seems, represents an apt metaphor for conveying the popular Darhad notion that the *taiga* zone is a vehicle of transformation, of mutation (of one's eyes, for example), and, above all, of multiplicity.

To the Darhads, I propose, the *taiga* zone constitutes an external homologue of what from the perspective of non-Darhads (and migrant Darhads), is internal to all Darhad persons. The steppe zone, on the other hand, is understood to provide the inhabitants of the Darhad Depression with what moving away from the Darhad homeland has accomplished for the migrant Darhads, namely a sort of spatial refuge from the "black side'" But how did this spatial refuge come into being? Why does the steppe constitute a "tame" safe haven beyond the reach of the dangerous "influence from nature"? To answer these questions, we need to consider how the Mongolian Buddhist church governed the Shishged Depression during the centuries up until the Communist revolution.

3. The Darhad Ih Shav'

Much has been written about Buddhist domestication projects and the search for hidden valleys in the borderlands of Tibet (see e. g. Blondeau & Steinkellner 1996; Huber 1999; Ehrhard 1999a, 1999b; Dienberger, H. & G. Hazod 1999). Less attention, however, has been paid to similar politico-religious agendas at play at the northernmost fringes of Inner Asia. Here, various representatives of the Mongolian Buddhist church were also engaged for centuries in an uneven struggle for domination with local shamanist practitioners, usually to the effect that the latter became marginalised in society or even disappeared altogether (Heissig 1980; Even 1991). The remote forests of northern Mongolia, for instance, seem to have represented a treasured pilgrimage spot for a certain class of *lamas*, the so-called *badarchid* ("itinerant *lamas*") (see Pozdneyev 1971, 343–44; Charleux 2002, 169). There is every reason to think that, as has been shown to be the case in Tibet's southern borderlands, these Mongolian *lamas* ventured out to discover "an untamed wilderness, one which awoke anxieties and terrors but also held the prospect of spiritual satisfaction." (Ehrhard 1999a, 228; see also Charleux 2002, 195–200).

It is well-known that "part of the aspiration of Tibetan religious ideology is to eliminate wilderness by subjugating it" (Ramble 1997, 133). The Darhad Ih Shav' is a case in point. Indeed, there is reason to believe this Buddhist estate entertained a privileged position within the total ecclesiastical estate (Shabi Yamen) of the Jebtsundamba Khutuktu. For one thing, the Darhad Ih Shav' was one of a few *shav'* situated inside the politically ambiguous borderland *hyazgaar* between the Qing and the Russian empire (Ewing 1981). Not only was the Darhad Ih Shav' uncharacteristically open to Russian influences, but its ecclesiastical subjects were also literally sealed off from the Outer Mongolian *aimags* located to the south of the *hyazgaar* (Badamhatan 1986, 26). Secondly, because of its location within the forest belt of Northern Mongolia, the Darhad Ih Shav' constituted the Shabi Yamen's primary source of fur and pelt revenues, obtained from the *shabinar* in the form of direct taxation as well as religious alms (Badamhatan 1986, 27–34). Finally, as Bawden notes (1986, 69), "the *Shabi yamen*, did not control actual territory, apart from the pastures of the Darkhat in the far north-west of Mongolia". So not only does the Darhad Ih Shav' seem to have comprised the largest and economically most vital population of *shabinar* within the Shabi Yamen estate, but it also comprised the only lands in Qing Outer Mongolia under *de facto* sovereignty of the Buddhist church (see also Ewing 1981; Pedersen 2011a).

On several levels, I propose, the Darhad Ih Shav' imposed a spatial dimension and political scale upon the Shishged Depression which had not been there before. The very entity known as the "Darhad people" is a case in point. For, even if one assumes that a grouping with this designation was found in north-western Mongolia before the Darhad Ih Shav's creation at the Khuren Belchir Assembly in 1668, it is still unquestionable that this original group subsequently were infused with a range of other ethnic or political groupings (Badamhatan 1986, 62–63). Indeed, the Darhad Ih Shav' was probably created by a close alliance between Mongolia's Buddhist church and its Halh nobility with the explicit aim of imposing a degree of political stability on the Hövsgöl region, which had suffered from civil war and general political unrest for several centuries (Ewing 1985; Bawden 1986; Pedersen 2011a). One could even argue that the very concept of "domestication" is inscribed into the Darhad ethnonym. The term *darhan* (pl. *darhad*) can denote something "sacred", "protected", and, most interestingly, "an area set aside for religious reasons or rites" (Hangin et. al. 1986).

It is not fully clear how many monasteries the Darhad Ih Shav' comprised at a given time during its more than two-hundred years long history. Several monasteries were relocated or perhaps even closed down due to the ongoing conflicts with local shamans as well as warlords from the White Army during the Russian Civil War (Pürev 1980, 46–48). But it is reasonably certain that the first monastery dates back to 1757, at which point the Darhads apparently resettled in their homeland following their forced migration to the Selenge region during the various Jungar invasions of the late 17th century (Badamhatan 1986, 25; 44–45; Badamhatan & Banzragch 1981, 13–15). This was the Züölöngiin Hüree (also known as the Darhadyn or Renchinlhümbe Hüree), which was to become the religious, administrative and commercial centre of the Darhad Ih Shav' for the next 175 years or so (see, for example, Sandschejew 1930). The location of this ecclesiastical centre was, however, changed several times. Initially, the Zöölön monastery was built at the mouth of the river Ivd, which is located in the present day Soyot sub-district of the Ulaan-Uul district. But, partly for occult reasons, and partly for practical ones, the Zöölön monastery was later relocated in two stages, eventually to find its permanent base East of the Shishged Depression's geographical centre, close to the present day Renchinlhümbe district centre (Dioszegi 1961).

At some point between 1821 and 1855, the Darhad Ih Shav' underwent an administrative reform. Until this point, this Buddhist estate had been organised into one *otog* (i. e. the ecclesiastical equivalent to the *hoshoo*, the main administrative unit in Qing Mongolia), but it was now reorganised into three *otogs*, known as the East, West and North *otog* respectively, each administered by an office headed by a secular *otog* leader (*otogyn darga*) ultimately answering to the ecclesiastical authorities (Badamhatan 1980, 26; see also Legrand 1976, 81–82; Vreeland 1962: 11–23). Not surprisingly, the introduction of this tripartite administrative structure was soon followed by the introduction of a tripartite monastic structure as well. Thus, in 1880, a second monastery was built at a place called Tsaram, but was soon relocated to another place called Burgaltai, after the river with the same name. Later, the monastery was moved once again finding its permanent location south of the Hög River in the present Ulaan-Uul district, though it kept its former name, the Burgaltai Hüree. In 1890, a third monastery was established, which in the early 1900's, was *also* moved to the aforementioned Ivdiin Am. Henceforth, this third monastery became known as the Ivdiin Hiid. Finally, two smaller prayer temples (*hural*) were at some point constructed to the north-east of the Zöölön monastery, namely the Töhiin Hural and the Mandalyn Hural respectively.

At the beginning of the 20th century, then, there seems to have been five (or possibly six) monastic sites in the Shishged Depression, all located within the spatial and political territory of the Darhad Ih Shav'. Now, if one plots these monastery locations onto a map, something resembling a star-shaped figure emerges. This could be taken to suggest that there was a deliberate design behind the construction of monastic sites in the Darhad Ih Shav'. Still, plenty of pragmatic reasons may account for the location of these sites. A variety of more contingent religious factors should also be taken into account, such as the possibility that the monastic locations were determined by divination activities carried out *in situ*, by the nature of the water flow of local streams and so forth.

If the strategic positioning of monasteries was one way in which the Mongolian Buddhist church sought to domesticate the Shishged Depression, then its appropriation of pre-existing sacred sites was another. Now there are many different kinds of sacred places in the Shishged Depression, some of which are (still today) distinctly shamanist in nature. Here I will focus solely on the so-called *ovoos* (i. e. sacred stone or wooden cairns constructed at prominent spots in the landscape to appease local "land-masters" (*gazaryn ezed*)). Both the written sources and my informants suggest that, by the mid-19 h century, the Darhad Ih Shav' had taken over the management of most prominent *ovoos* within its territory. Badamhatan (1980, 24) writes that, in 1855, the Darhad Ih Shav' comprised 26 *züünij ovoo*s and 24 *hilnii ovoo*s, where the former apparently served 'monastic' purposes, and the latter marking the borders of the non-ecclesiastical administrative units in the vicinity of the Darhad Ih Shav'. It is not known whether the 26 *züünij ovoos* existed before the creation of the Darhad Ih Shav', but it is likely that the majority did so, as these must have played an important politico-religious role for the indigenous patri-clans of the region (c.f. Humphrey 1995). So, it is reasonable to assume, as these and other clan migrant groupings became incorporated into the administrative structure of the Darhad Ih Shav', the corresponding *ovoos* and their 'resident' *gazaryn ezed* underwent a similar domestication, as local *lamas* gradually took over the ceremonial roles previously performed by shamanist clan-elders at these *ovoo* sites.

Nevertheless, the *ovoos* did not lose their pre-Buddhist significance with the increasing institutionalisation of the Darhad Ih Shav'. My data suggest rather that the former clannish cult-sites became subsumed under standardised liturgical forms imposed by the Buddhist church, and by equally standardised bureaucratic interventions instituted by the Darhad Ih Shav's secular arm. For example, a high-ranking *lama* from the Ivd monastery each year would preside over the *ovoo* ceremonies of the North Otog. The latter's territory, like that of the two other *otogs*, is likely to have comprised several *ovoos* as well several clan groupings. So, if the *otog* leader of the North Otog encompassed the diversity of human groupings (households, clans) within this *otog's* territory, the high-ranking *lama* assumed the leadership of the diversity of non-human entities ("land-masters" etc.) within the same *otog*. In that sense, the North Otog and its corresponding Ivd monastery seems to have fulfilled the same encompassing role towards the clans and *ovoos* within their territory as did the larger Züölön monastery towards them.

We are reminded here of Stanley Tambiah's notion of the "galactic polity" (1985). As Samuel (1993, 62–63) notes, this model only partly fits the Tibetan case, since the latter was often characterised by several politico-religious polities in conflict with one other. A similar objection might be raised with respect to long periods in Mongolian history, though the Qing colonial polity does, in fact, fit Tambiah's model (Bawden 1986, 108; Humphrey 1996, 275). Indeed, when it comes to the Darhad Ih Shav', Tambiah's model seems to work very well. For it is clear that the Züülön monastery was the centre of this micro-cosmos, and it is now also clear that its four adjacent monastic sites performed the role of "satellites" in Tambiah's sense, replicating (if to a higher or lesser degree) the political, economic and religious properties of the former, but on a smaller scale.

4. A magic circle?

Are we able to conclude, then, that the different politico-religious interventions of the Buddhist church outlined above gradually turned the Shishged Depression into "an enclosure, not necessarily circular, which separates a sacred area from the ..... profane world" ? (Snellgrove 1987, 198). In my view, this would imply the existence of a particular point from where this encompassment could be *seen* (or at least imagined); that is, a spatial vantage point from where the *entirety* of the Darhad Ih Shav' could be apprehended in a *mandala*-like way. In fact, the so-called Jargalantyn Ovoo (also called the Züülöngiin Ovoo) seems to have constituted precisely such a spatial vantage. This *ovoo* was – and still is – located on a hill top near the geographical centre of the Shishged Depression. Indeed, as the proud locals seldom fail to tell you, Jargalantyn Ovoo is "the only place from where it is possible to see the whole Darhad Depression". At this site, I was told, seven lamas from the Züülön monastery used to perform an important annual ritual.

First, the seven lamas would visit a sacred lake, at whose midst there was – and apparently still is – a tiny island with seven Siberian Larch trees (the lake is probably Deed Tsagaan Nuur). Here, the lamas made offerings and read prayers to the seven trees, which were named after the Great Bear Constellation (*Doloon Burhan Od*).[4] Following this, the

4 Pegg (2001, 117) presents a short Darhad wish-prayer for this star constellation, which is worth quoting due to its explicit Lamaist connotations: "Risen above //

lamas would climb the nearby hill, at whose peak the Jargalantyn Ovoo is located. Only the lamas were allowed to ascend all the way up to the *ovoo* itself. The laymen assembled for the ritual were left behind on the hill below the *ovoo* site (women were not allowed to participate at all). The seven lamas would then perform a full-blown *ovoo* sacrificial rite (*ovoony tahilga*), in which sacrifices (*tahil*), sutra-readings and beckonings (*dallaga*) were made for the local 'land-masters' / gods of the Great Bear. Finally, a big celebration of games (*naadam*) was held at the mountain pass below, in which ecclesiastical subjects from all over the Darhad Ih Shav' participated.

In discussing a comparable (Bonpo) case from the Tibetan context, Ramble suggests that "the pattern that formed around the site were, like a mandala in the most general sense, a magic circle that changed everything that came within its perimeter. The changes are thus not uniform but in accordance with *possibilities of form* offered by the nature of the quantities concerned: divinities are ranked hierarchically, stray events find themselves drawn into a unifying narrative, rocks are accorded resemblances to suitable subjects, and wildlife becomes tame" (1997, 134; emphasis added). This observation is highly pertinent for our present purposes, for it emphasises that a given project of domestication does not give rise to a perfect real-world instantiation of the *mandala* shape. Indeed, it is to some degree beside the point whether the Darhad Depression was meant to be transformed into a *mandala* or not. What matters is that the this landscape – with its unique hollow shape – must have offered the Mongolian Buddhist church a perfect "possibility of form" through which it could carry out its agenda of subjugation. Indeed, the term *hotgor* – which I elsewhere in this article have translated as "depression" – also means "concave", "a cavity" and "hollow" in the Mongolian language. (Incidentally, Darhads informally refer to their homeland simply as "The Hotgor").

But what was the outer perimeter of the sacred enclosure delineated by the Darhad Ih Shav'? It is relatively certain that the Buddhist sphere of influence did not reach into the depths of the *taiga*. Rather, the border zone between the steppe and the *taiga* appears to have been the major battlefront in the lamas' continuous struggle to subjugate the Darhads and

The Seven Gods of the firmament // Guard [us] like a hat // Guard us as a shadow // Please banish misfortune and evil spirits. (*Oroid mandsan // Ogtarguin Doloon Burhan // Malgai met mana // Süüder met sah' // Gai bartsat gamshig totgoryg arilgaj hairla*"; translation original).

their landscape. The frequent relocation of monasteries within the Darhad Ih Shav' testified to this. Clearly, had the Buddhists not met so fierce resistance from the local shamans, all these relocations might not have been necessary. We may therefore view the vectors delineated by the monastic relocations as indices of the changing power balance between shamanism and Buddhism in the course of Darhad history. A famous narrative published by Dioszegi (1961), for instance, refers to the earliest days of the Darhad Ih Shav', namely to a time when the only Buddhist site in the Darhad Depression was a temple inhabited by a sole lama; and to a time when the Buddhist church was forced to move this temple further into the steppe zone, because "shamanism was flourishing in the region" (1961, 202). Then, as the ecclesiastical estate gained more strength, its activities gradually expanded outwards towards the *taiga* zone, such that, eventually, the Ivdiin Hiid was erected at the very spot from where the temple originally had been removed, as explained above.

But this liminal zone between the steppe and the taiga also came to define the outer range of the Darhad Ih Shav's sphere of influence. The *taiga* zone proper, with its "too many different" animal, plant, and spirit entities, was left for the shamans to deal with (Pürev 1999, 342–44; Pedersen 2011a). But a more general observation also springs to mind. Just as in the comparable cases from Tibet (see MacDonald 1997), the Buddhist subjugation of the Darhad Depression never was – and nor could it ever have been – completed. Susan Mackinnon has demonstrated how, in Island Melanesia, "unified, fixed, weighted centers [are contrasted] to multiple, dispersed, moving, and weightless peripheries" (1991, 33). Were these centers to exist on their own, their solidity would be become void, for they would have no labile other in whose asymmetrical refraction they could imagine themselves as composed by congruent, commensurable entities that add up into one another. In much the same way, I suggest, for the weighty concept of a centred steppe to work in the Shishged context, it requires the weightless counter-image of a marginal *taiga*. The *taiga*, then, emerges as a zone of pure multiplicity and metamorphosis in opposition to a zone of pure sameness and singularity, the steppe, and this centre–periphery spectacle has come into being through a combination of consecutive political and religious interventions pertaining to the Mongolian empire, the Mongolian Buddhist church, and, as I have accounted for elsewhere (Pedersen 2011a), the Mongolian communist party.

We are now better equipped, ethnographically and historically, to account for the analogies that many Darhads make between the topography of their minds, and the topography of the Shishged landscape. Even on the assumption that shamans were already associated with the *taiga* before the creation of the Ih Shav' (see Pedersen 2011a), the interventions of this ecclesiastical polity unquestionably pushed the shamans, and the spirits, further into the *taiga.* In this light, the main reason why Darhads are today keen on mapping their minds onto the landscape is that, in doing so, they are following in the footsteps of their formerly Buddhist masters in attaining a perspective from which the *taiga* zone appears as a residual repository standing in asymmetrical opposition to the purified enclosure demarcated by the natural contours of the steppe zone.

5. Tame from Within

Several scholars have remarked that it was somewhat paradoxical that "the Darkhad region, [as a] stronghold of shamanic traditions . . . was controlled by the Buddhist church" (Even 1991, 200) until the revolution. But perhaps it was not so surprising after all. For it should now be clear that, with the continual institutionalization of the Darhad Ih Shav' from 1757 to 1921, an entire shadow world – the taiga zone – came into being that was filled up with all the "black" , "wild" and shamanic stuff which could not, and oftentimes would not, fit into the "pure" and "yellow" cosmological form imposed by the Buddhist church. In that sense, as I have argued, the lamas from the Ih Shav' *needed* the Darhad shamans to resist them; for had these shamans not done so, then it would have been hard for the lamas to justify, to themselves and to others, their self-acclaimed role as protectors of "wild and savage" Darhads of the north. Thus, for the Buddhist monks, flying demons and unruly shamans presented not just a problem to be solved, but, above all, an opportunity to be grabbed, for without an black and savage side to these people and their land, why would they need any outsiders to turn them yellow and pure?

Yet, on the Darhads' own understanding, what took place was subtly different. For it was not simply that they asked to be protected by the Jebtsundamba Khutuktu. It was also that the latter was irresistibly attracted to the Shishged and its people. Consider the following story, told to me in several different versions during fieldworks in the Shishged Depression:

Many years ago, a Tibetan *lama* came to our land: he had been sent by the Bogd Khan [the 8th Jebtsundamba Khutuktu]. As the *lama* reached to top of Öliin Davaa he was breath-taken: so beautiful was the sight that met him.[5] He exclaimed out loud: "This is a land of happiness. It is full of white merit (*tsagaan buyan*). This is because of the three Darhad White Animals: the Darhad White Sheep, the Darhad White Fish, and the Darhad White Horse'". The merit of the Darhad White Sheep comes from its special tail-bone, which is bigger than that of Halh sheep. The merit of the Darhad White Fish has to do with the healing powers of the water in the Shishged River. When one takes out a White Fish from the river, it will shine like gold. Also, the White Fish is very rare in the world.

This is the story about the Darhad White Horse. Once, there were two friends, and one was very ill. Shamans had been called in, but to no avail. Aware that his friend was about to go to Urga [present-day Ulaanbaatar], the sick man sent for him. "Take this gold and buy me medicine", he said, handing his friend a muddy stone. The friend took the stone and went off to Urga. But no one wanted to sell him any medicine, for all he had to pay with was the stone. Desperate, he went to see the Bogd Khan. The Bogd Khan weighed the stone in his hands, knocked it onto the wall, and said: "yes, this *is* gold. I will now tell you the cure. Back in your country there is the Darhad White Horse. Your friend must drink a cup of mare's milk three times a day and eat the mutton from the Darhad White Sheep. Then he will be cured".

The friend gave the stone to the Bogd Khan and travelled back home. Arriving empty-handed, the sick man scolded him for not having brought the medicine. He told of the Bogd Khan's advice, only to be met with the angry reply that the ailing person already had been drinking large amounts of fermented mare's milk (*airag*), and to no effect. Some time now passed, and, as the sick man was feeling worse and worse, he decided to heed the advice. And indeed, after nine days he began feeling better, and after one month he had completely regained his health. His companion, meanwhile, had been worrying: might his friend be dead? One day, as he was sitting inside his *ger*, his children shouted from the outside: "three horsemen are arriving". "Oh no", he thought, "could they be coming after me?". As he stepped outside his fears increased as he saw three proud men dismount and approach him. But, behold, it was his friend who was running up him, shouting: "the cure worked!".

Note that neither the Tibetan *lama* nor Bogd Khan are bringing anything new to the Darhad people. The *lama* is only saying out loud what can already be found in the Shishged Depression, namely the sacred blessing (*buyan*) of the three Darhad White Animals. Similarly, the Bogd Khan is pointing to a cure that is already available, namely the healing power of

5 The Öliin Davaa is the most important mountain pass in the Shishged, and home to a very prominent *ovoo*.

the milk from the Darhad White Horse. The stone/gold contrast seems to carry the same message. Only the sick man (because he is near to death?) can apparently glimpse its hidden golden quality; his friend and everyone else needs the Bogd Khan to weigh it and knock it against the wall, and thus make its inner purity visible to the world.

How are we to interpret this? I would suggest that the core idea is one of attraction. The Shishged Depression was attracting the Buddhist church to come to it, for it contained a superior "whiteness" that proved irresistible to the Tibetan *lama* (and his many Mongolian successors). But the attractor may not be aware of its own attraction: it may need someone or something to bring about its irresistible appeal, like when a fish requires to be pulled out of the water in order to shine golden, or when a piece of gold is revealed behind its outer shelter only if adequately treated. Still, everything was there from the beginning. The "white blessing" *was* already in the animals, and the muddy stone *was* already made of gold; it was simply that these hidden qualities needed to be extracted.

In more general terms, we are here reminded about the association made by many Darhads between the steppe zone and various pure and healing life forms. It is true that, by its very nature, the White Fish is not confined to the steppe zone. In fact, according to Dioszegi (1993, 78–80), the Shishged River is also said to be the genius loci of Zönög Aav, one of the most prominent Darhad shamanic spirits. There is, however, an alternative conception according to which the Shishged river has healing properties. These powers are deemed especially strong around Tsagaan Nuur Rashaan, a mineral spring located near the Jargalant Ovoo (see also Bat-Erdene 1992). Indeed, some people told me, the spring is the very source of the Shishged River. It is likely, then, that the Darhad White Fish embodies the healing power of Tsagaan Nuur Rashaan in the above narrative.

As domestic animals, the blessing (*buyan*) of the White Horse and the White Sheep correspond univocally to the steppe zone. But the connection between purity and the nomadic landscape goes one step further. Thus the *airag* from Renchinlhümbe – the most centrally located of the three districts in the Shishged – is widely considered the best. This is so, I was repeatedly told, because it is the only place where the right grasses grow. Indeed, people went on, the Shishged is a like a "small Mongolia with Renchinlhümbe at the centre". "Just take a quick look around", a man eagerly explained, "All varieties of landscape can be found here. There are steppe and salt marshes. There are rivers, freshwater lakes, and salt lakes. There is even desert and sand dunes. Even here, this far north, there is desert [*gov*]'. Interesting!"

The overarching idea to be discerned here, it seems, is one of self-containment. Unlike what is commonly believed by non-Darhads, the Shishged contains everything one needs to be a "real Mongolian" – and then some. For this land also contains the Darhad White Fish (with its rare healing powers); the unique Darhad White Horse (with its tasty and purifying milk); treasure-like stones that are full of gold; and (most important of all!) unusually tasty sheep with bigger and fatter tails than the Halh sheep. Crucially, all these qualities are associated with the steppe zone; in fact, they are especially associated with the most central parts of this landscape. This is where Tsagaan Nuur Rashaan is located; this is where the right grass is growing for the White Darhad Horse; and this is where one finds all the "interesting" deserts and lakes. And, indeed, it is where one finds the Jargalant Ovoo – the panoptic centre of all Shishged centers.

What, then, happened to that "untamed wilderness" which needed to be "domesticated" by the Buddhist church? It is unquestionable that the aforementioned legend is concerned with the historical existence of the Darhad Ih Shav'. But, and crucially, according to this and other narratives, the Buddhist church did not bring anything new the Darhads and their land; it merely brought out a hidden quality which had always been there. On this perspective, then, the Darhads were not actually domesticated by the Buddhist church, for, a bit like the natural refuge demarcated by the steppe zone, the Darhads were already *tame from within.* Like the muddy stone with the golden cavity inside, or the potentially hardworking Darhads who become successful outside their homeland, the Darhads' "white" or "yellow" side just needed to be extracted by a person, or a group of persons, who could sense this sacred attraction, and who had the unique capacity to make it visible. Invariably, in the different stories I have collected, this person was the Jebtsundamba Khutuktu (or various representatives of him).

6. Conclusion

In this chapter, I have explored certain forms which Darhad persons assume when seen from particular points of view, or, to put it the other way around, certain perspectives which Darhad persons require to be seen from. My point has thus been that Darhad subjectivities are grounded in concrete spatial vantages afforded by the topography of the Shishged Depression. The shamanic *taiga* is perceived as a zone of pure multiplicity and perpetual metamorphosis, which is asymmetrically contrasted to the Buddhist steppe as a zone of pure singularity and eternal sameness, and this centre–periphery spectacle has come into being through consecutive politico-religious interventions by first the Manchu imperial polity, later the pre-revolutionary Mongolian Buddhist church, and finally the socialist state. Thus the yellow and black side are immanent to all Shishged life: through a complex historical interplay between internal and external forces, and intended and unintended effects, they have become the two substances from which everything Darhad is made. This, of course, is not to say that this environment determines how the Darhads see themselves. It is only to suggest that the Shishged Depression, like any landscape, only offers certain "possibilities of form" (Ramble 1997, 134 and that it largely is *from within* these that Darhad concepts of themselves come into being.

In that sense, the Shishged is not just a landscape, and nor is its study only about how its people are perceiving their environment. For it is saturated with historically-cum-topographically generated perspectives, which elicit both individual bodies (persons) and social bodies (communities). Thus understood, Darhad identity is a "fractal" or "self-scaling" (Wagner 1991; Strathern 2004) phenomenon. Instead of being a single thing, or, conversely, a relative term constructed in opposition to a significant other (cf. Barth 1969), a Darhad person is an internally differentiated, "post-plural entity" (Holbraad & Pedersen 2009), whose paradoxical essence is to be both self and other at the same time. For, as we have seen, Darhadness is an uneasy and inherently unstable cocktail between the wild and the domesticated; between the most marginal, shamanic and least genuinely (Halh) Mongolian, and the most centered, Buddhist and most genuinely Mongolian. Darhad personhood, that is, amounts to a bifurcated inner topography comprised by wild, black, shamanic, and taiga-like components on the one hand, and domesticated, yellow, Buddhist, and steppe-like components on the other hand.

It is exactly this mutual folding of Darhad worlds, where the visible and the invisible are subject to constant figure-ground reversals so that concepts swim in and out of one another, palimpsest-like, which allows for a reconsideration of the seemingly obsolete binary between the wild and the tame within anthropology and cognate disciplines. For is the more general lesson that may be heeded from the present case study not that the concept of wilderness, for it to be theoretically progressive, must itself be conceived of in a fractal or "post-plural" manner? Thus understood, the wild and the domesticated are "internally" as opposed as "externally" related, just as the relations between "the black side" and "the yellow side" were found to be in the Northern Mongolian context. Wilderness, that is to say, is neither an intrinsic property pertaining to some phenomena as opposed to others (*viz.* essentialism), but nor is it a social, cultural or political category that is arbitrarily ascribed to certain phenomena and not others (*viz.* constructivism). Rather, wilderness is an intensive capacity that is potentially present within all things, but which is only elicited or actualized via certain social practices and material forms, ranging from inter-ethnic banter to mountain rituals.[6]

7. References

Badamhatan, S. BNMAU-in ündestnii ba ugsaatni högjliin asuuldad. *Tüühiin Sudlal, 9.* Ulaanbaatar 1980.

Badamhatan, S. Les chamanistes du Bouddha vivant. *Études Mongoles...et sibériennes* 17, 1986.

Badamhatan, S. & Banzragch. *Hövsgöl Aimagiin Tovch Tüüh.* Mörön 1981.

Barth, Frederik, ed. *Ethnic Groups and Boundaries.* Boston: Little Brown 1969.

Bat-Erdene, Baldangiin. *Hövsgöliin Rashaan Us.* Mörön 1992.

Bawden, Charles R. *The Modern History of Mongolia.* London: Kegan Paul International 1986.

Blondeau, Anne-Marie & Steinkellner, Ernst, eds. *Reflections of the Mountain.* Vienna: Verlag der OeAW 1996.

Bulag, Uradyn E. *Nationalism and Hybridity in Mongolia.* Oxford: Clarendon Press 1998.

Charleux, I. "Padmasambhava's Travel to the North. The Pilgrimage to the Monastery of the Caves and the Old Schools of Tibetan Buddhism in Mongolia". *Central Asiatic Journal* 46, 2 (2002): 168–232.

Dienberger, Hildegard. "Introduction: Mongols and Tibetans". *Inner Asia* 4 (2002), 171–180.

---

6 It is important to emphasize that this "radically essentialist" (Henare et al 2007) or "post-plural" (Holbraad and Pedersen 2009) stance differs from other branches of contemporary social and cultural theory in a number of ways. On the one hand, it represents an empirical ambition of turning anthropology itself into a "science of the concrete" in Levi-Strauss' famous sense by forging analytical concepts out of the ethnographic material generated from long-term fieldwork. On the other hand, it also represents a theoretical vision of adopting Boas' equally famous definition of anthropology as "philosophy with people in it" by making ontological claims of the sort some might expect to be the sole responibility of philosophers.

Dioszegi, Vilmos. "Problems of Mongolian Shamanism". *Acta Ethnographica* X (1–2) (1961), 195–206.
Dulam, S. "Darhad böögiin ulamjlal. Ulaanbaatar: MUIS-iin Hevlel 1992.
Even, Marie-Dominique. "Chants de chamanes de mongols". *Études Mongoles…et sibériennes* (1988–89), 19–20.
Even, Marie-Dominique. "The Shamanism of the Mongols". *In* S. Akiner, ed. *Mongolia Today.* London: Kegan Paul International 1991, 183–205.
Ewing, Thomas E. "The Forgotten Frontier: South Siberia (Tuva) in Chinese and Russian History, 1600–1920". *Central Asiatic Survey* 15 (3–4) (1981), 174–212.
Hangin, John et al. *A Modern Mongolian-English Dictionary.* Indiana: Research Institute for Inner Asian Studies 1986.
Heissig, Walther. *The Religions of Mongolia.* London: Routledge and Kegan Paul 1980.
Henare, Amira, Holbraad, S. and Wastell, Sari. *Thinking Through Things. Theorizing Artefacts Ethnographically,* London: Routledge 2007.
Holbraad, Martin & Morten Axel Pedersen. "Planet M. The Intense Abstraction of Marilyn Strathern". *Anthropological Theory* 9 (4) (2009): 371–394.
Huber, Toni, ed.: *Sacred Spaces and Powerful Places In Tibetan Culture.* Dharamsala: Library of Tibetan Works and Archives 1999.
Humphrey, Caroline. "Chiefly and Shamanist Landscapes in Mongolia". In E. Hirsch & M. O'Hanlon (eds) *The Anthropology of Landscape. Perspectives in Place and Space.* Oxford: Clarendon Press 1995, 135–162.
Humphrey, C. *Shamans and Elders. Experience, Knowledge, and Power among the Daur Mongols.* With U. Onon. Oxford: Clarendon Press 1996.
Jagchid, Sechin & Hyer, Paul. *Mongolia's Culture and Society.* Boulder: Westview Press 1979.
Lacaze, Gaelle. *Représentations et Techniques du Corps chez les Peuples Mongols.* Ph.D. Thesis, Université de Paris-X 2000.
Legrand, Jacques. *L'Administration dans la Domination Sino-Mandchoue en Mongolie Qalq-a.* Memoires de L'institut des Hautes Etudes Chinises, Vol. 2. Paris: Collège de France 1976.
Levi-Strauss, Claude. *The Savage Mind.* Chicago: Chicago University Press 1962.
Macdonald, Alexander W. Foreword. *In* A. W. Macdonald (ed.) *Mandala and Landscape.* New Delhi: D.K. Printworld, V-XI 1997.
Pedersen, Morten Axel. "At home away from homes. Navigating the *taiga* in Northern Mongolia". In P. Kirby, ed., *Boundless Worlds. An Anthropological Approach to Movement,* pp. 135–152. Oxford: Berghahn 2009.
Pedersen, Morten Axel. "The Virtual Temple. The power of relics in Darhad Mongolian Buddhism". In I. Charleux, G. Delaplace and R. Hamayon (eds.) *Representing Power in Modern Inner Asia: Conventions, alternatives and oppositions,* pp. 245–258. Bellingham: Western Washington University Press 2010.
Pedersen, Morten Axel. *Not Quite Shamans. Spirit Worlds and Political Lives in Northern Mongolia.* Ithaca, NY: Cornell University Press 2011a.
Pedersen, Morten Axel. "Non-identity Politics". *Common Knowledge* 17 (1) (2011(b)), 117–22.

Pedersen, Morten Axel. "The Task of Anthropology is to Invent Relations: For the Motion" *Critique of Anthropology* 32, 1 (2012).
Pegg, Carl. *Mongolian Music, Dance, and Oral Narrative.* Seattle: University of Washington Press 2001.
Pozdneyev, Aleksei M.. *Mongolia and the Mongols*, Vol. 1. Bloomington: Indiana University 1971 [1892].
Pürev, O. *Hövsgöl Aimgiin Ulaan-Uul Sum, 'Jargalant-Am'dral Negdel* (Tüühen nairuulal). Mörön 1980.
Pürev, O. *Mongol Böögiin Shashin.* Ulaanbaatar: The Mongolian Academy of Science 1999.
Ramble, Charles. "The Creation of the Bon Mountain Kongpo". *In* A. W. Macdonald (ed.) *Mandala and Landscape.* New Delhi: D.K. Printworld 1997, 133–232.
Samuel, Geoffrey. *Civilized Shamans – Buddhism in Tibetan Societies.* Washington: Smithsonian Institution Press 1993.
Sandschejew, G. D. *Darkhaty.* Leningrad: Ak. Nauk SSSR 1930.
Sneath, David. *Changing Inner Mongolia – Pastoral Mongolian Society and the Chinese State.* Oxford: Oxford University Press 2000.
Snellgrove, David L. *Indo-Tibetan Buddhism – Indian Buddhists and Their Tibetan Successors.* London: Serindia Publications 1987.
Strathern, Marilyn. "No Nature, No Culture: the Hagen case". In Nature, Culture, and Gender, edited by C. MacCormack & M. Strathern, pp. 174–222. Cambridge: Cambridge University Press 1980.
Strathern, Marilyn. *The Gender of the Gift. Problems with Women and Problems with Society in Melanesia.* Berkeley: University of California Press 1988.
Strathern, Marilyn. *Partial Connections* (Updated Edition). Walnut Creek, CA: Altamira 2004.
Tambiah, Stanley J. "The Galactic Polity in Southeast Asia". *In Culture, Thought, and Social Action – An Anthropological Perspective.* Cambridge: Harvard University Press 1985, 252–286.
Tatar, M. "Two Mongol texts concerning the cult of the mountains". *Acta Orientalia Academiae Scientiarum Hung.* XXX, 1 (1976), 1–58.
Viveiros de Castro, Eduardo. "Cosmological deixis and Amerindian perspectivism". *Journal of the Royal Anthropological Institute* 4, 3 (1998), 469–88.
Vreeland, Herbert H. *Mongol Community and Kinship Structure.* New Haven: Human Relations Area Files 1962.
Wagner, Roy. "The Fractal Person. In *Big Men and Great Men. Personifications of power in Melanesia*, M. Godelier & M. Strathern (eds.), pp. 159–173. Cambridge: Cambridge University Press 1991.
Wheeler, W. Alan. *Lords of the Mongolian Taiga: An Ethnohistory of the Dukha Reindeer Herders.* MA thesis, Dep. of Central Eurasian Studies, Indiana University 2000.
Zhamtsarano, Ts. "Ethnography and Geography of the Darkhat and other Mongolian Minorities". *The Mongolia Society, Special Papers* (8) (1979) [1934].

12. Ritual is Etiquette in the Larger than Human World: the two wildernesses of contemporary Eco-Paganism

Graham Harvey

1. What to expect

Wilderness, according to Russell in Disney-Pixar's *Up*, "isn't quite what I expected". When Carl asks him, "Yeah, how so?", Russell replies "Its kind of… wild. I mean its not how they made it sound in my book" (*Up* 2009). Many people share his experience of surprise, bafflement and even disappointment. Perhaps they ought not to be surprised when wild places do not seem particularly welcoming or easy. Nonetheless, wilderness can be sought and celebrated as much as it can be avoided or denigrated. Wilderness, as place and as category, turns out to be unruly, refusing to be confined by fixed expectations. Sometimes, surprisingly often in fact, what some people identify as wilderness turns out to be other people's homes. This seems definitive of the colonial exercise of invasion, expropriation and settlement, especially when based on the Doctrine of Christian Discovery and assertions about *terra nullius.* Wilderness might not only be the location of various kinds of work (herding, hunting, guiding or logging, for example) but may also be the product of considerable labour by people who dwell and/or work there. The rhetorics and rituals surrounding wilderness not only conflict but they often shift in ways that resist analysis from too secure a perspective. That, however, is part of the point of this book: different perspectives encourage another look, and then another. By approaching diverse forms of wilderness in various ways, we test what happens when we allow our comfortably familiar, civilised notions of the world, of religion and of humanity to become destabilised and re-examined. Tame words like "nature" become pallid when confronted by the unruly "wilderness" (whether inflected positively or negatively). There is a chance, however, that they might wake from their slumber and insist on playing a critical role in our deliberations.

This discussion arises from guesthood – i. e. a form of dialogical fieldwork research (Harvey 2003a, 2005a) – among some exemplary members of the diverse Eco-Pagan community, i. e. a network of environmental educators / shamanic practitioners. While I name only a couple of individuals (Gordon MacLellan and Barry Patterson) who are actually employed as environmental educators and practice something that could be called Pagan shamanism, it is their embeddedness in larger networks of like-minded and like-active people that justifies focussing on them. Sometimes, or in some places, it is environmentalism (and its dissemination) rather than shamanic practice that is foregrounded, or that brings groups of people together with these and similar core individuals. Sometimes the shamanic practice dominates. But these are different occasions and, often, different locations. It is crucial for these people (the individuals and their networks of friends, companions and collaborators) that their environmentalism and their shamanism inform and reinforce each other. This is what makes them distinctive among other environmentalists or shamans. However, the subject which MacLellan, Patterson and other Eco-Pagan environmental educators / shamanic practitioners teach is environmentalism. They are not in the business of promoting shamanism or eco-Paganism under the guise of lessons about "the natural world". The integration of these aspects of their lives is vital, but occurs in different contexts. They are interested in a wider population learning to engage with the larger-than-human world respectfully, responsibly and with joy but not necessarily as shamans or Pagans.

In this chapter, I show how two contrasting notions of what constitutes wilderness are important in understanding the practices and ideas of networks of these British Eco-Pagan environmental educators / shamanic practitioners. First I offer a brief orientation to the larger scene in which wilderness might be some kind of "nature" and might be found somewhere in or beyond the non-urban "countryside". Then I introduce the two understandings or experiences of wilderness, both of which are at least implicit in the discourse and practice of the people discussed here, but also occur in scholarly literature interested in other cultural contexts (as we'll see). Eco-Pagan environmental educators / shamanic practitioners, and their perceptions and activities in relation to both kinds of wilderness, and thus to the larger-than-human world (as an even more diverse community), are the subject of larger portions of this chapter.

In line with the now commonplace recognition that the nature/culture duality is misleading in many ways, this chapter concerns the social or cultural nature of nature and the constructed nature of wilderness. The chapter frequently plays riffs on Ronald Grimes' performance/essay (2002), much as he riffs on Gary Snyder's work (1990). The conclusion drawn here is that Eco-Pagan environmental educators / shamanic practitioners find employment and direction in the world because their relational participation with the larger-than-human world / community inspires and requires them to confront the threat posed by a (negative) kind of wilderness: anthropogenic monoculture and its ecocidal consequences. In turn, the threat of monoculture wilderness encourages such people to attempt to disseminate the urgent need for human conversation with the larger-than-human world / community. Learning the etiquette and protocols for human re-entry into such inter-species conversations (verbal and non-verbal) is perceived to eased by time spent ritualising in the other kind of wilderness: the less human dominated bio-diverse community of "wild(er)" places. Thus, the friction between these two understandings and experiences of wilderness is what explains the environmentalist, shamanistic, educative and ritualistic ideas and acts of the Eco-Pagans presented here.

What is most important in my argument in relation to Pagans and to religious environmentalism is the ongoing dynamic tension between these two value-laden uses of the word "wilderness". It becomes clear that these two wildernesses cannot be collapsed into one another, neither does the existence of one negate the other. It is the presence of both, and the contrast between them, that inspires and seems to require the kind of things that some people do and think. What is most important about my argument in relation to "wilderness" is its contribution to critical debates about alternatives to the nearly exhausted modernist rhetorical and impractical separation interposed between "culture" and "nature" (also see, Latour 1993, 2002; Plumwood 1993, 2002, 2009). This and other studies of the range of uses to which the word "wilderness" are put in particular contexts may, therefore, do more than describe interesting cultural diversities. They may contest the currently globally dominant culture's imperious dualisms (e.g. mind/matter, humans/animals, science/religion, rationality/relationality) and thereby advance the scholarly project of understanding the world more fully.

2. Nature and countryside

A deeply rooted modern romanticism in Britain equates countryside with rest, recreation and virtue. Since at least the mid-nineteenth century, and especially following the development of railway networks, people have taken breaks from cities to find purity and nobility in the countryside. In mountainous or upland areas, people have sought to encounter transcendence or "the sublime", and there to experience awe and ennoblement. Putatively wild nature has been positively valued as a means, a tool for re-creating humans as better people (see Hay 2002; Milton 2002; Mischi 2009; Harvey 2005, 283–99).

Despite the evidence of farming, forestry, quarrying and other labours, many locations have been deemed "wild". Even historical and archaeological knowledge that the British uplands are as they are now because of human actions, uses and abuses, over millennia, has not significantly damaged the notion that these are somehow pristine (and only recently endangered) "natural" environments. It is conveniently ignored that the relatively barren hills of the Lake District or the deliberately de-forested Scottish mountains are as they are now because of human assaults on previously biodiverse communities – and of communities comprised of both humans and other-than-humans (Hogan 2011; McIntosh 2004, 78). How could this not be so when even the easily accessible agricultural and pastoral landscapes of southern England are labelled "nature" and understood to be of a different category and character than the seemingly more constructed urban realms? Information boards offer visitors to wild-flower meadows beside the river Thames a mixed message: these water-meadows (i. e. fields deliberately flooded in winter) are hay-meadows (i. e. land that produces food for farmed animals), but it is not appropriate for human visitors to pick or tread on the "wild" flowers that, in fact, only grow because of human land-usage. Popular celebration of "wild nature" is encouraged in a thoroughly managed landscape, and particular ways of behaving in the world are cultivated as "environmentally responsible".

The term "wilderness" is not as common in British discourses about the world as terms like "countryside" or "nature". While references to "wild" animals, birds, habitats and locations (again, often more remote uplands or sea coasts) proliferate, these are less commonly associated with or identified as "wilderness" than they are elsewhere. Nonetheless, the same dubious dualities (nature/culture; humans/world; people/other species; species/environments) are now at play globally. The "wildernesses" of the Americas are places in which many humans live and work. They were home to even more humans prior to the European invasions. Many places now labelled "wilderness" are the result of the activities of both human and other-than-human populations who, often deliberately, shaped such communal places (or emplaced communities) over many millennia. The interlinked practices of genocide and ecocide in the Euro-American project (destroying biotic communities simultaneously with the human cultures that contributed to them) have eventually been seen, by some at least, to require efforts to preserve wilderness and to set aside areas of nature for protection from exploitation (Oelschlaeger 1991; Philippon 2005). The processes of preservation and protection have commonly involved the removal of yet more humans and conflict with particular species. The control of cervid and canid species in American national parks illustrates the continuing construction and cultivation of nature (Glass 2005; Bekoff 2010). Such wildernesses are not left to be natural but are the locations of continuous interventions. Notions of wilderness emerge from discourses of human actions in relation to the larger-than-human world.

In Britain, as elsewhere, distance from urban construction is deemed emblematic or definitive of "nature", but nature wildly refuses to recognise artificial boundaries. Foxes and other wild animals transgressively move into town and hunt in gardens. Wild birds are now more common in suburban gardens than they are out among the fields and (surviving) hedges of the countryside. They adopt people who feed them at bird-tables with seeds imported from China. Attempts are made at protecting them from only-partly tamed cats (whose consumption of fish, according to De Silva and Turchini 2010, now ranks them among the dominant marine predators). Contrary to the popular romantic confusion between "rural" and "nature", the animals and plants of the fields are largely products of human experimentation. Modern cattle, pigs and sheep can only survive in fields when fed a cocktail of chemicals. Modern grains only survive their unnatural environments by continuous human efforts to eradicate competition (i. e. natural evolutionary biodiversity) and the injection of manufactured nutrients. Nonetheless, people still go "out there" into the countryside to see wild creatures. TV programmes further encourage the valorisation of the countryside and the imagination of farmland as more natural and more biodiverse than human homes and gardens (especially urban ones). Conversely, TV adverts for various biocidal cleaning agents (for example, antibacterials and weed-killers) should reveal the truth that humans are far from alone in their dwellings. Simultaneously, the experiment attempts to impose human controlled order on the messiness and random profligacy of wild nature while imposing civilised habits on human behaviour.

In this messy context, I am interested in the ideas and activities of some British Eco-Pagan environmental educators / shamanic practitioners. Their movements between urban centres and relatively remote uplands (and, if we adopt their perspective, the sense that at least sometimes the uplands are central and cities are peripheral) involve them in efforts to challenge one kind of wilderness with another.

3. Two wildernesses

Two interpretations of wilderness interest me. There are other discourses that feed into or conflict with these ideas (such as the influential heritage of the interpretation of biblical, classical, and colonial words and motifs) but I do not intend to offer an exhaustive account of their origins and implications. A general outline, with reference to another cultural context, will provide an orientation to the wider context of the ideas, performances, labours and lives of the Eco-Pagans whose particular engagement with both kinds of wilderness will be focused on in later sections.

The first use of the term wilderness is the most common one, the first definition offered by most dictionaries, what any native speaker would assume was intended. This "wilderness" refers to spaces that are unmarked by humanity, nature untouched by and remote from human culture. It is not our environment, not our domain, not our place. The places it labels are not made by us but may be unmade by our presence and even more fatefully by our activity. Human absence and distance is definitive of this view of wilderness. This wilderness has long been denigrated but is now more commonly celebrated. It was once the domain of that which opposed and deserved to be opposed by civilised, cultured humans, or that which was an as yet untouched resource. It is now more often viewed as a domain that has been fortunate to survive. It is to be protected and honoured, either for its potential to save over-civilised, overly urban humanity, or for its preservation of biological, scenic or other forms of positively valued diversity (Oelschlaeger 2005, Philippon 2005). While members of some religions may continue to treat such wildernesses as negative or irrelevant to their spiritual concerns, Pagans – who positively value nature in general – often exalt this kind of wilderness (Harris 2008, Taylor 2010).

The second notion is that wildernesses are places fatally damaged by too much activity of inappropriate kinds, usually by humans who decline to dwell harmoniously and work cooperatively alongside other species. Put another way, wilderness is the result of human disruption. Human activities create rather than destroy this wilderness. This use of the term wilderness is most strongly expressed in a conversation between Debbie Rose and Daly Pulkara (about which more will be said below). Such wildernesses are desertified or severely degraded places, bereft of diversity and faced with either bare survival or further decline. They are places where over-consumption and/or overproduction have exhausted the ability of the soil to support anything but the most limited of ecosystems. An extreme example might be monocultural fields of chemical- and subsidy-dependent grains and grasses, which further exacerbate water pollution, soil collapse, and species extinction (Harvey 2001, 320–33). The increasing prevalence of intensive industrial farming means that the farmed countryside is now the primary location of (this second, anthropogenic kind of) wilderness.

In relation to both these understandings of wilderness, refinements and nuances are important. For instance, wildernesses might be *largely* rather than completely untouched by humanity, or *largely* rather than completely the product of human mistreatment or mismanagement. Both notions are pregnant with proliferating possibilities for understanding and acting towards the world – and thus, for us, now, for understanding and engaging with those who see and treat the world in such ways. Both uses of the term carry normative, behavioural, ethical and relational implications. They are not neutral, natural or merely descriptive. When used in the midst of conversation between people who share the same ways of being in the world, the referent and implications of each term can largely be taken-for-granted as if they were more simply descriptive. Wilderness *are* natural places out there beyond invasive human intervention. Or wilderness *are* the places damaged by humans. People should either leave such places alone or should do something to restore their health and vitality. And so on.

The first notion of wilderness (as a reference to relatively remote, largely natural places) is commonplace. Its expressions and manifestations are easily recognisable to most speakers of English and of other European languages. Any dictionary will indicate that this is what "wilderness" means and any encyclopedia that includes entries about "nature" or "wilderness" will reinforce the interpretation (e.g. Oelschlaeger 2005). Its exact outworkings and enactments among Eco-Pagan environmental educators / shamanic practitioners require some consideration, but these too will seem familiar to those of a similar cultural and linguistic heritage. The second notion of wilderness (as a reference to places damaged by human activities) requires some more introduction. This is best achieved by reference to an eloquent conversation between Daly Pulkara, an Aboriginal Australian culture-teacher, and Deborah Rose, an Ecological Humanities scholar. Rose writes that

> We had stopped to look at some of the extreme erosion on Humbert River Station, and I asked Daly what he called this country. He just looked at it long and heavily before he said: "It's the wild. Just the wild." Daly went on to speak of quiet country – the country in which all the care of generations of people is evident to those who know how to see it. Quiet country stands in contrast to the wild: we were looking at a wilderness, man-made and cattle-made. This "wild" was a place where the life of the country was falling down into the gullies and washing away with the rains. (Rose 2004, 4).

Pulkara and Rose speak not only of Euro-Australian caused erosion, nor only of anthropogenic ecological disaster, but of the wild assault by "whitefellas" imposing themselves and their alien system on other peoples and other places. The invaders think Aboriginal people are wild and that the land is un-cultivated. They do not heed Aboriginal law. They do not heed the law of the country or listen to its lore to know what is appropriate and expected of all beings who comprise its community (also see Rose 1996, 1998, 2009, 2011). They do not see "all the care of generations" that has resulted in viable, sustainable, equitable and vibrant lifeways not only for humans but also for other-than-human co-inhabitants and co-creators of these places, these countries. Mistakenly thinking that they are seeing wilderness (wild untamed land full of wild undomesticated animals and wild uncivilised people), "whitefellas" embarked on the eradication of varied multi-species communities and created a wilderness of loss, devastation, disconnection, irresponsibility and cruelty. The wilderness of this "wild" where there had previously been "quiet country" was established on the basis of a culture, with laws and customs, that divided, separated, cut and severed human participation in and belonging to, within or among larger-than-human communities.

It is a commonplace of indigenous, anti-colonial, environmentalist and other writings that engage with modernity (e.g. Latour 2002; Plumwood 2009; Franke and Latour 2010) that at the heart of colonialism and modernism are allied understandings of nature and culture. The former has been seen as a resource to be taken, bought, sold, developed, improved and disposed of. It had neither rights nor voice – except metaphorically. For Locke and other ideologues of the West's modernising reality the rights and voice of nature were those of owners, those who hold title deeds. Culture is the practice of humans utilising inanimate nature. Wilderness, in this context, is where humans have not yet imposed recognisably human-originated and human-benefitting order. In another frame, however, wilderness is the monoculture that results as the endgame of this degrading self-imposed alienation of humanity from the larger-than-human moral community.

For all the flaws hinted at in the sneering tones often used when people speak of Romanticism, there is something potentially initiatory and radical in this alternative wing of the modern project. By seeking awe, truth, beauty, majesty, alterity, reality, nobility, redemption or justification, Romantics placed themselves within reach of other kinds of relationship with the world. They grasped for the possibility that ownership, production, commodification and consumption were not the only ways to relate to the larger-than-human world. Romanticism was an experiment, as yet unfinished, to see if humans might find a more appropriate way of being in the world if they occasionally bowed to the grandeur of that which stands over and even against human pretensions to control and dominate without challenge. Nonetheless, perhaps Romantics could not grasp the full implications of the Aboriginal Australian social project in which humans and other species were and remain full participants in the enabling of "quiet country", expected by virtue of co-inhabitation to take on the responsibility of mutual care (Rose 1998).

While these two notions of wilderness, and the performances and cultures that energise and express them, are distinct and opposed, it is their simultaneous co-existence that is significant to my case study. It is the imperative felt by Eco-Pagans to do something in relation to *both kinds* of wilderness because of the existence of the other that generates the phenomena I identify as Eco-Pagan environmental education / shamanic practice. A brief note about the terms "Pagans" and "Eco-Pagans" and then an elaboration of the wild solidus that links and separates "environmental education" and "shamanic practice" will aid a more complete appreciation of exemplary members of this diverse community of shared practice and discourse.

4. Pagans and Eco-Pagans

People have been valorising Paganism as a positive term for over a century now, but increasingly since the 1950s some have self-identified as Pagans. What was once a largely derogatory accusation made against people for not being Christian, or for not being Christian enough, gradually came to be used as a term for celebrants of nature. Coterminous with the slow shift from the negative to positive valuation of terms like "nature" and "wilderness", the term Pagan acquired positive resonances. This trajectory has been traced by scholars interested in Paganism (e.g. Hutton 1991, 1994, 1996, 1999; Clifton 2006), including, in summary, by myself (Harvey 2006a, 2011). Briefly stated, it involves a heady brew of classicism, romanticism, celticism, naturalism, naturism, anti-clericalism, ritualism, individualism, communitarianism, evolutionism, esotericism, feminism, libertinism, rationalism, and much more. Evidently, some of these conflict with each other and their mixtures do not always result in experiences or systems that seem coherent to insiders let alone to outsiders. It is unlikely that anyone involved in these wider cultural currents intended or could have predicted the current situation or style of the miscellaneous forms of self-identified Paganism. However, within the proliferating diversity of groups and practices, a common theme and thread is regularly reiterated by Pagans and their scholarly observers. This is the notion that Paganism is a "nature religion," a "religion of nature" or a "religion centred on the celebration of nature". Quite what "nature" means, beyond the evidently religiously positive or even sacred tone with which it is invoked, is regularly debated (in addition to the works already cited, see Berger, Leach and Schaffer 2003; Taylor 1995, 2010).

"Nature religion" is portrayed by Catherine Albanese (2002) as a pervasive, unsystematic, and largely implicit strand in American and other religiosities. It might provide Pagans with a larger cultural context in which some of their ideas and practices resonate for others. But Albanese's "nature religion" is even less organised and institutionalised than Paganism – which now has recognisable subdivisions, authorities, shared ritual and social patterns and so on. It is a diffuse or implicit affirmation that human wholeness and/or wellbeing may be best found in communion with "wild" or "natural" places. The path from Romanticism to "nature religion" is easy to follow, albeit that it winds through forests and among mountains (not only metaphorically). Some Pagans certainly identify with the sense of being "at home" and "spiritually enlivened" in forests, among mountains, on wild coasts or simply outdoors in gardens and city parks. But some Pagans mean more or other than this. For some, Paganism is a "nature religion" because it is not about anticipating an afterlife in a transcendent, supernatural, other-worldly realm. It is a this-worldly religion, encouraging the celebration of physical well-being and encounter with the essential, praise-worthy materiality of the world and cosmos. For others, or at other times for the same people, Paganism is a "nature religion" when it provokes environmental activism or ecological responsibility. It is not enough, for some, to feel at one with nature only when watching the midsummer sunrise in a place of ancestral significance. The threat to the natural world requires that veneration should be expressed in the defence of threatened places and ecosystems by magical and mundane means. These are only some of the themes elicited from enquiries about the meaning of "nature" among Pagans.

Because the majority of Pagans turn out to be neither less nor more active in environmental campaigns than their neighbours (see Harris 2008, Jamison 2011), some Pagans or observers of Pagans sometimes prefix "Eco-" to "Paganism" when they wish to refer to eco-activist Pagans and/or Pagans who devote and direct their magical and practical efforts to radical environmentalism (see in particular Letcher 2000, 2004, 2005). Most Pagans celebrate seasonal festivals, seek to "tread lightly on the earth" (reducing consumption and not seeking inordinate wealth), and largely share (light) "green" values (recycling, joining pressure groups like Greenpeace and so on). But "dark green" (Taylor 2010) front-line activism, confronting land-developers or multi-national corporations, is not on their agenda. They may respect Eco-Pagans as a wing of the broader Pagan movement or they may object strongly to any suggestion that Paganism ought to require eco-activism. Some Pagans, however, think more is required of them and reject what they see as the quietism of many co-religionists.

It should be noted that the label "Eco-Pagan" is not always one that people apply to themselves. It can be used as a means by which observers distinguish among Pagans, Paganisms or possible Pagan activities. People who are so identified might prefer to name themselves simply "Pagans" or to insist that they need no name for the worldview or lifestyle that seems right to them. Nonetheless, broad common themes and practices suggest the value (to observers at least) of seeing Eco-Paganism as one of the many flavours of Paganism. Eco-activism emerged in confrontations with road builders, quarriers and land-developers in the 1990s. It made use of earlier forms of largely non-violent direct action but blended it with more-or-less elaborate, carnivalesque and/or transgressive forms of magical ritual. Binding rituals (aiming to prevent negatively perceived actions) and blessing rituals (aiming to energise positively desired outcomes) could be performed, publically or privately, for the benefit of threatened woodlands, hillsides, rivers and other "natural"-communities. Celebrations as well as protests blended recognisably Pagan and ritual-magical ideas and acts, and resulted in the evolution of shared tastes in music, costume, habitation, transportation, ideology, narrative and self-representation. For example, away from the frontlines, many Eco-Pagans became keen organic gardeners, devotees of all forms of sustainability, leading contributors to localisation movements (such as the Transition Town movement), and committed disseminators of environmentalist knowledges and practices. Among them, some identified with practices that can be labelled shamanic.

5. Environmental educators / shamanic practitioners

In introducing the religious people of interest in this chapter (about whom I'll say more in the following section) as "some British Eco-Pagan environmental educators / shamanic practitioners", I use a wild solidus, an un-solid or porous slash symbol, a permeable or transgressable line that links as it separates. The unpronounceable "/" attempts to convey the sense that while these people are educators *and* shamans, the performance of both roles (if, for now, we imagine these to be distinguishable) makes each different, more wild, than they are when (dis-)articulated by others who are *either* environmental educators *or* shamanic practitioners. Thus, it is not only true that the label "Eco-Pagans" is sometimes resisted and sometimes only playfully mobilised by those I discuss, it is also true that such people can be either hesitant or bold in identifying themselves and people like them as environmental educators and/or shamans.

Many different careers, roles, practices or performances might be identified as environmental education. The people who interest me here earn at least part of their living by offering educational opportunities in schools, libraries, country-parks, and other venues to groups of children and/or adults. They are trained experts in some discipline in the life sciences (especially biology, botany and/or zoology) and/or the environmental sciences (geology, geography, ecology). They are also voracious readers, passionate learners and skilled communicators about extensive overlapping fields of enquiry and knowledge. The sessions they run involve teaching and learning about habitats and their multi-species inhabitants. When working with school groups they refer to nationally defined curricula appropriate to the age group or to the subject studied by the pupils. Sometimes they work in heritage, country- or nature-parks. Then they introduce people to the environment and aid the increase of understanding of what makes living places flourish or decline. Such events might include lessons about wild creatures found by pond-dipping or about geology enhanced by handling rocks from red-sandstone hills. Participants are enabled to engage more knowledgeably and more enthusiastically with their local or regional environments. Frequently, and most characteristically, these lessons are creative, imaginative and dramatic. For instance, children or adults might make masks of the creatures they have encountered in woodlands, pools, rocky valleys or disused quarries. Educators might appear in the guise of folklore characters and facilitate imaginative shared story-telling. Groups may work together to produce entertaining dramas in order to show others what they have gained from their opportunity to engage afresh with some aspect of the larger-than-human world.

Similarly, the term "shaman" has been employed in diverse and often conflicting ways. Derived from the Tungus-speaking Evenki of Siberia, the term is now applied to a number of different kinds of social and religious experts worldwide. While this globalisation of the term is sometimes resisted as leading to vast over-generalisations and facile over-simplifications, it can also be valued as permitting enlightening critical comparisons and reflections. The long history of Western use and abuse of the term has been surveyed by Ronald Hutton (2001) and Andrei Znamenski (2007), while Robert Wallis and I have presented an A to Z overview of its contemporary applications and resonances (Harvey and Wallis 2010). Although the commonly emphasised "altered states of consciousness" or "ecstasy" (Eliade 1964), play a role in the shamanic practices of some Eco-Pagans, it is the more animistic "adjusted styles of communication" (Harvey and Wallis 2010,14) that justify the linkage between Eco-Pagan environmental education and shamanic practice. The use of a shared acronym "ASC" should hint at some playfulness and some polemic in the proffer of an alternative interpretation of what is central to the pursuits of shamans: not individual, interior consciousness but communal performance of interpersonal conversation and mediation. Note too that "animism" here is categorically not a reference to Edward Tylor's "belief in spirits" but to the "new" animism of more recent ethnography – contrast Tylor 1913 with Harvey 2005b.

Briefly put, the shaman's job is a necessity because of the nature of the world as understood and encountered by animists. Understanding the world as a multi-species community of persons, not all of whom are human, but all of whom deserve respect as these Eco-Pagans do (Harvey 2005b, xi), then the everyday necessities of eating and seeking not to be eaten may require the employment of shamans as well as of farmers, herders, hunters, soldiers, cooks, herbalists and others. Potential relationships entailed in living in a multi-species world include consuming and/or being consumed by others. Different forms of predation result from the different perspectives of particular kinds of persons, human or other-than-human (Viveiros de Castro 1998; Willerslev 2007). Predators and prey are not individuals but members of families, clans, societies and species. The injury, death and/or consumption of one member of a group may offend their relatives. Shamans are then required to mediate between the offending and offended communities to resolve, if they can, the situation to the satisfaction and future benefit of all concerned. They might, of course, be expected to pursue the well-being of their close kin and community above that of potential prey or predators. Nonetheless, to continue to obtain food or avoid predation they are likely to have to persuade the other parties to the dispute to compromise, at least.

The Eco-Pagan shamans discussed here do not live in hunter-gather societies in the Siberian taiga or tundra. They obtain food from supermarkets, shops, farmers' markets and gardens supplied by processes common to consumerist modernity. Nonetheless, they approach the world from an animistic perspective and it, in turn, invites animistic etiquette and shamanic acts. Not only is their world a community of persons, it is rife with competing desires and needs that lead to misunderstandings and conflicts of interest. Sometimes these seem best dealt with shamanically (MacLellan 1999). That is, rituals of negotiation with significant other-than-human persons are necessitated by human participation in an animate, multi-species world (see Grimes 2002). This is all the more so because of the increasing prevalence of modern practices that insult and assault other-than-human persons and create the pollutions, disruptions, extinctions, erosions and other features of the anthropogenic wilderness.

Some Eco-Pagans, then, understand themselves to perform shamanic acts both when they do things that others might identify as shamanic and when they do things that others might identify as environmentalist. Both when mediating between humans and other-than-humans to resolve tensions *and* when attempting to inspire more positive human participation in the larger-than-human world, these Eco-Pagans shamanise.

6. Eco-Pagan shamanic eco-educators

Among Pagans there is a subgroup who are identifiable as Eco-Pagans. They are likely to be animistic in the way they live towards the world. Among them, some are employed, at least part-time, as environmental educators. Among this group some understand themselves, or can be understood to be, shamans. While other British people may claim or be given the title "shaman" because of their (psycho)therapeutic practices, the people of interest here shamanise as part of their environmental activities or engage in environmental activities (especially educationally) as part of their shamanising. As noted above, my "guesthood" research has primarily been with and among the network of friends and companions who constellate around Gordon MacLellan who runs the "Creeping Toad" organisation which offers "environmental art and celebration", "education, training and workshops" (see Creeping Toad website) and Barry Patterson who offers environmentally educative events utilising drama, song and story (see Song and Story website).

For these people, it is the thoroughly relational nature of the world that requires both aspects of this complex performance and identity. It is the fact, for them, that what is commonly labelled "nature" is the realm of cultural interaction between persons of the human and other-than-human kinds that demands animistic relational engagement. Such engagements occur not only in wild places, or even only out in the non-urban countryside. Human homes too are locations of cross-species encounter. Nonetheless, for educational purposes these environmental educators run events in nature reserves, country parks, or seek to encourage school groups and other learners to learn in the "outdoors" world. Then, the fact that inter-species relationships (in all locations) can be damaged or undermined by varied forms of aggression (by or against humans) is what requires the efforts of shamanic mediators skilled at inter-species communication. The aggressive imbalance of modern consumerism and industrial land-(ab)use (i. e. the wilful construction of monocultural wilderness) is among the most dramatic of contemporary assaults of humanity against the larger-than-human world.

According to shamans from many cultures, including the Eco-Pagan shamans of interest here, the causes of conflict or misunderstanding are not all one-sided (e.g. Harvey 2003b; Whitehead and Wright 2004). Whether as revenge, defence or predation, various elements of the multi-species community aggress against humans. The causes of illness and injury can be sought (though they might not always be found) in personal acts of violence committed by particular other-than-human persons (including disease-persons and others). While most of the predators of humanity have been eradicated (or replaced by other humans) in Britain, folklore and ancestral narratives introduce the possibility that there are other beings around who, sometimes, may prey on humans. Despite the efforts of some Fantasy Fiction writers, a sinister tradition about elves and faerie survives and is being found to be meaningful among Pagans (Harvey 2006b). These are not the noble but tragic beings of Tolkien's epic, but the child-stealing, trout-disturbing, cherry-thieving denizens of W.B. Yeats' "The Stolen Child" (Yeats 1886). That they (still) exist, according to at least some Pagans, can be a cause of story-telling, celebration and rituals of protection: the world is not entirely safe and dull, and there are powerful beings who may, if Yeats was right, dance by moonlight in wild places. That they are reputed to endanger human lives as casually as humans endanger the lives of hedgehogs (for example) may be a sufficient cause for the existence and practice of shamans.

Whether or not Eco-Pagans take seriously the idea that the elves might be out there, all kinds of environmental education involve inculcating the knowledge that humans are not alone. With Darwin, eco-educators can insist on the kinship of humans and other species. Given all the signs of growing ecological global disaster, they may also insist that the full implications of Darwin's emphasis on human cross-species relatedness now requires major efforts to alter dominant human lifestyles and behaviours. The point here is not that these or other Eco-Pagans mask their religious practices in their performance of reputable environmental education. Nor is the point, contrary-wise, that they confuse environmentalism for shamanism. While decidedly "dark green" (Taylor 2010), they encourage all shades of green or all depths of greening (Kohák 2000) and seem not in the least interested in religious proselytism. Rather, what motivates them is that in a world in which two kinds of wilderness confront each other, and confront all living beings, they consider that environmental education is necessarily shamanic and shamanism is necessarily an education about the larger-than-human world. It is about mediation between what humans need and what other species need. Thus, although I may seem to have distinguished between therapeutic shamans and environmentalist shamans, the motivating factor for the people of interest here is that, at this point in time, human relationship with and belonging in the world needs healing. This is most clearly seen in the efforts of Eco-Pagan environmental educators / shamanic practitioners to engage with the two kinds of wilderness. First let us go with them into the "wilderness" of the larger-than-human world, and then consider their efforts to contest the growth of the anthropogenic monocultural "wilderness".

7. Retreating to advance

The positively valued wilderness where humans are less dominant and other-than-human species have some room to proliferate with less human interference is celebrated by Eco-Pagan environmental educators / shamanic practitioners. They visit to restore their energies, renew their commitment to the larger-than-human community, to be reminded that there are places where "wild" life can still flourish, and to listen and participate in conversations among multi-species participants in the larger-than-human community. Sometimes they go alone, sometimes they meet together in events that can be called, possibly with some irony, "retreats". There is a playful ambiguity here. These busy people can become exhausted by the continuous effort to be creative and to engage with the needs of clients and others. They acknowledge that sometimes they need to withdraw, to remove themselves from the locations of their busy-ness. Like other people, they find the (larger than human) "wilderness", relatively remote from the noise of humanity, to be a good place to contemplate, meditate, and seek visions of a reality alternate to that of modernity's ceaseless consumerism and competition. Sometimes they seek to retreat temporarily from being at the forefront of a movement attempting to create a radically different future for humans and other species. They explain that some aspects of this "wilderness" have survived from an earlier era when human domination was less evident or where human participation in the world was celebrated.

However, these people are not naïve. Their environmentalist and naturalist knowledges make them keenly aware that the places to which they "retreat" are not at all pristine natural environments. The uplands of the Peak District, the Lake District, Northumbria and the Scottish Highlands and Islands (to name just some favoured "wild" places) are the result of a long history of human intervention. The hills of the Lake District were denuded of trees by the earliest post-Ice Age farmers, and over the millennia they have become almost as species-poor as deserts. The Scottish mountains were cleared of humans and farmers by eighteenth century landlords seeking easier profits by means of other land-uses. All these locations are home to rural communities who farm, quarry or otherwise make significant use of the land. Although some rare species and habitats survive in these hills, these have not been venues of rich biodiversity for a long time. The continuing chemical "improvement" of the land may benefit sheep and their herders but massively threatens the enclaves of wild creatures (from ravens to mushrooms). None of this is lost on those who journey into the hills. Nonetheless, in these places it *is* still possible to hear the cronking of ravens rather than grumble of engines, to see starry skies rather than the glare of city lights, to discover moths in deep caves, and other experiences not immediately tied into human production systems. Because they are one place where humans can encounter the larger-than-human world, these wild uplands are valued. An edgy blend of sharp realism (about the true nature of these places) and colourful romanticism (about their restorative possibilities) emerges in conversations and activities during "retreats".

As individuals or as members of small groups of friends, then, our Eco-Pagan environmental educators / shamanic practitioners sometimes journey into the (positively valued) "wilderness" to commune with like-minded people (human or other-than-human), to dance and ritualise. They may appear to be only rambling in the hills, observing the movements of birds or rivers, dancing with friends (albeit in a somewhat trance-like state), feeling the textures of rocks or lichens, chatting about folklore or exchanging news about improvements or diminishments to ecological diversity in the places they have been. However, experience (including participative observation and polite questioning) reveals that these and other activities that place them in valued larger-than-human "wild" places allow Eco-Pagans to re-member what they take to be the true nature of the world: a multi-species moral and conversational community. Sometimes they perform more-or-less elaborate rituals: greeting the cardinal directions, making offerings of food, smoke or liquid to significant other-than-human beings. Sometimes they show respect more casually, but no less significantly, by lightly touching a leaf as they walk by a tree. Their purpose is to build more intimate relationships with the world, understood as a community.

It is not only conversation with the kind of "wildlife" listed in botanical and other recognition-guidebooks that take place in these events. They can also involve attempts to seek, become aware of, or acknowledge the presence and company of ancestral and otherworld beings. These latter might include the faery and elven people, and if they are sought out it may be with considerable care. They are wild, as "other" to humans as humans are to hedgehogs. A popular tradition claims that such beings have withdrawn as human domination of the world has increased. There are, though, places that are identified as locations in which humans have been known to encounter such elusive and ambiguous beings (Britain's indigenous tricksters). In the nearly forgotten or marginalised spaces of the world, it is possible to show respect to those whose aid in confronting overwhelming threats to the world caused by the wider culture's normative anthropocentricism may be sought. At some road-protest camps, some people sought to collaborate (mischievously) with fairies, trolls and pixies to advance the cause of increased well-being for all life (Letcher 2001). The Eco-Pagans of interest here make offerings or gestures of respect towards otherworldly beings, of the same kind as those made towards beings that others might consider "natural" rather than "metaphysical". That is, they may set aside a plate of food at each meal and put it out as a gift. They may sit quietly listening by rivers, rocks or trees, hoping to gain access to the ongoing conversation of the larger-than-human world. Sometimes they may (in isolation or as groups) vigorously chant to request visions and communion with other beings.

Eco-Pagan environmental educators / shamanic practitioners assert the truth that there is wildness not only in the extremes of confrontation and retreat, but in the middle space of their homes too. Not only are carpets and kitchens home to myriad "mini-beasts" and bacteria, urban gardens are full of creatures that do not contribute to human economic well-being, and some (like slugs) are rarely welcome. Similarly, ancestors and otherworld beings can be honoured, for example with gifts and the willingness to listen, in the more intimate surroundings of urban homes and pubs. Folklore records knowledges of how otherworld persons may be entertained (by appropriate gift giving and respectful behaviour that, in particular, recognises the autonomy and authority of such beings) in human dwelling places. In these and other ways, there is a wilderness (non-modern, non-consumerist, non-productive, non-development, non-separated) that reaches into places that are only imagined to be under human control. Thus, for all that people might retreat into relatively remote upland areas (which in actuality may be only a short car journey out of town), they may also celebrate the wilderness of multi-species community by refusing to use biocides. They may make offerings to the larger-than-human world at home as much as anywhere else. Barry Patterson (author of *The Art of Conversation with the Genius Loci*, 2005) told Adrian Harris that paying attention to the "threshold brook" (a reference to John Keats' evocation of "unheeded" places) even at the margins of one's garden or allotment can be not only healing but restorative of human respect for the world (Harris 2008; 2012). Wilderness, in this sense of relatively less human dominated places, can be found wherever an effort towards human participation or a practice of human restraint restores such possibilities. Wilderness is not so remote as ignored, marginalised or overlooked until efforts are made to pay it attention and to immerse oneself in its potential.

Taking another more careful look at "wilderness", then, we see that Eco-Pagan environmental educators / shamanic practitioners do not retreat into a romantic escapism of an imagined past. In the valued wilderness of larger-than-human community they advance into the world that is not and cannot be completely remodelled to suit only human needs.

8. Silent Spring in the Isle Full of Noises

The second kind of wilderness, the damaged world of over-consumption and soil exhaustion, is ever present in modernity. It shows how disastrous human domination would be if it ever became a complete success, a reality rather than a dystopian fantasy. In monocultural fields of grain or grass, or in land given over to overconsumption by dairy and meat bio-machines (engineered out of creatures that were once wild animals) the wilderness of the anthropogenic "wild" is expanding. It intrudes further into the world in such practices as genetic modification and the patenting of plants and plant derivatives. Rather than attempt to enrich the soils from which plants and animals might benefit, the agro-industry seeks ways to make soil redundant except as an inert base in which plant roots can seek their artificial fertilisers. Farmland fast becomes desert: deadly places for insects, birds, animals and anything beyond the desired monoculture. Humans have created and proliferated (monocultural) wildernesses.

The threat of what Rachel Carson (1962) called the "Silent Spring" (the emblematic silencing of bird song that results from over-use of "pesticides") motivates Eco-Pagan environmental educators / shamanic practitioners in their efforts. They seek to inspire change, to demonstrate the possibility and benefit of living as responsible citizens of this small but diverse community planet. For these Eco-Pagans, a world that should be "an isle full of noises" (Shakespeare's *Tempest* 3.2.148–9) as diverse beings talk with kin, neighbours and strangers may well be restored or enhanced by shamanic-environmentalist participation in those conversations. Etiquette demonstrating respect may be the foundation of shamanic rites that energise Eco-Pagan efforts to confront those ongoing activities that seem perversely designed to eradicate living species and eliminate biodiversity. While environmental activism is evidently inspired by the desire to confront those constructing an even more human dominated world (full of noisy roads, airports and quarries but empty of bird-song and insect buzz), to put an end to ecocide, and to enable sustainable and equitable living for human and other-than-human beings, shamanism may be a means to collaborate with the threatened wilder creatures.

Eco-Pagan environmental educators / shamanic practitioners respectfully greet wild beings (ravens as much as fairies) in weird rituals and almost casual acknowledgements of presence and participation, so that one kind of wilderness will not be the end-game of human cultural evolution and, concomitantly, of this long phase of the evolution of life on Earth. However, the chief effort of these people in their confrontation with the threat of monocultural wilderness enlists the aid of other humans. They acknowledge that a turning round is required, away from over-consumption and exploitative misuse, towards some more restrained version of human living in the world. This cannot succeed as a minority practice, one associated with strange labels like "shaman" or "Pagan", Therefore they offer forms of environmental education that are intended to enthuse a larger population about belonging within a larger-than-human community, and to inspire everyone to act differently from the consumerist norm of modernity. Shamanic rituals may get the attention of other-than-human persons (see Grimes 2002) but the rituals of fairly ordinary (but well delivered) environmental education will, these people hope, inspire the etiquette of respectful multi-species co-dwelling that may yet save the planet.

9. Ritualizing in combination with scientific education is etiquette

Nature is an overworked term. Lewis Carroll's character, Humpty Dumpty, would have told Alice that it deserves to be "paid extra". In Carroll's *Through the Looking Glass* Humpty Dumpty scornfully asserts "when *I* use a word … it means just what I choose it to mean – neither more nor less" (1962, 274–5). Although his use of words such as "glory" and "impenetrability" is certainly eccentric, his declaration carries some force: words can be used to mean "many different things". "Nature" certainly means many things, perhaps too many. However, in contrast with Humpty Dumpty's claim that "When I make a word do a lot of work … I always pay it extra" few users of the term "nature" pay it back. Rather, they lazily assume that readers will share their assumptions about its resonances and references. They do not polish it with explanatory glosses or elevate it by paying attention to other terms that constellate around it. The end result is that "nature" has come to mean too much about nothing specific. People who imagine that it is meaningful to separate humanity from the rest of the cosmos have also taken the term "nature" to refer to a realm that has, in reality, never really been separate or different from "culture". Given the possibility of speaking about "human nature" and of treating "natural" as a synonym of "material" or "physical", in modernist discourse the word even had to refer to everything except the minds of dualists. It is, in short, a term that should now be retired unless it can be reinvigorated by the more careful use of expansive and debateable terms like "wilderness".

In this book "wilderness" is paid handsomely. In each chapter it is given pride of place in reference to specific understandings and practices. In this chapter it refers to two domains, both created by humans and open to varied valuations. There is a wilderness that is relatively remote or generally unheeded. While there may be almost pristine wildernesses in some places (for example, those in the USA, such as Alaska's Tongass National Forest, protected by the Clinton and Obama administrations under the "Roadless Rule"), in Britain what passes for "the wild" is a range of marginal places. The uplands denuded of trees in the Neolithic or the Clearances, traditional hedgerows in family-farms, the "threshold brooks" at the edges of gardens, and woodlands endangered by developers, are some of the places that may, according to Eco-Pagan educator/shamans be discovered to be venues for respectful inter-species encounters. There is another "wilderness" of more recent origin that they engage: the "wild" created by the biocidal practices of industrial and consumerist global modernism. This is evident in monocultural fields and "forests", in swathes of tarmac that ease human travel (by road or air) while endangering all other species, and in other arenas utterly controlled only by humans.

Eco-Pagan environmental and shamanic ideas and practices are founded on the notion that humans cannot separate themselves from the world. However, the attempt to do so, beginning with the fantasy that a realm called "nature" exists over against human "culture", has created wild places. These include both the barren, monocultural "wildernesses" under complete human control and also the more biodiverse "wildernesses" (whether these are the Romantic fiction of pristine purity or the more messy marginal territories in which multi-species diversity survives). The British Eco-Pagan environmental educators / shamanic practitioners of interest here delight in and are concerned for the larger-than-human community (their positively valued wilderness) and they see their practices as offering inspiration to do something about and against the global disaster of proliferating the negatively valued wilderness and they stress the urgency of so doing. These efforts are, they say, not undertaken alone. There are two sides to the equation of their life and work. On the one side is environmental education in which they seek to inculcate knowledgeable and enjoyable engagement with other-than-human species. On the other side is shamanic practice in which they seek to persuade humans and other-than-humans to work together to address the perilous state of the world. The two sides inform each other. They are part of the same work. Nonetheless, a line can be drawn between them so that participants in their educational events are not conned into practicing animistic shamanic rituals. The "natural sciences" knowledge conveyed is solidly rooted in empirical and experimental science. The understanding of these Eco-Pagans, however, is that the world accessed by the best of modern science is one in which many species co-evolve and co-create the world, minute by minute. In that world, to be a shaman is to seek to understand other species. In the same world, one now threatened by pretensions of human control and consumption, to be an environmental educator is to be a mediator between humans and other-than-humans. Eco-Pagans can therefore be both environmental educators and/as shamanic practitioners, and vice versa, precisely because there are two kinds of wilderness.

The "Wild" and the "Silent Spring" are resisted by shamanic-eco-educators who, hearing conversations between wild creatures (ravens, mushrooms, elves), seek to demonstrate that some humans wish to collaborate in the project of enriching the larger-than-human community. In this project, the counter-modern practice of ritual *and* the thoroughly contemporary practice of scientific education have become, for some people, the chief form(s) of etiquette communicating respect and requesting partnership across species boundaries, restoring one wilderness to challenge the other.

10. References

Albanese, Catherine L. *Reconsidering Nature Religion.* Harrisburg: Trinity Press International, 2002.

Bekoff, Marc. "First Do No Harm", *New Scientist* | 28 August (2010): 24–5.

Berger, Helen, Evan Leach, and Leigh S. Shaffer. *Voices from the Pagan Census.* Columbia: University of South Carolina Press, 2003.

Carroll, Lewis. *Alice's Adventures in Wonderland and Through the Looking Glass.* London: Puffin, 1962 (1872).

Carson, Rachel. *Silent Spring.* New York: Horton Mifflin Company, 1962.

Clifton, Chas S. *Her Hidden Children: The Rise of Wicca and Paganism in America.* Lanham: Altamira Press, 2006.

Creeping Toad website. URL: www.creepingtoad.org.uk (last accessed 4 August 2011).

De Silva, Sena S. and Turchini, Giovanni M. "Towards understanding the impacts of the pet food industry on world fish and seafood supplies", *Journal of Agricultural and Environmental Ethics* 21.5 (2008): 459–67.

Eliade, Mircea. *Shamanism: Archaic Techniques of Ecstasy.* New York: Pantheon, 1964 (1951).

Franke, Anselm, and Bruno Latour. "Angels Without Wings: A Conversation Between Bruno Latour and Anselm Franke" in Anselm Franke (ed.), Sternberg Press, 2010. pp. 86–96.

Glass, Matthew. "National Parks and Monuments" in Bron Taylor (ed.) *Encyclopedia of Religion and Nature.* London: Continuum, 2005. pp. 1152–8.

Grimes, Ronald. "Performance Is Currency in the Deep World's Gift Economy: An Incantatory Riff for a Global Medicine Show", *Interdisciplinary Studies in Literature and Environment* 9.1 (2002): 149–64.

Harris, Adrian. *The Wisdom of the Body: Embodied knowledge in the practice of Nature Religions.* Unpublished PhD thesis, University of Winchester, 2008.

Harris, Adrian. "Embodied Eco-Paganism" in Graham Harvey (ed.) *The Handbook of Contemporary Animism.* Leiden: E.J. Brill, 2012.

Harvey, Graham.[1] *The Forgiveness of Nature: The Story of Grass.* London: Vintage, 2001.

Harvey, Graham. "Guesthood as ethical decolonising research method", *Numen* 50.2 (2003a): 125–46.

Harvey, Graham (ed.) *Shamanism: A Reader.* London: Routledge, 2003b.

Harvey, Graham. "Performing and Constructing Research as Guesthood" in Lynne Hume and Jane Mulcock (eds), *Anthropologists in the Field.* New York: Columbia University Press. 2005a. pp. 168–82.

Harvey, Graham. *Animism: Respecting the Living World.* London: C. Hurst & Co, 2005b.

Harvey, Graham. *Listening People, Speaking Earth: Contemporary Paganism.* 2nd edition. London: Hurst & Co, 2006a.

Harvey, Graham. "Discworld and Otherworld: The popularisation of Pagan fantasy literature" in Lynne Hume and Kathleen McPhillips (eds), *Popular Spi-*

1 Please note that this first Graham Harvey is a different Graham Harvey.

ritualities: The Politics of Contemporary Enchantment, Aldershot: Ashgate, 2006b. pp. 41–51.
Harvey, Graham. "Paganism: Negotiating between Esotericism and Animism under the Influence of Kabbalah" in Boaz Huss (ed.) *Kabbalah and Contemporary Spiritual Revival.* Beer Sheva: Ben-Gurion University of the Negev Press, 2011. pp. 267–84.
Harvey, Graham, and Wallis, Robert. *The A to Z of Shamanism.* Lanham: Scarecrow Press, 2010.
Hay, Peter. *A Companion to Environmental Thought.* Edinburgh: Edinburgh University Press, 2002.
Hogan, C. Michael. "Deforestation" in Encyclopedia of Earth, 2010. URL: http://www.eoearth.org/article/Deforestation?topic=58071 (accessed 3 August 2011).
Hutton, Ronald. *The Pagan Religions of the Ancient British Isles.* London: Blackwell, 1991.
Hutton, Ronald. *The Rise and Fall of Merry England: The Ritual Year 1400–1700.* Oxford: Oxford University Press, 1994.
Hutton, Ronald. "The Roots of Modern Paganism" in Graham Harvey and Charlotte Hardman (eds) *Paganism Today.* London: Thorsons, 1996. pp. 3–15.
Hutton, Ronald. *The Triumph of the Moon: A History of Modern Pagan Witchcraft.* Oxford: Oxford University Press 1999.
Hutton, Ronald. *Shamans: Siberian Spirituality and the Western Imagination.* London: Hambledon and London, 2001.
Jamison, Ian. *Embodied Ethics and Contemporary Paganism.* Unpublished PhD thesis submitted to the Open University, 2011.
Kohák, Erazim, *The Green Halo: A Bird's-Eye View of Ecological Ethics*, Chicago: Open Court, 2000.
Latour, Bruno. *We Have Never Been Modern.* New York: Harvester Wheatsheaf, 1993.
Latour, Bruno. *War of the Worlds: What about Peace?* Prickly Paradigm Press, Chicago, 2002.
Letcher, Andy. "'Virtual Paganism' or Direct Action? The Implications of Road Protesting for Modern Paganism," *Diskus* 6 (2000). online at http://www.basr.ac.uk/diskus/diskus1- 6/letcher6.txt (accessed 16 May 2011)
Letcher, Andy. "The Scouring of the Shires: Fairies, Trolls and Pixies in Eco-Protest Culture", *Folklore* 112 (2001): 147–61.
Letcher, Andy. "Raising the Dragon; Folklore and the Development of Contemporary British Eco-Paganism," *The Pomegranate: The International Journal of Pagan Studies* 6.2 (2004): 175–98.
Letcher, Andy. "Eco-paganism", in Bron Taylor (ed.) *Encyclopedia of Religion, Culture and Nature.* London: Continuum, 2005. pp.556–7.
MacLellan, Gordon. *Shamanism.* London: Piatkus, 1999.
McIntosh, Alastair. *Soil and Soul.* London: Aurum, 2004.
Milton, Kay. *Loving Nature: Towards an Ecology of Emotion.* London: Routledge, 2002.

Mischi, Julian. "Englishness and the Countryside. How British Rural Studies Address the Issue of National Identity" in Floriane Reviron-Piegay (ed.). Englishness Revisited. Newcastle upon Tyne: Cambridge Scholars Press, 2009. pp. 109–25.
Oelschlaeger, Max. *The Idea of Wilderness: From Prehistory to the Age of Ecology.* New Haven: Yale University Press, 1991.
Oelschlaeger, Max. "Wilderness Religion" in Bron Taylor (ed.) *Encyclopedia of Religion and Nature.* London: Continuum, 2005. pp. 1745–8.
Patterson, Barry. *The Art of Conversation with the Genius Loci.* Milverton: Capall Bann, 2005.
Philippon, Daniel J. "Wilderness Society" in Bron Taylor (ed.) *Encyclopedia of Religion and Nature.* London: Continuum, 2005. pp. 1750–1.
Plumwood, Val. *Feminism and the Mastery of Nature.* London: Routledge, 1993.
Plumwood, Val. *Environmental Culture: The Ecological Crisis of Reason.* London: Routledge, 2002.
Plumwood, Val. "Nature in the Active Voice", *Australian Humanities Review* 46 (2009): 113–29. Reprint forthcoming in Graham Harvey (ed.) *Handbook of Contemporary Animism.* Leiden: E.J. Brill, 2012.
Rose, Deborah B. *Nourishing Terrains; Australian Aboriginal views of Landscape and Wilderness.* Canberra: Australian Heritage Commission, 1996.
Rose, Deborah B. "Totemism, Regions, and Co-management in Aboriginal Australia", draft paper for the Conference of the International Association for the Study of Common Property, 1998. On-line at http://dlc.dlib.indiana.edu/dlc/bitstream/handle/10535/1187/rose.pdf?sequence=1 (accessed 16 May 2011).
Rose, Deborah B. *Reports from a Wild Country: Ethics for Decolonisation.* Sydney: University of New South Wales Press, 2004.
Rose, Deborah B. *Dingo Makes Us Human. Life and land in an Australian Aboriginal Culture.* Cambridge: Cambridge University Press, 2009.
Rose, Deborah B. *Wild Dog Dreaming: Love and Extinction.* Charlottesville: University of Virginia Press, 2011.
Song and Story website. URL: http://www.songandstory.co.uk/ (accessed 4 August 2011).
Snyder, Gary. *The Practice of the Wild.* New York: North Point Press, 1990.
Taylor, Bron. "Resacralizing Earth: Pagan Environmentalism and the Restoration of Turtle Island." In David Chidester and Edward T. Linenthal (eds) *American Sacred Space.* Bloomington: Indiana University Press, 1995. pp. 97–151.
Taylor, Bron. *Dark Green Religion.* Berkeley: University of California Press, 2010.
Tylor, Edward. *Primitive Culture*, 2 vols, London: John Murray, 1913 (1871).
Up. film, directed by Pete Docter and Bob Peterson. Emeryville: Disney Pixar, 2009.
Viveiros de Castro, Eduardo. "Cosmological Deixis and Amerindian Perspectivism." *Journal of the Royal Anthropological Institute* (N.S.) 4 (1998): 469–88.

Whitehead, Neil L., and Robin Wright (eds) *In Darkness and Secrecy: The Anthropology of Assault Sorcery and Witchcraft in Amazonia.* Durham: Duke University Press, 2004.

Willerslev, R. *Soul Hunters: Hunting, Animism and Personhood among the Siberian Yukaghirs.* Berkeley: University of California Press, 2007.

Yeats, William B. "The Stolen Child" in Daniel Albright (ed.) 1990. *W.B. Yeats: The Poems.* London: J.M. Dent, 1886. pp.44–5.

Znamenski, Andrei A. *The Beauty of the Primitive: Shamanism and the Western Imagination.* Oxford: Oxford University Press, 2007.

13. Wilderness, Spirituality and Biodiversity in North America – tracing an environmental history from Occidental roots to Earth Day

Bron Taylor

1. Introduction

Exploring competing perceptions of wilderness, and disputes over places said to be wilderness, opens an illuminating window into environmental history. Understanding such cultural evolution requires special attention to religion. My analysis begins with some generalizations about the idea of wilderness in the Occidental world and then focuses on the history of the nexus of wilderness and religion in North America from European contact to the first Earth Day in 1970. This history demonstrates how deeply the idea of wilderness and wilderness-related spirituality are related to nature conservation movements in U.S. culture. It also reveals how the idea of wilderness and the rationale for protecting it has shifted since the Darwinian revolution, from the preservation of natural beauty and its various spiritual values, to the notion that biological diversity is intrinsically valuable and sacred, and thus, worthy of reverence and defense.

2. "Wilderness" as idea and place in the Occidental World

In *Forests: The Shadow of Civilization*, Robert Pogue Harrison wrote:

> From the family tree to the tree of knowledge, from the tree of life to the tree of memory, forests have provided an indispensable resource of symbolization in the cultural evolution of humankind (Harrison 1992, 8).

The same could be said for the term "wilderness." Contemporary dictionary definitions generally agree that wilderness refers to environmental systems where natural processes occur with little or no significant influence by human beings. Wilderness definitions often also generally indicate that the place is uninhabited and uncultivated, and sometimes carry a connotation that such places are deserted, barren, wastelands. But the term's meaning is contested in ways that dictionaries do not reflect. There is, for example, no consensus whether the places the term stands for are sacred or profane; whether such places are needed for human physical and spiritual health, or in contrast, are actually physically and spiritually dangerous; whether such places reflect democratic values or are a preserve of an imperial elite. For some, therefore, the meaning of the term, and the purposes it has served, are troubling. Then there are the places to which the word supposedly refers. These places are in their own ways troubled, ever declining in size, biodiversity, and ecological resilience. So while there is at least vague agreement about the term's meaning, ever since the first significant wave of Europeans began to arrive in North America during the 16th century, the idea of wilderness and the places to which the term refers have both been ideological, spiritual, and physical battlegrounds. And by the late 20th century, with the spread of American-style conservation to many countries around the world, especially through National Parks and wilderness reserves, contention over the ideas and practices related to "wilderness" went global.

Before we can focus on wilderness ideas and places in North America we must begin with antecedent notions in the Occidental world, for as Roderick Nash explained in his seminal *Wilderness in the American Mind* (and as many ethnographers have noted) there is no cognate for the term wilderness in the languages of the peoples who were already inhabiting the continent before Europeans arrived (Nash 2001 [1967], xiii). Nash suggested that in English, "the root seems to have been 'will' with a descriptive meaning of self-willed, willful, or uncontrollable", adding that as expressed in early Teutonic and Norse languages, the term was initially applied to human beings; only later, in Old English, was it extended to other organisms and to wild, dangerous places and forested lands (Nash 2001 [1967], 1–2). Peering back deeper into Occidental history involves more conjecture, but in *The Idea of Wilderness* (1991), Max Oelschlaeger made a strong case that, while there are appreciative themes toward perceived beauties of nature in Judaism, Christianity and Islam, and ancient Western philosophy, the more nature-reverencing forms were likely remnants from earlier foraging cultures, which tended, to use contemporary terminology, to be animistic and pantheistic. Nash and many other scholars have noted that in the Hebrew Bible (the textual root of all Abrahamic traditions) wilderness (e.g., the Sinai Peninsula) was seen as a place to escape. Judaism, Christianity and Islam, therefore, have favored agricultural places replete with "milk and honey," and often derided wild habitats as deserts and wastelands, even if such habitats also at times were seen as places of refuge from moral corruption and other evils (Nash 1967, 13, 16, 18). The general scholarly picture painted by Nash and Oelschlager was that Occidental agricultures, including the religions accompanying them, while sometimes expressing appreciation for undomesticated habitats as places of spiritual trial and opportunity, and other times as a reflection of the power and beneficence of God, generally viewed the sacred not as the earthly world we inhabit but as somehow above and beyond it.[1]

It is equally important to recognize that agriculture is the basis of civilization, and that the now dominant, so-called "world religions" lend it spiritual and ethical legitimacy. As Harrison demonstrated during his examination of classical cultures as well as Christianity:

> The governing institutions of the West – religion, law, family, city – originally established themselves in opposition to the forests, which in this respect have been, from the beginning, the first and last victims of civic expansion (Harrison 1992, ix).

The same could be said of wildlands of all kinds, including wetlands and deserts.

It should not be missed that religion-infused agricultures are inherently and necessarily expansionistic (not to mention often violently imperial). Steven Stoll put it simply: "Agrarian societies also generated people – more than hunting and gathering societies did – so they continually created their own necessity for expansion, as sons and daughters sought to reproduce the material world of their parents" (Stoll 2007, 56). Since the land that agricultures need for expansion are almost always already inhabited, agricultures have been imperial; killing or displacing through force already-present inhabitants or converting them, either through example, persuasion, coercion, or threat, to their own agricultural and religious lifeways. This process, which began 10,000 years ago when humans began to domesticate plants and animals, has thus precipitated the dra-

---

1 When reading this section, Laura Feldt noted that it is difficult to assess the role that ancient Judaism, Christianity and Islam played in agricultural expansion and bio-simplification. She added in her comment to me that the difficulty is compounded by the differences between these religions and the distance in time and perception between them and the Hebrew Bible that they share as a sacred text, and the present time. As she noted, such interpretive difficulties should be kept in mind when considering the historical background presented in this section, which is necessarily brief and thus may involve some overgeneralization, in the interest of setting the stage for a consideration of wilderness in North America.

matic, global decline of both cultural and biological diversity (Marsh 1979 [1874]; Diamond 1987, 1997, Shepard 1992, 1998, Oelschaleger 1992, Mason 1993, Lockwood and McKinney 2001, Williams 2003, S. Stoll 2007, Pointing 2007). Agricultures exercised little restraint in this long process. Despite the expression of nature-related insights, such as of the interdependence of all life, or of ethics enjoining compassion for sentient creatures, by some religious figures in Occidental or Asian agricultures, the necessity and inertia of expansion in agriculture has been a powerful variable in producing biodiversity decline. Focusing on Europe, Harrison put provocatively the general pattern:

> The Christian church that sought to unify Europe under the sign of the cross was essentially hostile toward the impassive frontier of unhumanized nature. Bestiality, fallenness, errancy, perdition – these are the associations that accrued around forests in the Christian mythology. In theological terms forests represented the anarchy of matter itself, with all the deprived darkness that went with this Neoplatonic concept adopted early on by the church fathers. As the underside of the ordained world, forests represented for the church last strongholds of pagan worship (Harrison 1992, 61).

But on the North American continent, in land that became the United States of America, something remarkable happened in 1872: The United States Congress created Yellowstone National Park. Commenting on its significance, Steven Stoll wrote that this was "the first time that Congress imposed legal limits on the spread of agricultural settlement. The park created an entirely new category of land use – protected wilderness – where no crops or domestic animals would be allowed" (Stoll 2007, 68–69). And as Donald Worster has argued, this wilderness was a democratic space, owned by the public at large (Worster 2007). This was in marked contrast to most countries where wildlands held back from agricultural development were the reserves of the nobility, and commoners were excluded. The remarkable contrary development in the United States is inconceivable without the peculiar history of religion and nature that unfolded in North America after first Europeans reached its shore.

3. "Wilderness" perceptions early in the colonial history of North America

The imposition of legal limits to the spread of agriculture and the creation of wilderness areas owned by the public at large is not a development that one would expect when considered in the light of the reaction to wildlands by the first Europeans who arrived in North America with the intention of settling. This was significantly later than when Christopher Columbus, who in 1492 landed in the Bahamas, and then several Caribbean Islands, thus inaugurating the period of European imperial exploration of the Americas. It was also later than the next Spanish expeditions, the next of which in North America began in 1508 with Juan Ponce de León, who claimed for the Spanish crown what later became the U.S. state of Florida.[2] During the 16th century, after gaining military advantage over the Spanish, the English began trolling the Atlantic coast, looking for appropriate places to establish colonial settlements.

Like the Spanish and other colonial powers pursuing riches, power, and Christian converts, in Eastern North America, the English found a landscape domesticated by large numbers of Native Americans, who had shaped the land to their benefit and liking through fire and agriculture (especially corn cultivation).[3] Nevertheless, the English conveniently imagined the land as a little-populated wilderness. Such a perception, while difficult to maintain initially due to the many aboriginal people they encountered, became easier and more accurate a few decades after initial contact, for the Europeans carried diseases for which the aboriginal populations had no evolution-inherited immunities. This led to epidemics that devastated Native American populations, by some estimates, upwards of 80 %.[4] Some English colonists who noticed the diseases and the following depopulation took the epidemics as evidence that it was God's will that they take possession of the land (Perreault 2007, 23). To such individuals, the wilderness was God's gift.

It was the kind of gift, however, about which they were both naturally and culturally ambivalent. For anyone unaccustomed to large and dense forests inhabited by dangerous animals, fear would be perfectly natural.

2 I sidestep, therefore, the earliest, Spanish, claims made in Florida, by Juan Ponce de León, in 1513, and the subsequent battles between the French, and later the English, for control of the region, as well as discussion of the country's first settlement on Florida's north Atlantic coast, named for St. Augustine in 1565. For a number of reasons, Florida and the southeast was settled later than the rest of North America, much of which remained what would typically be called wilderness well into the 20th century, but this is not a story that can be told here. It is also a story largely ignored by the standard wilderness histories.

3 On population estimates see Denevan 1992a, Thornton 1987, and on Native American environmental impacts, Denevan 1992b.

4 This brief overview draws especially on Perreault 2007, Nash 1967, Denevan 1992b, Diamond 1997.

The fear was also of the aboriginal inhabitants who were considered "wild" or "savage" (Perreault 2007, 27–30). Such fears were magnified by Christian religious beliefs that, while viewing wild habitats as places of spiritual and commercial opportunity, also considered them to be places of religious and moral peril. As recalled by William Bradford, who arrived on the now-famous *Mayflower* in America in 1620, along with several dozen Puritan dissidents who were seeking religious freedom and economic opportunity, and who later became the second governor of the Plymouth Colony, the continent was a terrifying, "hideous and desolate wilderness, full of wild beasts and men" (Bradford, in Nash 23–24; Albanese 34; Perry 1956). As if that were not bad enough, for many pious European Christians, including the Spanish explorers and friars who founded missions and settlements in what would become Mexico and the American Southwest, as well as Florida, the American wildlands were also a haunt of Satan (Taylor and Van Horn 2006, 167).

Given these dangers, to tame Satan and wilderness, both the land and its people would need to be subjugated. The new arrivals generally believed that a religiously righteous culture could only be achieved by domesticating wildlands and subduing their inhabitants. Steven Stoll put it simply, "American expansion was agricultural expansion, and it came at the expense of Indians and wilderness" (Stoll 2007, 55). According to such historiography, the pattern resembled that of Europe, where the expanding agriculture, which often displaced people and usually reduced biological diversity, was deeply religious.

This sort of interpretation coheres with the history of the idea of wilderness in North America, which was famously advanced by Nash in 1967 (and reinforced by Oelschlaeger in 1991). Nash's book became as nearly canonical among scholars as histories can be, and in the main, it has held up well. Much of what follows draws on it but Nash's account is supplemented by more recent scholarship that has qualified and challenged some of it.

4. Wilderness in the 17th and 18th centuries in Europe & North America

For Nash and many of his scholarly progeny, European settlers had a twofold imperative to subjugate wilderness: First, because pioneer settlers "lived too close to wilderness for appreciation" for it "constituted a formidable threat to his very survival." Second, because these settlers felt obliged to battle "wild country ... in the name of nation, race, and God" (Nash 1967, 24). Even when early European Americans expressed appreciation for "nature" in early American history, "it was not for its wildness but because it resembled a 'Garden or Orchard in England'" (Nash 1967, 33). For Nash, John Bunyan's *Pilgrims Progress from this World to That which is to come*, which was published in London in 1681, well expressed "the prevailing viewpoint of wilderness as the symbol of anarchy and evil to which the Christian was unalterably opposed" (1967, 34). The Puritans, moreover, contrasted the city on the hill they wished to build with the wilderness, which first had to be transformed. Moreover, "their Bibles contained all they needed to know in order to hate wilderness," which showed, according to Nash, that "the colonist's conception of the wilderness was more a product of the Old world than the New" (Nash 1967, 35). One early puritan colonist even explicitly noted that towns and churches were established precisely where the "wilderness was subdued" and where previously there had been nothing but "*Heathenism, Idolatry,* and *Devil-worship*" (in Nash 1967, 37; see also Miller 1956). Nash noted that these New England colonists called themselves Christ's soldiers and thought they were engaged in "in a war against wilderness" (1967, 37). Nash's exposition analyzed the views of a variety of literary figures and politicians, who in the 18th and 19th centuries extolled agrarian civilization while contrasting it with untamed wilderness, including most notably Thomas Jefferson, who drafted the Declaration of Independence and became the republic's third president.

Nash noted, however, that antipathy to wilderness was not universal among the European arrivals: "A handful of mountain men and voyageurs were literally absorbed by the forest, ... in some cases even joining Indian tribes. These exceptions regarded civilization with the antipathy most pioneers reserved for wilderness" (Nash 1967, 43).[5] Moreover, Nash also acknowledged that in the 18th century Jonathan Edwards, the Reformed (Calvinist) theologian, and famous "fire and brimstone" preacher, found "spiritual joy" and beauty in "natural objects such as clouds, flowers, and fields" (Nash 1967,39). Nevertheless, Nash concluded, for Edwards, "wilderness was still beyond the pale," and as evidence, he quoted a passage from Edwards that considered wilderness an obstacle

5 Voyagers were the explorers, fur trappers, and guides who paddled interior waterways of present day Canada and the northern United States, often developing commercial and sometimes intimate relations with native peoples.

on the path to Heaven (Nash 1967, 39). Some other scholars, however, have found greater appreciation for nature among the Puritans and within the Reformed (Calvinist) Protestant tradition.

John Gatta and Mark Stoll, for example, have argued that the Reformed tradition was actually quite sympathetic to wild nature. As evidence, they cited the reverent attitude toward nature as early as in the 17th century poetry of Anne Bradstreet (who is sometimes called America's first poet), as well as in other passages by Edwards (Stoll 2007, 39; Gatta 2004, 40–48). To explain such views Stoll noted that if one believes God communicates with human beings through nature (which fits with the Platonic doctrine of correspondence that has often been embedded in Christian thought) then as John Calvin (the founder of the Reformed Tradition) once conceded, "the expression 'Nature is God,' may be piously used, if dictated by a pious mind" (in Stoll 2007, 39). Stoll buttressed his argument by citing a variety of wilderness proponents, all the way up through the 20th century, who had deep roots in the reformed tradition. Stoll also contended that Catholics, non-Reformed tradition Protestants such as Methodists and Baptists, new religious sects such as the Mormons (properly known as members of the Church of Latter Day Saints), and some ethnic groups, such as African Americans, were less prone to nature appreciation than those with roots in the Reformed tradition. Stoll concluded, "particularly for children of the Reformed tradition, wilderness has been irradiated with [positive] spiritual meaning" (Stoll 2007, 50). In a similar way, Gatta also concluded that, after their initial hostility and fear, the attitudes of some Puritans, and some Christian sects, such as the Society of Friends (Quakers), began to shift toward an appreciation of the American land as sacred. My own view is that these revisionist efforts rely, however, on scant evidence; they are based on cases that might more accurately be taken as exceptions that prove the rule.

Nash concluded his analysis of Early America by agreeing with Alexis de Tocqueville, who contended that those living in the wilds were biased against them, for, with only "a few exceptions, American Frontiersmen rarely judged wilderness with criteria other than the utilitarian or spoke of their relation to it other than a military metaphor" (in Nash 1967: 43). According to Nash, it would be as subsequent generations of European Americans were "removed from a wilderness condition" that they would begin "to sense its spiritual and aesthetic values… Appreciation of the wilderness began in the cities. The literary gentleman wielding a pen, not the pioneer with his axe, made the first gestures of resistance against the strong currents of antipathy" (Nash 1967, 43, 44). This is surely an overstatement, as the ability to appreciate the beauties of nature and the joys of existence within is part of the human emotional repertoire and known to people of all ethnic, class, and cultural backgrounds. Moreover, as Karl Jacoby has shown with regard to rural people, and other scholars are bringing forward with regard to African Americans, these groups often have their own environmental appreciation and ethics that are unknown or unseen by literary classes (Jacoby 2001, Glave and Stoll 2006, Smith-Cavros 2007a, 2007b, Smith 2007, Ruffin 2010). Nevertheless, it is true that in environmental history, it is often urban, not rural people, who have led conservationist battles. So there is significant truth to Nash's observation.

For Nash and other historians, and contrary to some of the views just mentioned, an appreciation of wild nature did not really begin until European Americans were primed for it by new intellectual trends in Europe, specifically, by "the flowering of Romanticism in the eighteenth and early nineteenth centuries" (Nash 1967, 45). And while the earlier fearful if not hostile attitudes remained dominant, by the middle of the eighteenth century, strong voices had emerged promoting an aesthetic and spiritual appreciation for wilderness, even, an "association of God and wilderness" (Nash 1967, 44). But the tributaries to it were not in orthodox Christianity, whether Protestant or Catholic. Nash saw this new trend as beginning with Enlightenment science, which heightened a sense of awe and wonder at the universe. Given the theistic cultural context in which such perceptions emerged, these wonders were often believed to derive from God. In this century, two new trends emerged, neither of which squared with traditional theism – Deism, which saw God as mysteriously ordering the universe in deep time, and Romanticism which, as Nash noted, "resists definition, but in general implies an enthusiasm for the strange, remote, solitary, and mysterious" (Nash 1967, 47). Romanticism also idealized wilderness and even primitivism, which according to Nash, is a belief "that man's happiness and well-being decreased in direct proportion to his greater degree of civilization" (Nash 1967, 47).

Both Deism and Romanticism, in their own ways, inculcated a perception that nature, including wild nature, was beautiful if not also sublime. Numerous books by theologians and philosophers began to appear that expressed the newfangled perceptions and beliefs.[6] Among the most important, which Nash duly noted, were Edmund Burke's *Philosophical Enquiry into the Origin of Our Ideas of the Sublime and Beautiful* (1757), Immanuel Kant's *Observations on the Feeling of the Beautiful and the Sublime* (1764), and most critically for Romanticism, Jean-Jacques Rousseau's writings, especially *Emile* (1762) and *Julie ou La Nouvelle Heloise* (1761). Rousseau contended that agriculture was the original human sin; that it led to private property, antagonism, and civilization, and that it destroyed the good life enjoyed by foraging peoples. For Rousseau, happiness and wellbeing depended on finding a way back to a natural existence (Oelschlaeger 1991, 110–111).

Those inspired by Rousseau, including poets and other artists, would soon travel to America to experience first hand the wild places and peoples there, playing some role in the spread of Romanticism (Nash 1967, 49). Among the most prominent Romantic poets was the English nobleman George Gordon "Lord" Byron (1788–1824). Byron was disenchanted with civilization but enchanted with nature, and even though he never traveled to North America, his poetry was much loved and cited by American nature enthusiasts, even many generations later. The following poem is perhaps the most often quoted one. Nash quoted it only partially, as do I, although I add two lines printed below that were omitted by Nash because I think they add significantly to the affective feeling of this poem, and because they are typical of Romantic poetry more generally.

> THERE is a pleasure in the pathless woods,
> There is a rapture on the lonely shore,
> There is society, where none intrudes,
> By the deep sea, and music in its roar:
> I love not man the less, but Nature more,
> To mingle with the Universe, and feel
> What I can ne'er express, yet cannot all conceal.[7]

Here are the feelings of belonging and connection to nature, and perceptions of its sacredness, that characterize earth and nature-based religions in diverse cultures and periods around the world (Taylor 2001a, 2001b,

6 By the late 18th century Romanticism was growing rapidly in Europe and finding expression in philosophy, poetry, visual art, and music. There are too many figures to review here, given the focus her on American wilderness, but they include Goethe, Shelling, Blake, and Wordsworth.

7 Lord Byron (George Gordon), from Childe Harold, Canto iv, Verse 178, available at http://www.theotherpages.org/poems/byron01.html

2010). And while there have long been people with such spiritual sentiments, it is nevertheless true that there are some times and places that provide fertile ground for the spread of such perceptions and nature-revering practices. Romanticism emerged and spread influentially from European soil, but American soil, especially during the 19th and 20th centuries, provided even more fertile ground for it, playing a leading role in the emergence of a unique wilderness protection movement. The movement would eventually spread to over 100 other countries, in large part because people in other parts of the world came to share the sentiments animating it.

There were hints of this emerging nature appreciation during the late 18th century, as Philip Freneau, a hermit known as "the Philosopher of the Forest," began publishing essays in 1781, which criticized civilization and celebrated wild forest life. Meanwhile, Daniel Boone, who was celebrated during his own time and, in the mid 20th century, became an icon in American popular culture due to Disney films celebrating his life, wrote an autobiography in 1784. In this book, Boone depicted himself not only a mountain man engaged in settling the American wilderness, but as a philosopher of nature and its beauties who was ambivalent about the expansion of human settlements that he, ironically, had helped make possible (Nash 1967, 63). Boone's popularity may in part reflect an aesthetic appreciation for wild places by the Americans who read him, as well as a nascent ambivalence about the course of "progress" and its increasingly obvious destructive impacts. His popularity might also be one of the early signs that some Americans felt something important about a man's strength and character depended on his connection to wilderness. As wilderness shrunk a corresponding anxiety would grow.[8]

8 The foremost exemplar of this anxiety was Frederick Jackson Turner, whose "Frontier Thesis" (1893) captured such anxiety, shortly after as the last of the tribes resisting the European invasion were suppressed and the frontier was officially declared closed by the U.S. Census Bureau, as the European presence then stretched coast to coast.

5. Wilderness in the 19th and early 20th century in North America

Possibly exemplifying this nascent trend of nature appreciation and anxiety about the loss of human wildness was the Philadelphia physician Benjamin Rush, who in 1800, "explicitly connected primitivism and wilderness by observing that 'man is naturally a wild animal, and … taken from the woods, he is never happy … 'till he returns to them again'" (Nash 1967, 56). This recognition of human animality was a remarkable acknowledgement for a person writing more than a half century before Darwin's *On the Origin of Species* was published. Soon after Rush, in 1818, the New Hampshire lawyer Estwick Evans put such feelings into practice. Wearing clothes and moccasins made of buffalo and bearskin, he headed west, writing, "I wish to acquire the simplicity, native feelings, and virtues of savage life; to divest myself of the factitious habits, prejudices and imperfections of civilization … and to find amidst the solitude and grandeur of the western wilds, more correct views of human nature and the true interest of man" (in Nash 1967, 56). Soon Evans would write what Nash called a "Romantic paean" to wilderness:

> How great are the advantages of solitude!
> How sublime is the silence of nature's ever-active energies!
> There is something in the very name of wilderness, which charms the ear, and soothes the spirit of man.
> There is religion in it (Evans 1819, in Nash 1967, 56).

Nash accurately commented: "In the sweep of Western thought, this was a relatively young idea, and one with revolutionary implications. If religion was identified with wilderness rather than opposed to it, as had traditionally been the case, the basis for appreciation, rather than hatred, was created" (Nash 1967, 56). Evans' statement also represents an example and early recognition of "nature religion", namely, the idea that nature itself can be the wellspring of religious belief, perception, and reverence.[9] It also represents the long stream in North American culture of appreciation of Native Americans for their supposed connection to nature,

---

9 This paper is rooted in an approach to the study of religion that is unconcerned with where the boundaries of "religion" lie, and more interested in what can be learned about social phenomena that have traits and characteristics typically associated with the term. For more on this "family resemblance" approach to religion studies, see Taylor (2007 or 2010, 1–4).

which often coexisted with fears and negative stereotypes (Pedersen 1995; Krech 1999; Kalland 2005).

Poets and other writers proliferated, expressing feelings of being spiritually connected to nature. William Cullen Bryant, for example, wrote "A Forest Hymn" in 1825 about his conviction that the Creator's hand could be found in the very forests that most European Americans had previously found perilous. Shortly afterward, James Fenimore Cooper began publishing five "Leatherstocking tales" novels, beginning with *The Last of Mohicans* in 1826. The tales focused on a Native American named Leatherstocking (and Natty Bumpo), who was a liminal figure navigating a perilous space between Native American, nature-revering cultures, and the dominant, expanding, Euro-American empire. Readers learned that good character depends on a rootedness in nature, nature is holy, deforestation is a desecrating act, and America's superiority over Europe is due to its wild nature (Nash 1967, 75–77). Cooper not only expressed reverence for the land but an appreciation for Native American lifeways and what he considered to be their closeness to nature, a perspective that would become an important feature typical of much environmentalist thinking. This perspective was also evident in the earliest calls for nature preservation, such as in 1832 when George Catlin, "an early student and painter of the American Indian," became the first to promote the idea of setting aside large national parks that would include both wild natural beauty as well as Indians (Nash 1967, 100–101). This would not be the course that the wilderness preservation movement would take, as has now been pointed out by numerous critics; Indian removal, even if not always immediate or violent, became the official policy in the management of National Parks and forests (Keller 1998, Spence 1999).

More dramatic expressions of Romanticism soon followed Cooper and Catlin, this time, in the landscapes of a group that became known as the Hudson River School painters. Led by Thomas Cole, who was directly influenced by Lord Byron, and whom Nash called a romantic pantheist, these artists generally depicted human beings as insignificant and nature as sublime (Nash 1967, 78–79, 81). Cole's famous "Course of Empire" series, five paintings he produced in 1836, repeated the romantic theme of a decline from a harmonious state of nature to a devastated world wrought by civilization. Nevertheless, in the final painting, nature was seen re-emerging, after the environmental apocalypse. Many environmental artists would revisit such themes in the 20th century. Yet another painter from the school, Edward Hicks, depicted "Noah's Ark" in 1849, perhaps implicitly expressing an affinity for the earth's biological diversity, shortly before advocates for wilderness would begin to make the case explicitly. The fledgling Christian environmental movement, in the late 20th century, used the painting when lobbying congress in the hopes of protecting endangered species (Kearns 1997).

The same year that Cole focused his paintings on empire, 1836, the great transcendentalist Ralph Waldo Emerson published his now-famous essay, *Nature*, arguing that all natural objects can awaken reverence, "when the mind is open to their influence" (Emerson 2000 [1836]). Emerson wrote in ways that sometimes sounded pantheistic, other times animistic, speaking of spiritual truths conveyed by nonhuman organisms. He did not advocate protection of wilderness, however, very likely because of his Platonic view (central to Transcendentalism) that nature is more "the pathway to spiritual truth than … a spiritual end"; nevertheless, he "contributed decisively to the dramatic rise in nature appreciation in the latter decades of the nineteenth century" (Taylor and Van Horn 2006, 169). It was an appreciation advanced famously in an often-quoted passage by Emerson's friend Walt Whitman, who wrote in *Leaves of Grass* (1855), "This is what you shall do: love the earth and sun and animals" (Whitman 2005 [1855], vi). Whitman seemed to be more clearly speaking about here-and-now nature than did Emerson, however, and he articulated a religious kinship ethic with non-human nature that easily fits with the then nascent ethics that values wilderness for its biological diversity.

One can certainly wonder whether Henry David Thoreau would have become the 19th century's most important philosopher promoting wilderness (and wildness) had he not forged his thinking with Emerson as his mentor and muse. But to do so, he had also to break decisively with Emerson's Platonic spiritualism in favor of a more naturalistic spirituality in which wild nature, including a wild human nature, could become the centerpiece. Indeed, Thoreau wrote of experiences and perceptions that were pantheistic and animistic, and he spoke sympathetically of paganism. He rejected Christianity, doing so in a way that reflected his lifelong interest in Native Americans and *their* worldviews, once stating that that he had much to learn from Indians (and implicitly their spirituality and life practices) but nothing to learn from missionaries or even Christ.[10] While in his most mature thinking he seemed to remain

10 "A snowstorm was more to him than Christ," according to Walter Harding 1965, 464, in Gould 2005, 1635.

convinced that there was some divine dimension to the universe, he also eschewed metaphysical speculation, and instead, articulated an ecological understanding of the interdependence, and mutual dependence, of all life. Moreover, he confessed a strong faith in nature (which was idolatrous in the eyes of the religious mainstream) and promoted kinship ethics and concern for biodiversity, a century before the term was even coined.[11] Indeed as Lawrence Buell noticed, Emerson implicitly recognized Thoreau's ecocentrism when he paid "tribute to him as the attorney of the indigenous plants" (Buell 1966, 363).

With Thoreau, a decisive shift began, only rarely hinted at previously, in which wilderness ecosystems would be understood holistically and valued for the biological diversity they harbor. Before Thoreau, appreciation for nature and wilderness, if present at all, was for anthropocentric reasons, either for aesthetic pleasures, spiritual enlightenment, or for its ability to buttress the national self-esteem among the citizens of the relatively young American republic by providing a positive, comparative reference point to Europe. Increasingly from this time forward, and this accounts for much of the material interests that drove wilderness protection movements, wilderness would also be valued for its commercial value as a tourist destination. All of these wilderness values would get wrapped up and exceptionally complicated in the American conservation movement, including among its wilderness advocates. But after Thoreau, there would always be some for whom the protection of the country's natural heritage, and all of its diverse ecosystem types and species, was the central objective (Taylor 2001a, 2001b, 2010).

The next seminal figure was John Muir, who was born in Scotland in 1838, and immigrated to rural Wisconsin as a youngster, after which he helped to convert wild woods into a farm, as demanded by his severe and ardently Christian father. In 1861 Muir enrolled at the University of Wisconsin, where he developed an interest in natural science, but he eventually drifted away on long journeys, first to the southeastern

11 David Takacs stated that while the importance of biological diversity had been known and discussed for several generations, the term was not coined until 1986 by someone involved in the 1986 "National Forum on BioDiversity," which "was sponsored by the National Academy of Sciences and the Smithsonian Institution" (Takacs 1996, 36). His book shows the strong, religion-resembling love of nature among many biologists who work to defend biodiversity, some of whom have also been ardent wilderness partisans, such as the conservation biologists Michael Soulé (see 1995) and Reed Noss (see 1994, 2002); see also Meine, Soulé and Noss (2006).

U.S., and eventually to California. Along the way, he rejected the anthropocentric and anti-nature views of the Christianity he knew growing up. He found such views to be incompatible with natural facts, especially, the "conceit" that nature and all creatures were created just for humankind (Muir 1997 [1911], 231]; see also Muir 1997[1916], Taylor 2010, 61–70). Although inspired early by the Transcendentalist writings of Emerson, he was more drawn to the German explorer and naturalist Alexander von Humboldt (1769–1859), whose path-breaking analyses help account for Muir's keen awareness of ecological interdependence, notions reinforced by his reading of Thoreau. Muir became an adventurer, naturalist, writer, and eventually, the driving force behind the Sierra Club, which he founded in 1892. His preservationist goals were incorporated into National Park Service's management philosophy in the United States, and later, the idea and model spread to over 100 countries beyond the U.S.

Muir's spirituality was rooted, first and foremost, in ecstatic experiences in nature that variously involved both animistic and pantheistic perceptions, as well as a deep understanding of ecological interdependence. While some Christian writers consider Muir as one of their own (Cartwright 1987), the most comprehensive studies, as well as my own research, portray Muir as more pagan than theistic (Fox 1981, Cohen 1984, Worster 1998, Taylor 2010, 61–70). Muir wrote in prose indebted variously to Transcendentalism and Romanticism; nature as both beautiful and "sublime" is woven into his descriptions of dramatic natural areas, as well as were descriptions of intimate encounters with plants and animals, and expansive feelings of belonging to the universe. And while he would sometimes use theistic language it is likely he did so because he thought it was politically useful to use language that, he hoped, would be compelling to the dominantly Christian public he sought to enlist in the wilderness cause. Certainly Muir's writing often viciously (and humorously) attacked the Christian thinking prevalent in his day. Even though the terms anthropocentrism or biocentrism had not been coined, without any doubt, he had contempt for the former and affinity with the latter. Indeed, Muir's biocentric moral sentiments were more consistently and clearly expressed than Thoreau's.

Muir understood that the anthropocentric Christianity predominant at his time was antithetical to wilderness preservation; but there were forms more conducive. Gifford Pinchot, the utilitarian forester and founder of the United States Forest Service, was influenced by the Social Gospel movement that flourished at the turn of the 20th century, and placed a priority on providing for the poor and needy (Naylor 1999, 2005). With his utilitarian and progressive approach, Pinchot wanted to conserve natural resources so that all people could benefit from them, not just the wealthy. Pinchot succeeded in establishing a utilitarian, "multiple use" management philosophy for most federal forests and other wildlands. This approach seeks to allow the extraction of "natural resources," as well as diverse forms of human recreation, so long as these resources would be taken in a way that would not deny them to future generations. In practice, however, the multiple use philosophy has seldom prevented serious degradation of natural ecosystems, a fact that preservationists are quick to point out. But preservationists have also, often, viewed wildlands as sacred places that should remain inviolate, set off from human extractive and commercial enterprise; anything less is desecration, they often feel. Indeed, for those who consider wilderness and wildlife as sacred, the terminology of "natural resources" was misleading and spiritually misguided. Muir certainly clearly felt this way: after a falling out with Pinchot (who he had met and with whom he initially felt great affinity), for pursuing dam building, unregulated grazing, and other commercial enterprises, he likened him and his allies to temple destroyers. For just one example, when in the early 20th century Pinchot sought to dam the Tuolumne river at Hetch Hetchy valley in California's Sierra Nevada mountains, Muir compared him to "Satan," and called him and his allies "mischief makers and robbers" (in Fox 1981, 141) – and with the fervor of a prophet, he declared "Dam Hetch Hetchy! As well dam for water-tanks the people's cathedrals and churches, for no holier temple has ever been consecrated by the heart of man" (in Cohen 1984, 330).

This strategy, to label perceived opponents of wilderness as desecrating agents, has periodically been deployed by wilderness advocates ever since Muir's time. It has at times been effective because many Americans have come to view their National Parks and wilderness areas as sacred places that should be off limits to commercial incursions (Graber 1976, Ross-Bryant 1990, Taylor 1995). Indeed, the breech between the anthropocentric and utilitarian conservationism of Pinchot, and the biocentric and spiritual preservationism of Muir, would, throughout the 20th and into the 21st century, characterize much of the tension among conservation-minded Americans.

One more figure deserves mention at this point, even though he was not as directly engaged in wilderness issues. John Burroughs was a contemporary of Muir and Pinchot, living and writing during the late 19th and early 20th centuries. He became a prominent literary naturalist who is best known for the decisive role he played in spurring back-to-the-land movements in North America. Even though he was rooted in a secular, evolutionary worldview, he was nevertheless an important voice for appreciating the wonders of nature, and he often expressed this appreciation using religious rhetoric, as in this remarkable statement, which captured a trend then well underway, and that fits many Americans in subsequent generations: "The forms and creeds of religion change, but the sentiment of religion – the wonder and reverence and love we feel in the presence of the inscrutable universe-persists. Indeed, these seem to be renewing their life today in this growing love for all natural objects and in this increasing tenderness toward all forms of life. If we do not go to church so much as did our fathers, we go to the woods much more, and are much more inclined to make a temple of them than they were" (Burroughs 2001 [1912], 246). The very effort to establish and protect wilderness areas can be understood to be a way of establishing such places as sacred spaces, as temples for those who have left behind conventional religions.

6. From the early 20th century to the Wilderness Act of 1964

The religious dimension of the diverse figures and movements that understand nature as sacred in some way, and that think all life deserves respect and reverent care, has not always been as obvious as they were in Burroughs' words. Over time, as increasing numbers of Americans came to see themselves as secular naturalists (either agnostics or atheists), the religious (or religion-resembling) perceptions and feelings that lead to passionate wilderness advocacy sometimes became more subtle. With the right lenses, however, these perceptions can be brought into view.

Just as Thoreau is properly considered the greatest environmental philosopher in North America during the 19th century, Aldo Leopold (1887–1948) deserves a similar title in the 20th. One of the country's first foresters with the United States Forest Service, and later, while a professor of game management at the University of Wisconsin, in 1935, Leopold co-founded the Wilderness Society. These experiences, as well as a great deal of time spent in close observation of natural systems, went into *A Sand County Almanac*, which was published posthumously in 1949, shortly after Leopold's untimely death. In this text, Leopold subtly expressed a deep emotional connection to and reverence for the earth, which has resonated with a great many readers. He also articulated an organicist understanding of ecological interdependence that presaged James Lovelock's Gaia theory (1979), and most importantly, set forth his "land ethic," which many consider to be the foremost expression of an ecocentric land ethic.[12] The *Almanac* has become a sacred text for many environmentalists and wilderness lovers, who commonly cite certain passages including:

> The land is one organism ... . If the land mechanism as a whole is good, then every part is good, whether we understand it or not... . All ethics so far evolved rest upon a single premise: that the individual is a member of a community of interdependent parts... . The land ethic simply enlarges the boundaries of the community to include soils, waters, plants, and animals, or collectively: the land... . A thing is right when it tends to preserve the integrity, stability, and beauty of the biotic community (Leopold 1949, 190, 239, 262).

Less well known is the influence on Leopold's thought by the Russian mystic Pyotr Demianovich Ouspensky, who was in turn influenced by other Russian mystical thinkers, including Georges Ivanovitch Gurdjieff, as well as by Theosophy. Leopold's holism, some scholars think, can be traced to these influences (Pecotic 2005; Pryor 2011). Like Thoreau and Muir, however, far more decisive for Leopold was his long personal observation of nature. Indeed, although Leopold was reticent to talk publicly about spiritual matters, he did confide to family members that his own religion came from nature, and that if he had to characterize it, he would probably say that he was pantheistic. His offspring were certain that he did not believe in a personal God, but nevertheless, felt that that the living world was sacred to him, and that he considered it to be valuable apart from human needs (Meine 1988, 506–07).

Although it took some time for Leopold's book to find its audience, Muir's Sierra Club, and the Wilderness Society Leopold helped to shape, and a host of environmental groups that would follow, drew Romantics who found wild habitats aesthetically pleasing and spiritually meaningful. Generally speaking and increasingly, such feelings were fused with ecological understandings of the interdependence, and corresponding values that put a premium on the retention of all of the living things that constitute and contribute to the resilience of environmental systems. Wilder-

12 Biocentrism and ecocentrism are often understood to be synonyms, but there are differences; the former refers to life-centered values, the latter to ecosystem-centered value priorities, which are more focused on the health of environmental systems than on individual species, and is thus a more holistic ethics than is biocentrism.

ness "purists", as the geographer Linda Graber (1976) labeled them in an early book about their perception of wilderness areas as sacred, ranged from those deeply suspicious of or hostile to civilization, to those who valued civilization but thought it must be constrained.

A good example of an individual who *valued* civilization but fought tenaciously to protect wilderness *from* it was the Minnesotan canoe outfitter, teacher, and writer, Sigurd Olson. Olson is best known for a series of books promoting wilderness as essential for human spiritual health and wellbeing, and also, for playing a major role in the protection of the multi-lake wilderness of north-eastern Minnesota, which eventually was designated the Boundary Waters Canoe Area Wilderness (Backes 1997). Olson was also one of the first persons to join the Wilderness Society, and he not only eventually served on its governing board and as its President, he also served as the President of the National Parks Association. His writings expressed many of the themes common among wilderness partisans, a deep feeling of belonging and connection to nature, animistic perceptions, and feelings of kinship with non-human organisms. Like John Muir, Olson had rejected the severe form of Christianity that he knew from his passionately evangelical father, partly because he so often had to battle such Christians in his efforts to save wilderness. Olson nevertheless seemed to retain a sense that God was in nature, or responsible somehow for it (Olson 2001). Unlike Leopold, however, the protection of biodiversity never seemed to be a central rationale behind Olson's wilderness advocacy. This shows, then as now, that there are many motivations animating wilderness activists.

One incident David Backes mentioned in his excellent biography of Olson is worth special mention. In 1963, John Carver, an assistant secretary in the U.S. Department of the Interior, which supervises the National Park Service, gave a blistering speech that offended many of the assembled National Park superintendents. He charged, "When all else fails the Park Service seems always able to fall back upon mysticism." Then, referring to a Park Service memorandum, Carver declared that it had "the mystic, quasi-religious sound of a manual for the Hitler Youth Movement"! He underscored that the Park Service was a branch of the U.S. government and insisted, "it isn't a religion, and it should not be thought of as such" (in Backes 1997, 303). Here, Carver gave voice to a critique of the wilderness preservation movement that would within a few decades become common, that wilderness fanatics were a spiritual and political danger; some went so far as to label environmental and wilderness advocates "ecofascists."[13]

Just as revealing, Olson saved the conference and the morale of its employees with a rousing defense of the spiritual dimension of their conservationist mission. According to Backes, he exhorted these Parks employees to be unashamed about their devotion to the cause of conservation because their ultimate sentiment in so doing was love, and a desire to protect the "silent sanctuaries and the eternal perspectives" that the Parks provide (in Backes 1997, 304). In essence, Olson was affirming that indeed, federal employees charged with protecting wilderness had a sacred calling and that they should stand firm and be proud of it.

Carver's frustrated critique was the earliest expression of what became a common criticism by critics of wilderness and of wilderness advocates, that they consider wilderness sacred and are motivated by religious (or at least religion-resembling) sentiments. Sometimes, to such criticisms are added charges that these are idolatrous and spiritually perilous beliefs that worship or trust in nature rather than God. Other times the charge has been that the U.S. Park or Forest Service, or at least some of its employees, publications, interpretive displays, films, and presentations, and even management decisions, promote nature-venerating religion in a way that violates the non-establishment clause of the United States constitution.[14] Some scholarly studies also have also noticed that Interior Department agencies (or personnel) sometimes promote nature spiritualities or otherwise take sides in religion-related public land disputes (Burton 2002; Mitchell 2007, Glass 2005, Taylor and Geffen 2003). A book edited by scholars associated with the Forest Service, including and especially the Foreword written by Jack Ward Thomas, Chief of the United States Forest Service during much of the 1990 s, titled *Nature and the Human*

13 For arguments that environmentalism has fascist tendencies see Bramwell 1989, Ferry 1995, Biehl 1995; for more nuanced assessments, see Zimmerman 1994, 1995, Clark 1996, 2005, Taylor 1998, 2004.

14 In October 1999 there was even a lawsuit filed by loggers against the U.S. Forest service and two environmental groups claiming they were in cahoots with environmental organizations to establish "deep ecology religion" on the public lands by restricting logging. Although the case was dismissed, and some of the legal premises were absurd, the notion that some Americans consider wilderness sacred and want to protect it for that reason was not. For a discussion, see Taylor and Geffen 2003, and for the case, *Associated Contract Loggers, Inc. and Olson Logging, Inc, Plaintiffs, versus United States Forest Service, Superior Wilderness Action Network, and Forest Guardians*, Case No.99–1485 JMR/RLE, United States District Court, District Of Minnesota.

Spirit, provides compelling evidence in this regard (Thomas 1996; see also Taylor and Geffen 2003). All this shows how deeply wilderness, and wilderness spirituality, has become rooted in U.S. culture, so much so that for the most part, Americans absorb and resonate with the subtle nature spirituality expressed by their resource agency rangers and other personnel; generally speaking, only the most astute individuals even notice, and only the most religiously conservative of those who do notice, object (Taylor 2010, 203–07).

It was only a year after the incident in which Carver expressed frustration with the religious sentiments animating and promoted in the National Parks, in 1964, that the U.S. Congress passed the Wilderness Act.[15] In its own words, the Act established procedures

> to assure that an increasing population, accompanied by expanding settlement and growing mechanization, does not occupy and modify all areas within the United States and its possessions, leaving no lands designated for preservation and protection in their natural condition.

Like the establishment of the National Park and Forest Service lands, this was another stunningly unusual, if not unique, example of restraint by a nation, given the previous 10,000 years of agricultural history. The Act then stated, "It is hereby declared to be the policy of the Congress to secure for the American people of present and future generations the benefits of an enduring resource of wilderness." It also provided a historically important definition of wilderness, as

> an area where the earth and its community of life are untrammeled by man, where man himself is a visitor who does not remain … land retaining its primeval character and influence, without permanent improvements or human habitation, which is protected and managed so as to preserve its natural conditions and which (1) generally appears to have been affected primarily by the forces of nature, with the imprint of man's work substantially unnoticeable; (2) has outstanding opportunities for solitude or a primitive and unconfined type of recreation; (3) has … land … of sufficient size as to make practicable its preservation and use in an unimpaired condition; and (4) may also contain ecological, geological, or other features of scientific, educational, scenic, or historical value.

It is noteworthy that, while the preservation of natural conditions and forces, and the protection of features that contain ecological value are mentioned, the preservation of biological diversity, which includes ecosys-

15 See http://www.wilderness.net/index.cfm?fuse=NWPS&sec=legisAct for the text of the Act.

tem variety and the genetic and species diversity of all organisms, was not mentioned, either as a rationale for the Act or as a central objective of it. That concern was not enshrined in law until the US Congress passed, and President Richard Nixon signed, the Endangered Species Act of 1973.

7. After the Wilderness and Endangered Species Acts

Space constraints preclude a detailed review of the cultural history of wilderness after the Wilderness Act, but let me briefly highlight key issues and trends.

In 1967, the historian Lynn White Jr. published a now-famous article blaming monotheistic religions, and Christianity especially, for promoting anthropocentric values and indifference if not also hostility toward nature. Such views were nothing new and had been expressed in various ways by Thoreau, Muir, and Leopold. But White's article was published at the dawning of the environmental age in the widely read journal *Science*; consequently it made sense to many readers and became higly influential. Moreover, for generations, many in the U.S. had been drawn to or developing spiritualities that were divorced from organized religions and rooted in nature. So White's critique fell on fertile cultural ground. Meanwhile, while some Christians dismissed his argument, others took it to heart, calling for an enriched appreciation and care for God's created order. Individuals from many other religious traditions, awakening to an increasingly obvious global environmental crisis, looked to their own traditions for spiritual resources to respond, sometimes being quite innovative in so doing, after finding not much in them that provided such resources. This is a process that has now been going on nearly a half century. Scholars have both participated in such revisioning and have been tracking it.[16] Some have mustered evidence and argued that the world's predominant religions are beginning to turn green (Tucker 2003, Gardner 2006, Gottlieb 2006a). I have contended that the most dramatic greening of religion has been outside of long-established religious traditions in "dark green religions," in which people feel a deep sense of belonging and connection to nature and a sense of its sacredness, which is accompanied by feelings of kinship with other organisms and a corre-

16 For a brief review, see Taylor 2005a. See also Gottlieb (2006a), and journals *Worldviews* and the *Journal for the Study of Religion, Nature and Culture*, where much of this ferment is recorded.

sponding belief in their intrinsic value, all of which redound in a passionate concern to protect the biological diversity of the planet, which generally depends on preserving all ecosystem types (Taylor 2010). I have also noted that at least some in the world's predominant religions are developing their own dark green forms. Since wilderness is seen as a good way to describe the large, relatively undisturbed habitats that are needed to ensure that all ecosystems and species survive, such religious evolution is providing at least some modest support for the preservation of wilderness areas today.

The truth is, however, that the cultural gestalt shifts that lead to religious wilderness activism are relatively new, and the future of such efforts, and these places themselves, remains obscure. We do not know with any confidence whether and to what extent the world's predominant religions *are* greening, nor do we know what impact unconventional nature religions will have, including with regard to wilderness preservation. Much more research is needed to more fully understand currently unfolding trends (Taylor 2011).[17] What is clear is that the rationale for preserving large scale habitats, which used to be first and foremost for the spiritual benefit and aesthetic pleasures of the human species, has taken a dramatic turn, and the arguments increasingly turn on preserving wilderness, both terrestrial and marine, because these are the habitats upon which biodiversity, and even the evolutionary process itself, depends. It is important to note, however, that even as the rationales for wilderness

17 While in the cited article I emphasized how little we know about the influence of the religion variable and call for further research, I think that, *with regard to wilderness protection movements*, that the weight of current evidence is that religion has been an important and even a decisive variable for some individuals and groups. My views about religion and environmental action in general are indebted both to Emile Durkheim and Max Weber. Durkheim saw religion as a reflection of society, and I think most environmental concern among religious individuals and groups, when it exists, reflects cultural trends rather than does a natural outgrowth from these traditions. Weber also understood religion as a reflection of the society in which it was situated, but he also postulated that sometimes, religious innovation shapes a culture in important and even decisive ways. My view is that it is in the naturalistic nature spiritualities that I discuss in *Dark Green Religion* where the most significant, culture-influencing, and environment-valuing, religious innovation is unfolding (Taylor 2010). Many of the figures and movements promoting wilderness preservation have been a part of this innovative religious (or religion-resembling) productivity, and so at least with regard to these forms, if my current view is correct, we could also conclude that sometimes religion significantly contributes to environmental concern and action, and may continue to do so for the foreseeable future.

turn more scientific, and even when the arguments insist, as Leopold argued, that saving all parts of nature is the essence of prudence, these arguments are usually infused with religious, or at least, religion-resembling sentiment.

8. Conclusions

Understandings of wilderness, and the places the term refers to, are diverse and contested, as are the policies that emerged that have designated wilderness areas, as are the agencies that have been established to protect them.[18] In the decades since the passage of the Wilderness Act, the criticisms of the idea of wilderness and the management of places so designated have proliferated. Put briefly:

- Some say the wilderness ideal is inherently dualistic and assumes that people are somehow not a part of nature, and that this leads them to care only for places they think, falsely, they can remove themselves from and leave unimpacted. The result is that the places people routinely inhabit, and the everyday practices that people need to change if they are to create environmentally sustainable livelihoods and lifeways, receive too little attention, even from environmentally concerned people.
- Others contend that the wilderness idea has been and remains elitist and exclusionary, that only affluent white Americans really benefit from it. They cite the deracination of indigenous and other long-term inhabitants from their lands when Parks and wilderness reserves are established, both in the United States and other countries, and conclude that the model is thus fundamentally unjust.

18 The sort of oversimplified binaries that fail to consider the wide range of wilderness-related ideas, spiritualities, and practices, as well as the diverse cultural and environmental habitats from which they emerge, have become all to prevalent in scholarly discourse about wilderness. Moreover, there has been a tendency to reify beliefs and practices about wilderness, as though criticisms and resistance to certain wilderness-related ideas and practices have had no effect on such ideas and practices. The articles in this volume, including the perceptive introduction by its editor, importantly call for much greater nuance in the study of "wilderness," its various cognates, and the bio-cultural changes influenced by it. Hopefully, more such scholarship will follow.

- Still others insist that wilderness preservation is rooted in a spiritually perilous paganism, or a politically dangerous authoritarianism and an economically pernicious collectivist or socialist impulse.[19]

Defenders of the idea of wilderness acknowledge that while some wilderness defenders may be guilty of dualistic myopia, only caring about the wildest of habitats, but they rejoinder that properly understood wilderness is precious because all life and environmental systems are interconnected, so a desire to protect wilderness should by no means lead to a lack of effort to transform human political systems and everyday practices. Indeed, one cannot save wilderness or the city without reforming human cultures in a very comprehensive way, the most astute of wilderness advocates would argue.

Historically-informed defenders of wilderness certainly recognize shortcomings among those who have advanced the ideal, as well as injustices in the way such areas have been established and managed. This does not, they would nevertheless respond, indicate that the ideal should be abandoned, but rather, that the model should adapt and change in response to compelling criticisms. One such adaptation is that wilderness advocates speak of biosphere and wilderness reserves, and acknowledge that some human habitation can be compatible with the preservation of the spiritual and biological values found in the earth's wildest places.

Defenders of wilderness could also accurately note that as America has become more religiously plural, people of many religious faiths, including some theologically conservative Christians, have come to value wilderness. So have people of other faiths, and no religious faith.

Finally, defenders of wilderness point out that by understanding environmental systems scientifically, we can come to value biological diversity by recognizing its importance to the flourishing of ecosystems and all who depend on them, including our own species. This in turn, provides a strong basis to argue for minimizing human impacts on every type of ecosystem, so we have them as a baseline for understanding how such systems work. This sort of scientific rationale has the potential to provide a prudence-based consensus for establishing and protecting significant areas of the earth as wild habitats. It may be, however, that by understanding such

19 Space constraints preclude reviewing the contending arguments here, but the most influential perspectives appear in these (mostly edited) works: Noss 1994, Noss and Cooperrider 1994; Soulé and Lease 1995, Cronon 1995, Callicott and Nelson 1998, Rothenberg and Ulvaeus 2001, Butler 2002, Meine, Soulé and Noss 2006, Smith 2007, Nelson and Callicott 2008.

efforts as ways that moderns have been establishing new sacred places, that those involved in efforts to celebrate and protect such places will find more compelling ways to do so. Perhaps greater attention to the ways that ritual focuses on that which is considered sacred, for example, could help biodiversity advocates to construct experiences that will shape people's desire to care for the world's remaining, relatively intact and wild habitats.

In 2009, a multi-part documentary aired on public television in the United States that was titled "The National Parks: America's best idea."[20] The establishment of National Parks and wilderness reserves in the United States was certainly not the first time that human beings have set biological systems aside, by labeling them sacred, and requiring special and reverent behavior of humans when present. There has, however, been a strong religious dimension to the wilderness and biodiversity preservation movements that have promoted the establishment and preservation of such reserves. These efforts on a large scale first bore fruit in the United States. The wilderness and biodiversity reverence movements might just prove to be ecologically and culturally adaptive, helping human beings to finally learn their planetary manners.

9. References[21]

Albanese, Catherine L. *Nature Religion in America: from the Algonkian Indians to the New Age.* Chicago: University of Chicago Press 1990.

Backes, David. *A Wilderness Within: the life of Sigurd F. Olson.* Minneapolis: University of Minnesota Press 1997.

Buell, Lawrence. *The Environmental Imagination: Thoreau, nature writing, and the formation of American culture.* Belknap/Harvard University Press 1996.

Burke, Edmund. *Philosophical enquiry into the origin of our ideas of the sublime and beautiful.* Edited by A. Phillips. Oxford: Oxford World Classics 1990 [1757].

Burroughs, John. *Time and Change.* Amsterdam, The Netherlands: Fredonia Books 2001 [1912].

20 Ken Burns, The National Parks: America's Best Idea," U.S. Public Brodcasting Service, 2009. See http://www.pbs.org/nationalparks/ and from the Internet Movie Database, http://www.imdb.com/title/tt1464482/.

21 Bron Taylor wishes to acknowledge the generous support of the Rachel Carson Center for Environment and Society which supported this research subsequent to my appointment as a Carson Fellow there in 2011.

Burton, Lloyd. *Worship and Wilderness: culture, religion, and law in the management of public lands and resources.* Madison: University of Wisconsin Press 2002.

Butler, Tom, ed. *Wild Earth: wild ideas for a world out of balance.* Minneapolis, MN: Milkweed 2002.

Callicott, J. Baird, and Michael P. Nelson, eds. *The Great New Wilderness Debate.* Athens: University of Georgia Press 1998.

Cartwright, Richard Austin. *Baptized into Wilderness: A Christian perspective on John Muir.* Louisville, KY John Knox Press 1987.

Clark, John. "The French take on environmentalism". *Terra Nova* 1 (1) (1996): 112–19.

Cohen, Michael P. *The Pathless Way: John Muir and American wilderness.* Madison: University of Wisconsin Press 1984.

Cronon, William, ed. *Uncommon Ground: toward reinventing nature.* New York: Norton 1995.

Denevan, William M. "The pristine myth: the landscape of the Americas in 1492". *Annals of the Association of American Geographers* 82 (3) (1992):369–385.

Denevan, William M., ed. "*The Native population of the Americas in 1492*". 2nd ed. Madison, Wis.: University of Wisconsin Press 1992.

Diamond, Jared. "The worst mistake in the history of the human race". *Discover* (May1987): 64–66.

Diamond, Jared. *Guns, Germs, and Steel: the fates of human societies.* New York: Norton 1997.

Emerson, Ralph Waldo. *Nature.* In *The Essential Writings of Ralph Waldo Emerson,* edited by B. Atkinson. New York: The Modern Library 2000 [1836].

Evans, Estwick. *A pedestrious tour of four thousand miles through the Western states and territories during the winter and spring of 1818.* Concord, New Hampshire 1819.

Ferry, Luc. *The New Ecological Order.* Chicago & London: University of Chicago Press 1995. Original edition, Bernard Graset; Paris; 1992.

Fox, Stephen. *The American Conservation Movement: John Muir and his legacy.* Madison: University of Wisconsin Press 1981.

Gardner, Gary. *Inspiring Progress: Religion's Contributions to Sustainable Development.* New York: Norton 2006.

Gatta, John. *Making Nature Sacred: literature, religion, and environment in America from the puritans to the present.* Oxford & Cambridge: Oxford University Press 2004.

Glass, Matthew. "National Parks and Monuments". In *Encyclopedia of Religion and Nature,* edited by B. Taylor. London & New York: Continuum 2005.

Glave, Dianne, and Mark Stoll, eds. *To Love the Wind and the Rain: African Americans and Environmental History.* Pittsburgh: University of Pittsburgh Press 2006.

Gottlieb, Roger S. *A Greener Faith: religious environmentalism and our planet's future.* Oxford & New York: Oxford University Press 2006.

Gottlieb, Roger S., ed. *Oxford Handbook of Religion and Ecology.* Oxford & New York: Oxford University Press 2006.

Gould, Rebecca Kneale. Thoreau, Henry David. In *Encyclopedia of Religion and Nature*, edited by B. Taylor. London & New York: Continuum International 2005.

Graber, Linda. *Wilderness as Sacred Space.* Washington, D.C.: Association of American Geographers 1976.

Harding, Walter. *The days of Henry Thoreau: A biography.* New York: Knopf 1965.

Harrison, Robert Pogue. *Forests: the shadow of civilization.* Chicago: University of Chicago Press 1992.

Jacoby, Karl. *Crimes against Nature: Squatters, Poachers, Thieves, and the Hidden History of American Conservation.* Berkeley, California: University of California Press 2001.

Kalland, Arne. "The Religious Environmentalist Paradigm". In *Encyclopedia of Religion and Nature*, edited by B. Taylor. London & New York: Continuum International 2005.

Kant, Immanuel. *Observations on the feeling of the beautiful and sublime.* 1960 [1764]. Second Paperback Edition 2003 ed. Berkeley, California: University of California Press.

Kearns, Laurel. "Noah's Ark Goes to Washington: a profile of evangelical environmentalism". *Social Compass* 44 (3) (1997):349–366.

Keller, Robert H., and Michael F. Turek. *American Indians and National Parks.* Tucson, Arizona: University of Arizona Press 1998.

Krech, Shepard. *The Ecological Indian: myth and history.* New York: Norton 1999.

Leopold, Aldo. *A Sand County Almanac with Essays on Conservation from Round River.* Enlarged ed. New York: Sierra Club and Ballantine Books 1966 [1949]. Original edition, Oxford University Press,1949.

Lewis, Michael L., ed. *American Wilderness: a new history.* Oxford & New York: Oxford University Press 2007.

Lockwood, Julie L., and Michael L. McKinney, eds. 2001. *Biotic Homogenization.* New York: Springer.

Lovelock, James. *Gaia, a new look at life on earth.* Oxford 7 New York: Oxford University Press 1979.

Marsh, George Perkins. *The Earth as Modified by Human Action.* New York City: Arno 1970 [1864]. Original edition, Scribner, Armstrong and Co.; New York; 1874.

Mason, Jim. *An Unnatural Order: uncovering the roots of our domination of nature and each other.* New York: Simon and Schuster 1993.

Meine, Curt. *Aldo Leopold: his life and work.* Madison: University of Wisconsin Press 1988.

Meine, Curt, Michael Soulé, and Noss Reed. "A mission-driven discipline: the growth of conservation biology". *Conservation Biology* 20 (2006): 631–651.

Miller, Perry. *Errand into the Wilderness.* Cambridge, Massachusetts: Harvard University Press 1956.

Mitchell, Kerry. "Managing spirituality: public religion and National Parks". *Journal for the Study of Religion, Nature and Culture* 1 (4) (2007).

Muir, John. Cedar Keys. In *Muir: nature writings*, edited by W. Cronon. New York: The Library of America 1997 [1916].
Muir, John. "My First Summer in the Sierra". In *Muir: nature writings*, edited by W. Cronon. New York: The Library of America 1997 [1911].
Nash, Roderick Frazier. *Wilderness and the American mind.* 4th ed. New Haven: Yale University Press 2001 [1967]. Original edition, Yale University Press, 1967.
Naylor, D. Keith. "Gifford Pinchot, the conservation movement, and the social gospel". In *Perspectives on the Social Gospel: papers from the inaugural social gospel conference at Colgate Rochester Divinity School*, edited by C. Evans: Edwin Mellon Press 1999.
Naylor, Keith. "Pinchot, Gifford". In *Encyclopedia of religion and nature*, edited by B. Taylor. London & New York: Continuum International 2005.
Nelson, Michael P., and J. Baird Callicott. *The Wilderness Debate Rages On: continuing the great new wilderness debate.* Athens, Ga.: University of Georgia Press 2008.
Noss, Reed F. "Wilderness–now more than ever: a response to Callicott". *Wild Earth* 4 (4) (1994): 60–63.
Noss, Reed F., and Allen Y. Cooperrider. *Saving Nature's Legacy: protecting and restoring biodiversity.* Washington, DC: Island Press 1994.
Oelschlaeger, Max. *The Idea of Wilderness: from prehistory to the age of ecology.* New Haven: Yale University Press 1991.
Oelschlaeger, Max, ed. *The Wilderness Condition: essays on environment and civilization.* San Francisco: Sierra Club Books 1992.
Olson, Sigurd F., and David Backes. *The Meaning of Wilderness: essential articles and speeches.* Minneapolis: University of Minnesota Press 2001.
Pecotic, David. "Ouspensky, Pyotr Demianovich". In *Encyclopedia of Religion and Nature*, edited by B. Taylor. London & New York: Continuum International 2005.
Pedersen, Poul. "Nature, Religion and Cultural Identity: The Religious Environmentalist Paradigm". In *Asian Perceptions of Nature: a critical approach*, edited by O. Bruun and A. Kalland: Curzon Press 1995, 258–276.
Perreault, Melanie. "American wilderness and first contact". In *American Wilderness: a new history*, edited by M. L. Lewis. Oxford & New York: Oxford University Press 2007.
Ponting, Clive. *A New Green History of the World: the environment and the collapse of great civilizations.* Rev. ed. New York: Penguin Books 2007.
Pryor, Ashley. "Thinking like a mystic: The unacknowledged legacy of P.D. Ouspenksy's Tertium Organum on the development of Leopold's, "Thinking Like a Mountain". *Journal for the Study of Religion, Nature and Culture* 5 (4 (2011): 465–490.
Quinn, Daniel. *Ishmael:* "A novel". New York: Turner/Bantam 1992.
Quinn, Daniel. *The Story of B.* New York: Bantam 1996.
Reed, Charles A. *Origins of Agriculture.* The Hague: Mouton 1977.
Ross-Bryant, Lynn. "The land in American religious experience". *Journal of the American Academy of Religion* 58, 3 (1990).

Roth, Dennis M. *The Wilderness Movement and the National Forests.* College Station, Texas: Intaglio 1990.
Rothenberg, David, and Marta Ulvaeus. *The World and the Wild.* Tucson: University of Arizona Press 2001.
Rousseau, Jean-Jacques. *Émile.* Phoenix, AZ: Everyman 1963 [1762].
Rousseau, Jean-Jacques. *Julie or the New Heloise (Juli ou la Nouvelle Héloïse).* Lebanon, NH: Darthmouth College Press 1997 [1761].
Ruffin, K. N. *Black on Earth: African American ecoliterary traditions.* Athens, University of Georgia Press 2010.
Shepard, Paul. "A post-historic primitivism". In *The Wilderness condition: essays on environment and civilization,* edited by M. Oelschlaeger. San Francisco: Island Press & Sierra Club Books 1992, 40–89.
Shepard, Paul. *Coming Home to the Pleistocene.* San Francisco: Island Press 1998.
Smith, Kimberly K. *African American Environmental Thought.* Lawrence, Kansas: University Press of Kansas 2007.
Smith-Cavros, Eileen M. "Modern Black Churchgoers in Miami-Dade County, Florida: Place, Nature, and Memory". *Journal for the Study of Religion, Nature and Culture* 1 (3) (2007): 351–370.
Smith-Cavros, Eileen M. "Black Churchgoers, Environmental Activism and the Preservation of Nature in Miami, Florida". *Journal of Ecological Anthropology* 10 (1) (2007): 33–44.
Soulé, Michael, and Gary Lease, eds. *Reinventing Nature?: responses to postmodern deconstruction.* Washington D.C.: Island Press 1995.
Spence, Mark David. *Dispossessing the Wilderness: indian removal and the making of the national parks.* Oxford University Press 1999.
Stoll, Mark. "Religion 'irradiates' the wilderness". In *American wilderness: a new history,* edited by M. L. Lewis. Oxford & New York: Oxford University Press 2007.
Stoll, Steven. "Farm against forest". In *American Wilderness: a new history,* edited by M. L. Lewis. Oxford& New York: Oxford University Press 2007.
Takacs, David. *The Idea of Biodiversity: philosophies of paradise.* Baltimore, Maryland: John Hopkins University Press 1996.
Taylor, Bron. "Resacralizing earth: pagan environmentalism and the restoration of Turtle Island. In *American Sacred Space,* edited by D. Chidester and E. T. Linenthal. Bloomington: Indiana University Press 1995, 97–151.
Taylor, Bron. "Religion, Violence, and Radical Environmentalism: From Earth First! to the Unabomber to the Earth Liberation Front. *Terrorism and Political Violence* 10 (4) (1998): 10–42.
Taylor, Bron. "Earth and nature-based spirituality (part II): from deep ecology and bioregionalism to scientific paganism and the New Age. *Religion* 31 (3) (2001b): 225–245.
Taylor, Bron. "Earth and nature-Based spirituality (part I): from deep ecology to radical environmentalism". *Religion* 31 (2) (2001a): 175–193.
Taylor, Bron. "Threat Assessments and Radical Environmentalism". *Terrorism and Political Violence* 15 (4) (2004): 172–183.

Taylor, Bron. "Religious studies and environmental concern". In *Encyclopedia of Religion and Nature*, edited by B. Taylor. London & New York: Continuum International 2005a.
Taylor, Bron, ed. *The Encyclopedia of Religion and Nature.* London & New York: Continuum 2005b.
Taylor, Bron. "Exploring religion, nature, and culture". *Journal for the Study of Religion, Nature and Culture* 1 (1) (2007): 5–24.
Taylor, Bron. *Dark Green Religion: nature spirituality and the planetary future.* Berkeley & Los Angeles: University of California Press 2010.
Taylor, Bron. "Toward a robust scientific investigation of the 'religion' variable in the quest for sustainability". *Journal for the Study of Religion, Nature and Culture* 5 (3) (2011): 253–62.
Taylor, Bron, and Joel Geffen. "Battling Religions in Parks and Forest Reserves: facing religion in conflicts over protected places". In *The Full Value of Parks and Protected Areas: from economics to the intangible*, edited by D. Harmon and A. Putney. Lanham, Maryland: Rowman & Littlefield 2003, 281–294.
Taylor, Bron, and Gavin Van Horn. "Nature religion and environmentalism in North America". In *Faith in America*, "Personal Spirituality Today" (v. 3), edited by C. H. Lippy. Westport, CT, London: Praeger 2006, 165–190.
Thomas, Jack Ward. "Foreword. In *Nature and the Human Spirit: toward an expanded land management ethic*", edited by B. L. Driver, D. Dustin, T. Baltic, G. Elsner and G. Peterson. State College, Pennsylvania: Venture Publishing 1996.
Thornton, Russell. *American Indian Holocaust and Survival: a population history since 1492.* 1st ed., The Civilization of the American Indian series v. 186. Norman: University of Oklahoma Press 1987.
Tucker, Mary Evelyn. *Worldly Wonder: religions enter their ecological phase*. LaSalle, Illinois: Open Court 2003.
Turner, Frederick Jackson. "The significance of the frontier in American history". In *The Frontier in American History* (collected works), edited by unknown. Melbourne, Florida: Krieger 1962 [1893].
White, Lynn. "The historic roots of our ecologic crisis". *Science* 155 (1967):1203–1207.
Whitman, Walt. *Leaves of Grass.* Oxford & New York: Oxford University Press 2005 [1855].
Williams, Michael. 2003. *Deforesting the Earth: from prehistory to global crisis.* Chicago: University of Chicago Press.
Worster, Donald. "Nature, Liberty, and Equality". In *American Wilderness: a new history*, edited by M. L. Lewis. Oxford & New York: Oxford University Press 2007, 263–272.
Worster, Donald. *A Passion for Nature: the life of John Muir.* Oxford & New York: Oxford University Press 2008.
Zimmerman, Michael E. "Ecofascism: threat to American environmentalism?" *Social Theory and Practice* 21, 2 (1995):207 f.
Zimmerman, Michael E. "Ecofascism". In *Encyclopedia of Religion and Nature*, edited by B. Taylor. London & New York: Continuum International 2005.

Contributor biographies

Bremmer, Jan N.
Jan N. Bremmer is Emeritus Professor of Religious Studies at the University of Groningen. He has published widely on the social and religious history of Greece, Rome and Early Christianity as well as on contemporary religion. His most recent books are: *Greek Religion & Culture, the Bible and the Ancient Near East* (2008), *The Rise of Christianity through the Eyes of Gibbon, Harnack and Rodney Stark* (2010) and, as co-editor, *Perpetua's Passions* (2012).

Endsjø, Dag Øistein
Dag Øistein Endsjø is Professor in Religious Studies, University of Bergen. He has published articles on a variety of subjects, among them religious change in antiquity, religious space, beliefs in physical immortality, religion and human rights, and religion and popular culture. He has also written the books *Primordial Landscapes, Incorruptible Bodies. Desert asceticism and the Christian appropriation of Greek ideas on geography, bodies, and immortality* (2008), *Greek Resurrection Beliefs and the Success of Christianity* (2009), and *Sex and Religion. Teaching and taboos in the history of world faiths* (2009/2011), the latter published in nine languages.

Feldt, Laura
Laura Feldt, PhD, is Assistant Professor in the Section of the Study of Religion, Department of Cross-Cultural and Regional Studies, University of Copenhagen. She has published articles on ambiguity and uncertainty in ancient religious narratives from Mesopotamia and the Hebrew Bible, on monsters, miracles, religion and contemporary fantasy literature, and the book *The Fantastic in Religious Narrative from Exodus to Elisha* (2012). She is currently engaged in a research project about wilderness conceptions in ancient religions.

Fibiger, Marianne C. Qvortrup
Marianne C. Qvortrup Fibiger, PhD, is Associate Professor at the Study of Religion, Department of Culture and Society, University of Aarhus, Denmark. She is especially interested in contemporary Hinduism and has done fieldwork different places in India as well as in Sri Lanka,

Kenya, England, Mauritius and Denmark. She has published articles on Hinduism in Diaspora, especially Sri Lankan Tamils in Denmark, on Śāktism, Religion and Pluralism and how tradition is transmitted and remembered, and she has written a book on Hindu women and their role as tradition keepers.

Gilhus, Ingvild S.
Ingvild Sælid Gilhus is Professor of Religion at the University of Bergen, Norway. She works in the areas of religion in late antiquity and new religious movements. Main publications include *Lauging Gods, Weeping Virgins: laughter in the history of religions* (1997) and *Animals, Gods and Humans: changing attitudes to animals in Greek, Roman and early Christian ideas* (2006). She is book review editor of *Numen* and is editorial board member of *Temenos.*

Harvey, Graham
Graham Harvey, PhD, is Reader in Religious Studies, Department of Religious Studies, The Open University, UK. He has published books and articles about religion among Pagans, indigenous peoples and Jews. He is particularly interested in animist rituals and everyday practices but has also written about the rhetorics of self identities.

Hoffmann, Thomas
Thomas Hoffmann, PhD, is Associate Professor in Arabic and Islamic Studies, Department of Culture and Society, Aarhus University. He has published a monograph on the Qur'an as a poetic text as well as articles dealing with Qur'anic and Islamic studies. He has recently embarked on research concerning early Islam and natural disasters.

Hornborg, Anne-Christine
Anne-Christine Hornborg, PhD, is Professor in the History of Religions, Department of Theology and Religious Studies, Lund University, Sweden. Hornborg has published articles and books concerning indigenous worldviews, the anthropologist in the field, shamanism, religion and ecology, ritual studies and neospiritual therapies and coaching practices.

Pedersen, Morten Axel
Morten Axel Pedersen is Associate Professor at the Department of Anthropology, University of Copenhagen. He is author of *Not Quite Shamans: Spirit Worlds and Political Lives in Northern Mongolia* (Cornell

University Press, 2011), and co-editor of *Inner Asian Perspectivism* (special issue of the journal *Inner Asia*, 2007), *Technologies of the Imagination* (special issue of the journal Ethnos, 2009), *Comparative Relativism* (special issue of the journal *Common Knowledge*, 2011), and *Times of Security: Ethnographies of Fear, Protest, and the Future* (forthcoming with Routledge).

Segal, Robert A.
Robert A. Segal is Sixth Century Chair in Religious Studies at the University of Aberdeen, where he is also Director of the *Centre for the Study of Myth.* He is the author of *The Poimandres as Myth* (1986), *Religion and the Social Sciences* (1989), *Joseph Campbell* (rev. ed. 1990), *Explaining and Interpreting Religion* (1992), *Theorizing about Myth* (1999), and *Myth: A Very Short Introduction* (2004). He is the editor of *The Gnostic Jung* (1992), *The Allure of Gnosticism* (1995), *The Myth and Ritual Theory* (1998), *Jung on Mythology* (1998), *The Blackwell Companion to the Study of Religion* (2006), and *Myth: Critical Concepts in Literary and Cultural Studies* (2007).

Schjødt, Jens-Peter
Jens-Peter Schjødt, Dr. Phil., is Professor in the Study of Religion, Department of Culture and Society, Aarhus University. He has published books and numerous articles on subjects from pre-Christian Scandinavian religion and on religio-phenomenological issues, such as myth, ritual religious diversity, etc.

Taylor, Bron
Bron Taylor is Professor of Religion and Environmental Ethics at the University of Florida, and a Carson Fellow of the *Rachel Carson Center for Environment and Society* in Munich, Germany. His research focuses on the emotional and spiritual dimensions of environmental movements, and he has led and participated in a variety of international initiatives promoting the conservation of biological and cultural diversity. His books include *Dark Green Religion: Nature Spirituality and the Planetary Future* (2010), the award winning *Encyclopedia of Religion and Nature* (2005), and *Ecological Resistance Movements: the Global Emergence of Radical and Popular Environmentalism* (1995). He is also the founder of the *International Society for the Study of Religion, Nature and Culture*, and editor of its affiliated *Journal for the Study of Religion, Nature and Culture.* For more information see www.brontaylor.com.

Index